Electrical Breakdown and Discharges in Gases

Part B
Macroscopic Processes and Discharges

NATO Advanced Science Institutes Series

A series of edited volumes comprising multifaceted studies of contemporary scientific issues by some of the best scientific minds in the world, assembled in cooperation with NATO Scientific Affairs Division.

This series is published by an international board of publishers in conjunction with NATO Scientific Affairs Division

A	Life Sciences	Plenum Publishing Corporation
B	Physics	New York and London
C	Mathematical and Physical Sciences	D. Reidel Publishing Company Dordrecht, Boston, and London
D	Behavioral and Social Sciences	Martinus Nijhoff Publishers The Hague, London, and Boston
E	Applied Sciences	
F	Computer and Systems Sciences	Springer Verlag Heidelberg
G	Ecological Sciences	

Recent Volumes in Series B: Physics

Volume 84— Physical Processes in Laser–Materials Interactions
edited by M. Bertolotti

Volume 85— Fundamental Interactions: *Cargèse 1981*
edited by Maurice Lévy, Jean-Louis Basdevant, David Speiser, Jacques Weyers, Maurice Jacob, and Raymond Gastmans

Volume 86— Surface Mobilities on Solid Materials: Fundamental Concepts and Applications
edited by Vu Thien Binh

Volume 87— Relativistic Effects in Atoms, Molecules, and Solids
edited by G. L. Malli

Volume 88— Collective Excitations in Solids
edited by Baldassare Di Bartolo

Volume 89a— Electrical Breakdown and Discharges in Gases: Fundamental Processes and Breakdown
edited by Erich E. Kunhardt and Lawrence H. Luessen

Volume 89b— Electrical Breakdown and Discharges in Gases: Macroscopic Processes and Discharges
edited by Erich E. Kunhardt and Lawrence H. Luessen

Electrical Breakdown and Discharges in Gases

Part B
Macroscopic Processes and Discharges

Edited by
Erich E. Kunhardt
Texas Tech University
Lubbock, Texas

and

Lawrence H. Luessen
Naval Surface Weapons Center
Dahlgren, Virginia

Plenum Press
New York and London
Published in cooperation with NATO Scientific Affairs Division

Proceedings of a NATO Advanced Study Institute on
Electrical Breakdown and Discharges in Gases,
held June 28 – July 10, 1981,
in Les Arcs, France

Library of Congress Cataloging in Publication Data

NATO Advanced Study Institute on Electrical Breakdown and Discharges in Gases (1981: Les Arcs, France)
Electrical breakdown and discharges in gases.

(NATO ASI series. Series B, Physics; v. 89)
"Proceedings of a NATO Advanced Study Institute on Electrical Breakdown and Discharges in Gases, held June 28 – July 10, 1981, in Les Arcs, France"—T.p. verso.
Includes bibliographical references and indexes.
Contents: pt. a. Fundamental processes and breakdown—pt. b. Macroscopic processes and discharges.
1. Electric discharges through gases—Congresses. 2. Breakdown (Electricity)—Congresses. 3. Plasma (Ionized gases)—Congresses. I. Kunhardt, Erich E. II. Luessen, Lawrence H. III. North Atlantic Treaty Organization. Scientific Affairs Division. IV. Title. V. Series.
QC710.N37 1981   537.5'2   82-19006
ISBN 0-306-41194-6 (v. 1)
ISBN 0-306-41195-4 (v. 2)

©1983 Plenum Press, New York
A Division of Plenum Publishing Corporation
233 Spring Street, New York, N.Y. 10013

All rights reserved. No part of this book may be reproduced, stored in a retrieval system, or transmitted in any form or by any means, electronic, mechanical, photocopying, microfilming, recording, or otherwise, without written permission from the Publisher

Printed in the United States of America

PREFACE

This volume contains the lectures and seminars on Coronas, Gas Discharges, Diagnostics, Plasma Chemistry, and Applications presented at the Advanced Study Institute on Breakdown and Discharges in Gases. The Institute was held in Les Arcs, France, during June 28 to July 10, 1981. The first volume has been devoted to the areas of Basic Discharge Processes, Kinetic Theory and Gas Breakdown.

This collection is both tutorial and representative of the state of the field. We hope it will be useful both to beginners and experienced researchers.

We are grateful to a number of organizations for providing financial assistance. The Scientific Affairs Division of NATO provided the major contribution for the Institute. The Office of Naval Research, the Naval Surface Weapons Center at Dahlgren, the Air Force Office of Scientific Research and the Air Force Weapons Laboratory provided additional support. The National Science Foundation financed the travel of one student.

We would like to acknowledge the many people who helped us organize and carry the Institute to its successful completion. Our thanks to the organizing committee, lecturers and participants pants for their support; to Dr. H. Davidson of the Army Research Office for his guidance, to W. P. Allis, A. V. Phelps and A. Rees' whose initial interest in the Institute encouraged us to pursue its organization; to Dr. Mario di Lullo, and Dr. T. Kester of the Scientific Affairs Division of NATO for their personal interest and advice, to J. P. Boeuf for reviewing manuscripts and indexing, to our wives Christine K. and Lynn L. and to Marie Byrd for their assistance in Les Arcs.

A special thanks to Elizabeth Orem of Associated Editors & Authors, whose editorial assistance was invaluable to the preparation of these proceedings.

E. E. Kunhardt
Texas Tech University
Lubbock, Texas

L. H. Luessen
U.S. Naval Surface Weapons Center
Dahlgren, Virginia

July 26, 1982

CONTENTS

CORONA PHENOMENA

Corona Discharge Physics and Applications.   1
   R. S. Sigmond and M. Goldman

DISCHARGES

Modeling of a Glow Discharge 65
   L. Vriens, A. H. M. Smeets, and
   H. J. Cornelissen

The Glow-To-Arc Transition 119
   E. Marode

Arc Discharge Electrode Phenomena. 167
   G. Ecker

Dynamics of Arc Discharges 181
   J. Hackmann

DIAGNOSTICS

Diagnostic Techniques for Discharges
      and Plasmas. 203
   R. T. Waters

Spectroscopic Diagnostics in Gas Discharges. 267
   F. Bastien

PLASMA CHEMISTRY

Plasma Chemistry 293
   A. Goldman and J. Amouroux

HIGH FREQUENCY DISCHARGES

Microwave Discharges. 347
    J. Marec, E. Bloyet, M. Chaker, P. Leprince,
    and P. Nghiem

APPLICATIONS

Plasma Applications 383
    M. Kristiansen and A. H. Guenther

SEMINARS

Diffuse Discharge Opening Switches. 415
    K. Schoenbach, G. Schaefer, M.
    Kristiansen, L. L. Hatfield, and
    A. H. Guenther

The Electroluminescence of Pure Noble
    Gases Below the Threshold for
    Electron Avalanche 429
    C. A. N. Conde, M. F. A. Ferreira,
    T. H. V. T. Dias, and A. D. Stauffer

Plasma Formation in a Tokamak 443
    R. D. Bengtson and J. F. Benesch

APPENDIX

Notes on Symbols, Units and Nomenclature
    in Gas Discharge Physics. 451
    R. S. Sigmond

PARTICIPANTS 459

INDEX 467

CORONA DISCHARGE PHYSICS AND APPLICATIONS

R. S. Sigmond* and M. Goldman**

*Electron and Ion Physics Group
Norwegian Institute of Technology
7034 Trondheim NT, Norway

**Laboratoire de Physique des Décharges "CNR"
Ecole Supérieure d'Electricité
911910 Gif-sur-Yuette, France

INTRODUCTION

Together with the lightning spark or arc, electrical coronas probably are the first electrical phenomenon observed by man. Certainly a phenomenon like the cold Fire of St. Elms playing around. mast-tops must have been considered as a thing apart from this world, with properties as unfathomable as the whims of the kings whose crowns probably should simulate it. Unfortunately, the name corona has also been given to other crown-like phenomena; thus let us warn you that the solar corona is hardly a corona discharge, and that corona equilibrium is the nonthermal transport equilibrium common for the solar corona, corona discharges, and most other low current, low density discharges.

Today, the electrical corona probably is the industrially most important type of gas discharge, even if the more spectacular arc discharges get most of the press. As will be elaborated later, the development potential of coronas is equally great because electrical energy may be utilized for reactions directly without unnecessary heating of nonparticipating components. Also surfaces seem to develop and keep surprising catalytic capabilities in coronas.

Another type of application, possibly more dear to scientists, is the use of coronas as testing grounds for gas discharge theories and computer models — very exacting because of the wide range of space and time scales encountered in a single discharge. However, for just this reason, coronas are not very suitable for accurate measurements of fundamental discharge coefficients.

It is perhaps unnecessary to mention that this chapter is written more as an introduction to corona physics than as a complete history or review of the latest developments. Fundamentals and phenomena that fit the undoubtedly preconceived corona picture of the authors are stressed, while some of the less comprehensible experiments (for us) and theories of others may be relegated to the reference list. The chapter relies heavily on the preceding chapters about fundamental gas discharge properties and general diagnostic techniques and is limited in scope by the following chapters on spark-over characteristics and processes of long discharge gaps.

The presentation is divided into main sections. Following this introduction, the next section will treat fundamental corona phenomena and corona diagnostics, containing subsections on ionization processes, preview of corona types, space charge effects in coronas, and corona stability. Having thus established a base and a vocabulary, we treat negative and positive coronas separately in the following sections and end with a survey of corona applications in the last section.

Before closing this introduction, we would like to pay tribute to J. S. Townsend (1914 a,b) who in the beginning of this century made gas discharges a science and gave coronas their first useful formulae and to L. B. Loeb and his school, who did an impressive number of detailed corona studies and wrote the first and still indispensable book about the subject (Loeb, 1965).

More recent reviews of coronas and of breakdown in inhomogeneous fields, apart from the later chapters in this book, will be found in Goldman and Goldman (1978), Sigmond (1978), Waters (1978, 1981) and Gallimberti (1979). The paper of Fieux and Boutteau (1970) gives an excellent and detailed description of the various corona forms in air as observed by one team of investigators. Finally, there is a recent review paper on "corona and insulation" by the present authors (Goldman and Sigmond, 1981 b). Needless to say, the authors have drawn heavily on the still current parts of their earlier reviews cited above.

FUNDAMENTAL CORONA PHENOMENA AND DIAGNOSTIC TECHNIQUES

Definitions

When pressed for a quick answer, an experimenter probably would describe a corona as a gas discharge emitting bluish light and crackling sounds from regions close to sharp-pointed or thin-wire electrodes. Electrically, it would be expected to have a rising current-

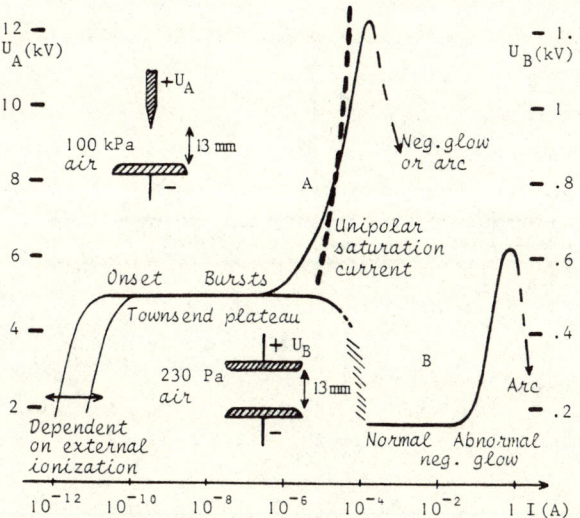

Fig. 1. Current-voltage characteristics for a 13 mm point-to-plane corona gap in 1 atm air (A) and a 13 mm parallel-plane discharge gap in 230 Pa air (B) (authors' laboratories). Note that the gap width of B is of the order of the width at the ionization region in A, when measured in mean free paths for the dominant collision processes. Also drawn (in heavy, dashed line): The unipolar saturation current limit for the corona, according to Eq. (24).

voltage characteristic over a substantial current range, like curve A of Fig. 1, so that it could be fed directly from a constant-voltage power supply without risk of current run-away into arc or spark.

A more precise, physical definition is: *A corona is a self-sustained electrical gas discharge where the Laplacian (geometrically determined) electric field confines the primary ionization processes to regions close to high-field electrodes or insulators*[*]. Thus, a corona discharge system will consist of high-field active electrodes or surfaces surrounded by ionization regions where free charges are produced; low-field drift regions where charged particles drift and react; and low-field passive electrodes mainly acting as charge collectors. This is illustrated in Fig. 2, showing the very widely used point-to-plane (rod-to-plane) corona test configuration. Another popular corona geometry is the coaxial cylinders (or wire-cylinders), shown and discussed later.

---

[*]Corona currents to insulators must necessarily be pulsed or oscillatory.

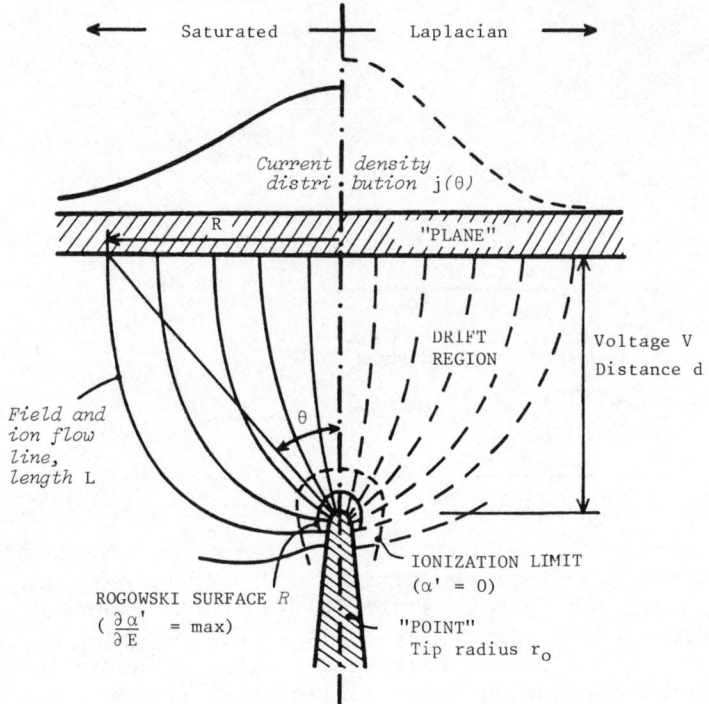

Fig. 2. Illustration of a point-to-plane corona gap, with:
- Laplacian and space charge saturated field line systems (Sigmond, 1981).
- Characteristic equipotential surfaces. $\alpha' \equiv \alpha - \eta$, net or effective ionization coefficient.
- Warburg current density distribution (see Fig. 6).

A DC corona is called positive, negative or bipolar according to the polarity of the active electrode(s). AC Coronas are traditionally power frequency fed, while HF coronas are characterized by HF supply voltage periods shorter than the electron transit time across the corona gap.

It is important to realize that the geometrically based corona definition lies outside and across the electrically based subdivision of DC discharges into Townsend, glow and arc discharges, as shown in Table 1.

Thus, a corona discharge may be a Townsend discharge, a (negative) glow discharge, or neither. (We never speak about corona arcs or sparks, as arcs and sparks seem quite independent of geometry, though not of gap length). It is also worth noting that the term "Townsend generation mechanism" refers to the mechanism of discharge

Table 1. Classical Classification of DC Gas Discharges.

Discharge type	Field and potential distribution	Feedback (γ-) mechanism and cathode fall voltage
Townsend	Essentially Laplacian	Feedback to cathode by photons (γ_p), pos. ions (γ_i), and metastables (γ_m), needing a high cathode fall voltage $V_c \gg V_i$
(Negative) glow	Space charge dominated, with a concentrated cathode fall and a non-LTE or LTE positive (plasma) column	
Arc, spark		Feedback to cathode via thermal or field electron emission, only needing $V_c \approx V_i$

LTE : Local Thermodynamic Equilibrium
$\gamma = \gamma_p + \gamma_i + \gamma_m$ = the secondary emission coefficient
  $\equiv$ number of secondary electrons from the cathode per ionizing collision in the gas.
V_c : Cathode fall voltage V_i : Gas ionization potential (5-20 V)
Spark : Arc of short duration (discharge of cable or capacitance)

self-sustainment whereby each electron avalanche on the average provides for its successor by feedback to the place of avalanche initiation (usually the cathode; see the definition of γ in Table 1). This mechanism is by no means exclusive to Townsend discharges.

Finally, we wish to stress that the real hallmark of coronas is the drift region. All discharges have ionization regions, but it is the drift region that gives coronas their unique intrinsic stability.

Fundamental Corona Processes

The building block of all DC and AC coronas, as of most other discharges, is the Townsend electron avalanche, dealt with in earlier papers. N_{eo} electrons released in a gas of density n in an electric field $\vec{E}(\vec{r})$ will drift with velocity $\vec{v} = -\mu_e \vec{E}$ up the electric field line, each producing α new electron-ion pairs and suffering η attachments per unit drift length. In addition, important amounts of neutrals will be excited to radiating of metastable states. Thus:

$$dN_e = N_e \cdot (\alpha - \eta) \cdot dr = N_e \cdot \alpha' \cdot dr \;, \quad \alpha' \equiv \alpha - \eta \tag{1}$$

which integrates directly to give:

$$N_e(r) = N_{eo} \cdot \exp \int_{r_o}^{r} \alpha' ds \equiv N_{eo} \cdot M \tag{2}$$

(Detachment neglected)

The net primary ionization coefficient α' is thus of fundamental importance and is here given for dry air and SF_6 by the curves of Fig. 3. Note the sharp zero crossings, especially for SF_6, marking the so-called critical fields E_c (at 1 atm: 2.8 MV/m for air, 8.3 MV/m for SF_6). Curves for other gases are shown in earlier chapters.

The exponential in Eq. (2), called the "electron multiplication M," can easily reach values 10^3-10^8 across an ionization region. However, feedback is necessary to make any amplifier self-excited, and we need secondary ionization (γ-) processes, replacing each electron which leaves its starting point with a new one, to get a self-sustained discharge. In the positive corona shown in Fig. 2, each electron appearing at the ionization region boundary must cause so many positive ions, photons, and metastables that at least one

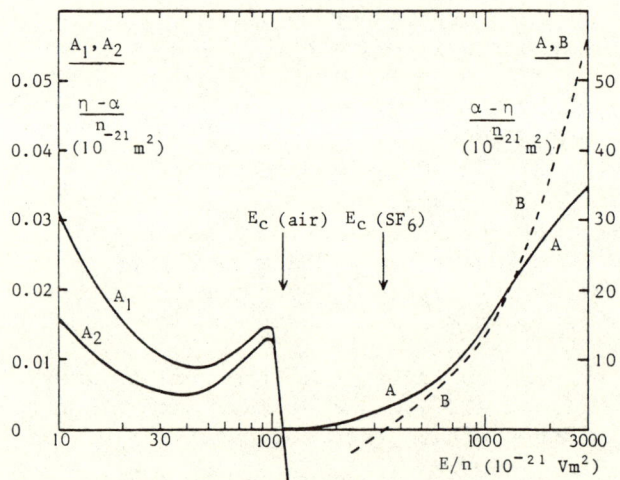

Fig. 3. Net reduced attachment and ionization coefficients for dry air (A) and SF_6 (B) at 20°C.
A_1, $n = 25040 \times 10^{21}$ m^{-3} e.g. 1 atm 20°C (Phelps, A. V., 1980, personal communication)
A_2, $n = 12520 \times 10^{21}$ m^{-3} e.g. 1 atm 310°C
A , $n = 25040 \times 10^{21}$ m^{-3} e.g. 1 atm 20°C (Dutton, 1975; Hartmann, 1980)
B , $E/n < 500$ (Bahalla and Craggs, 1962)
B , $E/n > 500$ (Teich, 1981).

CORONA DISCHARGE PHYSICS

of these succeeds in providing a new electron at the ionization boundary, if the discharge shall not die. These secondary processes are usually summarized by the secondary ionization coefficient γ, defined as the number of replacement electrons produced per ionizing collision in the ionization region. As indicated in Table 1, $\gamma = \gamma_p + \gamma_i + \gamma_m$. Further, we will use γ_c for feedback to the cathode, γ_g, for feedback to the gas. This brings us to Townsend's criterion for self-sustainment:

$$\mu_r \equiv \gamma \int_0^d \alpha\, dx \cdot \exp \int_0^x \alpha'\, ds = \begin{matrix} <1 & i \to 0 \\ =1 & i = \text{const} \\ >1 & i \to \infty \end{matrix} \qquad (3)$$

where μ_r is the reproduction factor and i is the discharge current.

In some form or another, this equation rules the realm of gas discharge physics, and coronas are no exception. Its simplicity, however, is deceptive:

(a) Although α and $\alpha' = \alpha - \eta$, for many gases are tolerably well-known functions of the electric field E (Fig. 3), the very concept of transport coefficients as functions only of the local electric field becomes dubious in very high or very inhomogeneous fields. Thus, in many coronas, ionization region multiplication can only be treated properly by Boltzmann equation or Monte-Carlo methods, or, more approximately by introducing some kind of field memory factors [see Morton (1946), Rajapandyan and Raju (1972), and other chapters in the present book].

(b) The electric field is everywhere the sum of the applied (Laplacian) field and the field from the space charges created through the strongly field-dependent processes of Eq. (3). In a corona geometry, accurate solutions are generally only possible by computer. Fortunately, the ionization region is the one least disturbed by its own space charge, as its Laplacian field and charged particle velocities are high. Thus, a classic approximation assumes a Laplacian field distribution across the ionization region of a strength determined by the geometry, the applied voltage, and the gross effect of the space charges in the drift region (as if the drift region was a series resistor).

(c) The approximation in (b) breaks down when the number of electrons in a single avalanche exceeds 10^6–10^8, as then the avalanche space charge field equals the applied electric field. A field-free region then tends to form between the electrons in the avalanche head and the trailing positive ions. In this region negative and positive charges accumulate in equal numbers, and a plasma filament called a streamer channel tends to form. In homogeneous fields, this signals the spark breakdown of the gap (see Schumann (1923) and Meek (1940) for breakdown criteria).

(d) As for the γ of Eq. (3), let us first note that the feedback processes in positive and negative coronas necessarily must be very different. In negative coronas, the cathode borders on the ionization region, and this makes the quite well-known cathode secondary mechanisms (γ_p, γ_i, γ_m) fast and efficient. In positive coronas, the cathode is insulated from the ionization region by the low-field drift region, which often absorbs photons and traps cathode secondary electrons by attachment. Thus, negative coronas become self-sustained by the Townsend cathode feedback mechanism, while photon feedback to the gas bordering on the ionization region (γ_{pg}) often predominates in positive coronas. The feedback coefficients γ_{pg} for the latter processes are still poorly known and a subject of debate.

When charged particles drift in an electric field, they continuously transfer momentum to the neutral gas. Indeed, transport equilibrium is only achieved when <u>all</u> the momentum gained from the field is immediately transferred to the gas. The effect of this transfer is most easily seen in coronas as an electric wind directed away from the active electrode. Up to now this wind has primarily been used to blow out candles and drive windmills in demonstration experiments. However, it carries a large population of excited and chemically very active species, which may account for up to 50% of the corona energy input (Goldman, 1974, 1976). If nothing else, it will mess-up all retarding-field or retarding-potential diagnostics in its way and disturb electrode field measurements as well. Its effects are better not neglected.

A Preview of DC Corona Types and Terminology

Up to now we have neglected the fundamental quantized and stochastic nature of electrons and electron multiplication, except by noting that an avalanche started by one electron might contain 10^6-10^9 charged particles, enough to cause a measurable current pulse, a streamer, or breakdown. The feedback processes are stochastic as well, and this makes it quite probable that trains of avalanches will die out, even at voltages a little above corona self-sustainment or onset.

Around onset, corona currents thus show very erratic behavior, with small current pulses due to single avalanche trains, mixed with larger pulses caused by initial enhancement and subsequent reduction of μ_r due to space charge accumulation and drift, after long current pauses while we wait for space charges to clear away and new electrons to appear in the ionization region. This has been called the "autostabilization," or, in positive coronas, the "burst pulse" regime. Positive bursts, merging into burst pulses at higher voltages, are identical to the pulses from a Geiger counter operated on the normal Geiger plateau.

Above the corona onset, after the appearance of the first successful seed electron, quite a variety of phenomena are observed as the corona current is increased, depending on the corona polarity and on the gas density and electron attaching properties.

In negative coronas in even weakly electron attaching gases, the interplay between positive and negative space charges leads to the current pulsations known as "Trichel pulses," which first at mA currents are damped to the continuous glow corona. In positive coronas, the burst pulses will either join together, becoming a regularly pulsating positive glow* covering the active electrode, or the largest will develop into positive streamers, which are plasma filaments that carry their own ionization region ahead and propagate rapidly towards the cathode.

All coronas reach their end when their current density causes sufficient thermal ionization across the gap to turn its differential resistance sharply negative; see Fig. 1. A spark or an arc results, depending upon the external circuit parameters.

As usual, the "time lag" is defined as the time delay between corona voltage turn-on and the establishment of a specified corona form or current. This is subdivided into the statistical time lag before the appearance of the first free seed electron, and the formative time lag after that. Measurements of statistical time lags give information about cathode properties in negative coronas, negative ion densities and detachment rates in positive coronas (Berger, 1981); while formative time lags tell about primary and secondary avalanche developments, as discussed in other papers in these proceedings.

Ionization and Related Effects in Space-Charge Perturbed Inhomogeneous Fields; Corona Stability

Given the necessary data for the fundamental corona processes discussed above, we are in principle able to simulate any form of corona by computer. For accurate predictions of corona behavior or comparisons with experiments, this is the only way. However, the task of modeling, for instance, a complete point-to-plane corona in dry atmospheric air is formidable:

Ionization region reduced field: $100 < E/n < 10\,000 \times 10^{-21} Vm^2$

Drift region reduced field: $1 < E/n < 100 \times 10^{-21} Vm^2$

Electron-ion-molecule ionization and conversion coefficients: About 40 for each polarity, to be known for all E/n in the range $(1 - 10^4) \times 10^{-21} Vm^2$

---

*Sometimes called Hermstein glow or ultracorona.

Time scales: Trichel pulse or streamer rise time ≃ 1 ns
                Trichel pulse or streamer duration   ≃ 1 μs
                Repetition periods   ≃ 1 ms

Space scales: Cathode fall thickness   ≃ 5 μm
                Streamer diameter   ≃ 20 μm
                Gap width d   1-100 cm

Thus, the modeling must cover a space range of meters and a time range of milliseconds with respectively micrometer and nanosecond resolutions, for an E/n span of 10^4. For each time step the Poisson equation must be <u>re-solved</u> to yield the new electric field distribution. At present, this means that a complete modeling will be found only for a few important - or "showcase" - coronas (Davies and Donne, 1980; Graf, 1980). For the corona that needs to be investigated, the much simpler, seminanalytical ways of treating coronas sketched below will always be necessary. In the present subsection, we will discuss:

- Corona space charge and the resulting field perturbation.

- The effect of this perturbation upon the corona ionization processes.

- The possibility of cumulative effects, leading to oscillations, pulses, or spark.

From the simple relation between current and charge densities $\rho = j/v_d = j/\mu E$, we see, for continuous current coronas:

- Drift region space charge will always be much larger than ionization region space charge.

- Electron current contribution to space charge will always be negligible, except in the drift region of negative coronas in nonattaching gases.

Further, as the feedback coefficient $\gamma \ll 1$:

- Space charge of the corona polarity will always dominate the drift region of unipolar coronas, and thus, always tend to reduce the ionization region field. (In negative coronas this is obvious. In positive coronas each electron entering the ionization region from the drift region will inject $1/\gamma \gg 1$ new positive ions into the drift region; see Eq. (3)).

The last point above is, of course, of great importance to corona stability. However, remember that space charge enhancement of ionization often is of decisive importance under pulse conditions, as in single avalanches (see below).

We now turn to the effect of space charge field perturbations on the corona reproduction factor μ_r, Eq. (3). Its somewhat complicated form is due to attachment processes, which make the net number of electrons produced less than the number of ionizing collisions. In the ionization region, where $\alpha \gg \eta$, we can write as a fair approximation:

$$\mu_r \simeq \gamma(M - 1) \simeq \gamma M = \gamma \exp \int_o^d \alpha' ds. \qquad (4)$$

Here the exponential is the main field-dependent factor. The effect of a field perturbation $\delta E(s)$ is then

$$\frac{\delta \mu_r}{\mu_r} \simeq \frac{\delta M}{M} \simeq \delta \int_o^d \alpha' ds = \int_o^d \frac{\partial \alpha'}{\partial E} \delta E(s) \, ds \, . \qquad (5)$$

Now, a thin layer of space charge extended over a Laplacian equipotential surface at a distance coordinate s_o from the active electrode will decrease the field on the side of the electrode of the same polarity as the layer and increase the field on the other side. From Eq. (5) one may suspect that the net effect on μ_r will depend on the location s_R of the Rogowski surface R where $\partial\alpha/\partial$ is a maximum: if s_R is on the increased-field side of the space charge layer $\partial \mu_r > 0$, and vice versa. This is verified in Sigmond's review (1978, pp. 329-32), based on the ideas of Rogowski (1936). The Rogowski point in air is at $E/n \simeq 700 \times 10^{-21} Vm^2$, while SF_6 has a practically constant $\partial\alpha'/\partial E$ from below the critical field E_c and past 3000×10^{-21} Vm^2 (difficult to see from Fig. 3 because of the logarithmic scale).

Thus, positive space charge created just outside a positive point electrode will actually increase the ionization if the Rogowski surface exists, i.e., at low pressures or small point radii. However, at 1 atm, normally the R surface is very close to or "inside" the active electrode. Then an extended layer of space charge of the corona polarity will <u>always</u> decrease the corona ionization, no matter the position of the layer. As the dominating, extended space charge in unipolar coronas always is of the corona polarity, we usually find that coronas have positive V(I) slopes, as shown in Fig. 1(A).

It is instructive to apply this reasoning to a parallel plate gap Townsend discharge, as shown in Fig. 1(B). The uniform Laplacian field is then either everywhere below or everywhere above the Rogowski value E_R. In the former case ("normal" in the sense of van Geel (1939)), any field redistribution will increase the avalanche multiplication M. Furthermore, as the space charge is positive, the cathode field, and thus γ, also will be increased by increasing the discharge current. The gap voltage U_B necessary for self-sustainment will decrease, and we obtain the negative-slope $U_B(I)$ characteristic shown in the figure. For electric fields above E_R (n.d well below the Paschen curve minimum) ionization will decrease with increasing current and space charge field redistribution. The $U_B(I)$ curve would have a positive slope, and the gap would need no series resistance for stabilization.

The considerations above are only valid for laterally extended space charge layers. Such a case is the stable positive glow corona illustrated in Fig. 4(a), consisting of a carpet of small avalanches covering the active electrode. A spurious current increase will increase the positive space charge layer close to the anode and thus weaken the field there and push the ionization outwards into weaker field regions. With the Rogowski surface at or inside the anode surface, ionization is reduced and the original current restored.

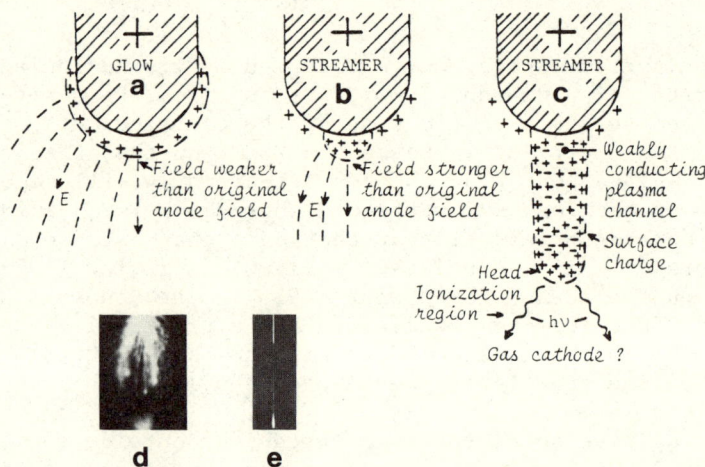

Fig. 4. Positive glow stabilization and positive streamer development. a) The positive glow is stabilized by large lateral avalanche diffusion. The field outside the positive space charge region is weaker than the anode field without positive space charge. b-c) Large concentrated avalanches will leave a positive space charge with a high field region in front, leading to a propagating plasma filament if electrons are available ahead of it (γ-process?) Note the similarities: Streamer   -Glow discharge
          Channel   -Positive column
          Streamer head   -Neg. glow region
          Ionization region -Cathode fall region
d) Photograph of single, nonrepetitive streamer in ambient air. e) Photograph of repetitive streamers in ambient air. Note: The photographs show the paths of the streamer heads. The post-streamer channel is visible only in frame (e).

For a gas which favors large and concentrated trains of avalanches, because of very low γ and small avalanche diffusion, a stochastic current increase may result in a concentrated positive space-charge accumulation at the anode [Fig. 4(b)]. This will reduce or cancel the electric field inside the lump of charge, thus simulating a spiked extension of the metal anode (in itself a plasma!). The field outside this spike is increased because of the geometry, provoking the formation of still larger avalanches in front and a weakly conducting plasma filament behind, forming a positive streamer [Fig. 4(c-e)]. This very constricted, bipolar type of instability might propagate straight through an otherwise perfectly stable extended positive glow drift region, make contact with the cathode, and then act as a time-dependent resistor parallel with the concurrently running diffuse unipolar corona current (the background or continuous current).

Apart from the streamer puncture instability, we should expect the positive glow corona in Fig. 4(a) to be stable. Nevertheless, it oscillates at MHz frequencies, as will be discussed later. The negative diffuse corona ionization region, on the other hand, should be unstable, indeed, at pressures not too low, as here positive space charge is generated outside a negative electrode and certainly will increase μ_r. The resulting Trichel pulses will be discussed in detail later.

Now we will look at a more general but qualitative model of corona small-signal stability (Fig. 5) suggested by similar calculations for parallel-plate Townsend discharges (Sigmond, 1961). First we lump the capacitances in Fig. 5(b) into one equivalent capacity C_i' across the ionization region. From Sigmond (1961), we see that we have two possibilities of instabilities

(1) <u>Catastrophic</u> $R_i + R_d + R_o < 0 \rightarrow$ Spark or arc   (6)

If C_o too large: $R_i + R_d < 0 \rightarrow$ Recurrent spark   (7)
                                             discharging of C_o

(2) <u>Oscillatory</u> $R_d > L_i/C_i'|R_i|$ $\rightarrow$ Oscillations at   (8)
              angular frequency $\simeq (L_i C_i')^{-1/2}$

The ionization region, negative differential resistance $- R_i$ (slope of the U-I curve) is usually constant at low currents, increasing at large. The self-inductance L_i, equal to the initial inverse current rate of rise $(dI/dt)^{-1}$ when a unit overvoltage step is applied across the region, will be about $\propto I^{-1}$. The resistance of the drift region can be estimated from Eq. (23): $R_d = (2 \mu \varepsilon_o I/d)^{-1/2} \alpha (\mu I)^{-1/2}$. Thus, in positive glow coronas, where the ion mobility μ is reasonably

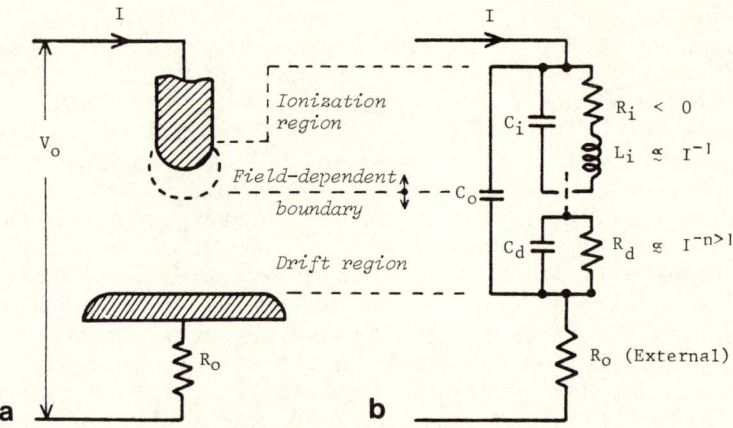

Fig. 5. Simplified, small-signal equivalent circuit of an unipolar corona, viewed as a parallel-plate ionization region discharge in series with a passive drift region (see Fig. 1). a) Corona gap, b) Equivalent circuit. R_i is constant at low I, increases at larger I. All C are approximately current independent. L_i and R_d vary inversely with the current.

constant, the inequality (8) should be increasingly fulfilled at increasing currents, and oscillations should increase in strength. The same should happen in negative Trichel coronas in electron attaching gases up to the currents where the effective mobility µ increases sharply because of increase in the electron population. As the effective mobility may be increased also by space charge redistribution of the electric field, R_d may even turn negative. Then the inequality (8) ceases to be satisfied, stopping the oscillations. Finally, in both types of coronas, inequality (6) or (7) is fulfilled by decrease of R_d and/or of R_i, and the coronas break down. In positive coronas, breakdown via streamers usually occurs before this gross negative resistance type of breakdown.

Finally, we wish to mention the possibility of attachment instability in gases like air, oxygen and carbon dioxide, that in streamer channels might be combined with thermal (density) instability. The former occurs in regions where the net electron attachment rate $(\eta-\alpha) \cdot v_d$ increases with increasing field, such as for $20 < E/n < 90 \times 10^{-21}$ Vm² in air (Fig. 3), a range typical both for negative corona drift regions and positive streamer channels. Here a homogeneous region of electrons and negative ions will tend to split into high-field, negative-ion-rich and low-field, electron-rich domains. In narrow channels, the high power dissipation in high field regions will heat the channel, causing weak shock wave formation and, later, a gas density reduction. The resulting increase of E/n may be substantial

CORONA DISCHARGE PHYSICS

and cause a run-away ionization effect leading to spark or arc. This subject will be treated by Marode.

Space-Charge Dominated Coronas: The Townsend Formula, the Warburg Distribution, and the Unipolar Saturation Current Limit

The following assumptions will be used in the subsequent discussion:

- An extended (nonstreamer) drift region is dominated by charged particles of one sign and constant mobility μ^*,

- Charged particle diffusion is negligible.

We will now derive simple formulas for:

- The unipolar charge density development $\rho(t)$

- The unipolar saturation current density j_s

- The current density distribution $j(\theta)$ in point-to-plane coronas (the Warburg distribution; see Fig. 2)

- The point-to-plane corona unipolar saturation current limit I_s.

The derivations are taken from Sigmond (1981), where also further examples and references will be found.

The stationary nondiffusive ion drift equations:

$$\vec{j} = \rho \vec{v} = \rho \mu \vec{E} \quad , \quad \nabla \cdot \vec{j} = 0 \qquad (9)$$

and the Poisson equation:

$$\nabla^2 V = - \rho/\epsilon_o \quad , \quad \vec{E} = - \nabla V \qquad (10)$$

give, by elimination of $\vec{j}$, ρ and $\vec{E}$:

$$(\nabla^2 V)^2 + (\nabla V) \cdot (\nabla \nabla^2 V) = 0 \qquad (11)$$

---

*In long rod-to-plane gaps at voltages well below the breakdown value, streamers will be confined to a region around the rod tip. If this is considered as an extended ionization region, the rest of the drift region can be treated by the methods of this subsection.

which, together with the boundary conditions, contains the complete stationary ion drift problem in a very hard nutshell. This we will not try to crack, though it is easy in mathematically one-dimensional cases. However, we will cite Townsend's (1914 b) well-known corona current formula:

$$I_T = C \cdot V(V - V_o) \, , \quad C_{cyl} = \frac{8 \pi \mu \varepsilon_o}{R^2 \ln(R/r)} \tag{12}$$

which is an approximate solution of (11) valid for <u>very</u> low currents in a coaxial-cylinder corona of radii r and R and onset voltage V_o (see Fig. 7). The ionization region is here considered to be very thin, with a current-independent burning voltage. As shown by the experimental (Waters and Stark, 1975) and more exact theoretical curves (Thomson and Thomson, 1933; Waters and Stark, 1975) of Fig. 7, one must not take the straight-line appearance of I/V vs. V curves as a proof for the validity of (12) with that value of C_{cyl}. However, with empirical values of C, Eq. (12) has a surprising range of validity for all corona geometries, and experimental current-voltage corona characteristics are often plotted as I/V vs. V for easy comparison.

Instead of (12), we will attack the time-dependent problem, where the ions are drifting along a (space-charge-perturbed) field line $\vec{E}(\vec{r},t)$ with density $\rho(\vec{r},t)$. We take the convective derivative:

$$\frac{d\rho}{dt} = \frac{\partial \rho}{\partial t} + \vec{v} \cdot \nabla \rho \, , \tag{13}$$

combine with the continuity equation and the Poisson equation:

$$\frac{\partial \rho}{\partial t} + \nabla \cdot (\vec{v}\rho) = 0, \quad \vec{v} = \mu \vec{E}, \quad \nabla \cdot \vec{E} = \rho/\varepsilon_o \tag{14}$$

and obtain the unipolar charge drift formula:

$$\frac{d\rho}{dt} = - \frac{\mu \rho^2}{\varepsilon_o} \, , \quad \frac{1}{\rho(t)} - \frac{1}{\rho_o} = \frac{\mu}{\varepsilon_o} (t-t_o) \, . \tag{15}$$

These formulas are exact in the absence of diffusion:

$$\text{Low } \rho \text{ limit}: \quad \rho(t) = \text{const}$$
$$\rho_o \gg \rho \text{ limit}: \quad \rho(t) \to \rho_s(t) = \varepsilon_o/\mu t \tag{16}$$

where ρ_s is the unipolar saturation charge density.

However, the unipolar charge drift formula is useful only if you know approximately the ion path $\vec{r}(t)$. This limits the applications to the two extreme cases:

CORONA DISCHARGE PHYSICS

(1) Ion drift and diffusion experiments, where space-charge effects are to be avoided: Relative space-charge broadening $\simeq \rho/\rho_s$. Thus, for a drift distance L in a field E we must keep the ion current density below j_s:

$$T = L/\mu E, \quad \rho_s = \varepsilon_o/\mu T = \varepsilon_o E/L, \quad j_s = \mu E \rho_s$$
$$j \ll j_s = \mu \varepsilon_o E^2/L \tag{17}$$

(2) Space-charge saturated coronas. The exact space-charge, saturated solutions for the electric field variations as function of the distance r from the unlimited ion source are:

Plane geometry : $E \propto r^{1/2}$ (Laplacian : E=const)

Cylindrical geometry : $E = $ const (Laplacian : $E \propto r^{-1}$)

Spherical geometry : $E \propto r^{-1/2}$ (Laplacian : $E \propto r^{-2}$)

Unipolar space charge tends to smooth all field variations.

Now, the field line geometry in point-to-plane coronas will start $\simeq$ spherical, continue $\simeq$ cylindrical, and end $\simeq$ plane. Thus, a reasonable assumption for space charge saturated point-to-plane coronas will be simply $E \simeq $ const $= V_o/L$ along any field line of length L. This choice also makes the ion transit time T a minimum, insensitive to deviations. Assuming unlimited ion density at the point, we get from Eq. (15):

$$\rho L = \varepsilon_o E/L, \quad J_L = \mu E \rho_L = \mu \varepsilon_o E^2/L = j_s$$

$$j_s = \mu \varepsilon_o V^2/L^3 \quad \text{the unipolar space charge saturation current density,} \tag{18}$$

an equation similar to Mott's for seminconductors.

Now, for a point-to-plane gap it has been shown that the field line shape does not vary much from the Laplacian to the saturated case. For a hyperboloid point-to-plane system, the field lines are ellipses of approximate length; see Fig. 2:

$$L \simeq \sqrt{R^2 + d^2/\cos^2\theta} = d\sqrt{2\,tg^2\theta + 1}. \tag{19}$$

Inserted in Eq. (18):

$$j_s(\theta) \simeq j_{so}(2\,tg^2\theta + 1)^{-3/2}, \quad j_{so} = \mu \varepsilon_o V^2/d^3 \tag{20}$$

which show as good a fit to experimental current density distributions as the well-known Warburg (1899, 1927) distribution

$$j(\Theta) = j_o \cos^5 \Theta \qquad (21)$$

(see Fig. 6). By integrating Eq. (20) over all Θ up to the experimentally observed limit of about 60°, we find:

$$I \simeq 2.0 \, d^2 j_o \, . \qquad (22)$$

Inserting j_s from (18) for j_o, with $L = d$, we finally obtain the unipolar point-to-plane saturation current limit:

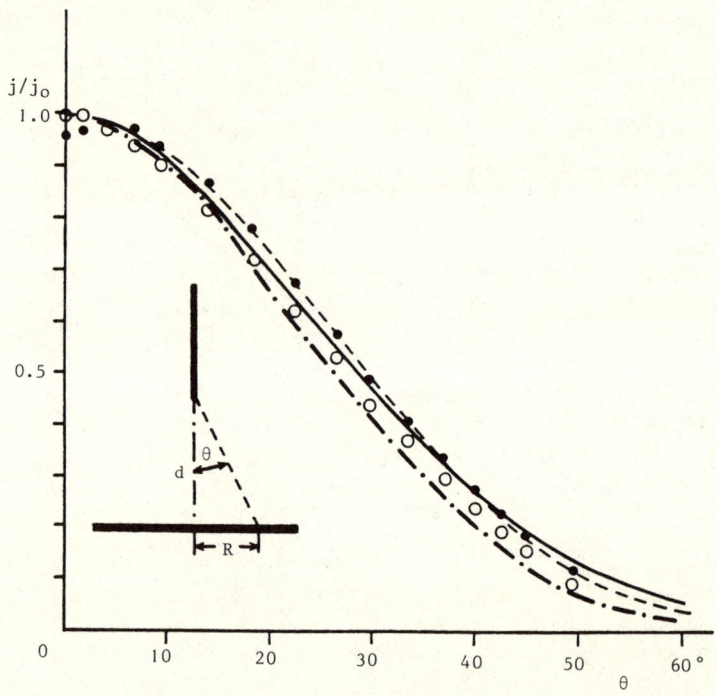

Fig. 6. Radial current density distributions at the plane in unipolar point-to-plane coronas (Sigmond, 1981). The experimental points are normalized to 1 near the axis. The negative Trichel pulse data show a dip in the center.
————— Formula (20)
------ $j/j_o = \cos^5 \Theta$ (Warburg 1899, 1927)
● Neg. Trichel ⎱ Ambient air, $d = 120$ mm, $V = 40$ kV
○ Pos. glow ⎰ (Goldman et al., 1978)
—·—·— Neg. pulseless glow, ambient air
$d = 12, 13$ and 14 mm, $V = 22$ kV (Kondo and Miyoshi, 1978).

$$I_s = 2.0 \, \mu\varepsilon_o V^2/d \, , \quad I_s/V^2 = 2.0 \, \mu\varepsilon_o/d, \tag{23}$$

or, for air at 1 atm with $\mu = 2.3 \times 10^{-4}$ m^2/Vs :

$$I_s(\mu A) = 4 \, V^2 \, (kV)^2/d(mm). \tag{24}$$

This formula, which also has been derived semiempirically by Selim and Waters (1980), is not any substitute for the Townsend corona current formula (12), despite obvious similarities. Its importance is as a definite and often quite close upper limit to the unipolar current in the given geometry. If we measure a corona current/(voltage)^2 in excess of this limit, we may be certain that either:

- Our ion mobility is wrong (in negative coronas, part of our ions are fast electrons), or

- Part of our current is bipolar (streamers, or a new ionization region has formed near the passive electrode).

Note that free electron current and bipolar streamers may well exist below the saturation current limit, but that they must exist above it.

In Figs. 7 and 8, we show the unipolar space charge saturation current limits for various geometries, compared to more exact formulas and to experimental curves. We stress that the curves of Fig. 8 depict only the unipolar part of the corona currents, while the much larger bipolar (streamer) component is given by numbers along the curves. The unipolar current density distribution obeys the Warburg law, while the bipolar streamer current is concentrated in short-lived, repetitive filaments of some 10 μm radius. Let us conclude this subsection by noting:

- The unipolar conduction currents seem to flow also in coronas where the main current flow is by bipolar streamer conduction. The unipolar current size and distribution seem remarkably uninfluenced by the streamers and probably dominate the field in which the streamers propagate.

- The unipolar current component is often called the background or continuous current, as its oscillations usually are of high frequencies and smeared out by the slow ion movement through the drift region.

<u>Corona Diagnostics</u>

The main diagnostic tools for corona studies are in principle not different from those used on any other discharge form. Thus, for general optical, spectroscopical, and electrical methods we refer to other chapters in this book. But the peculiarities that

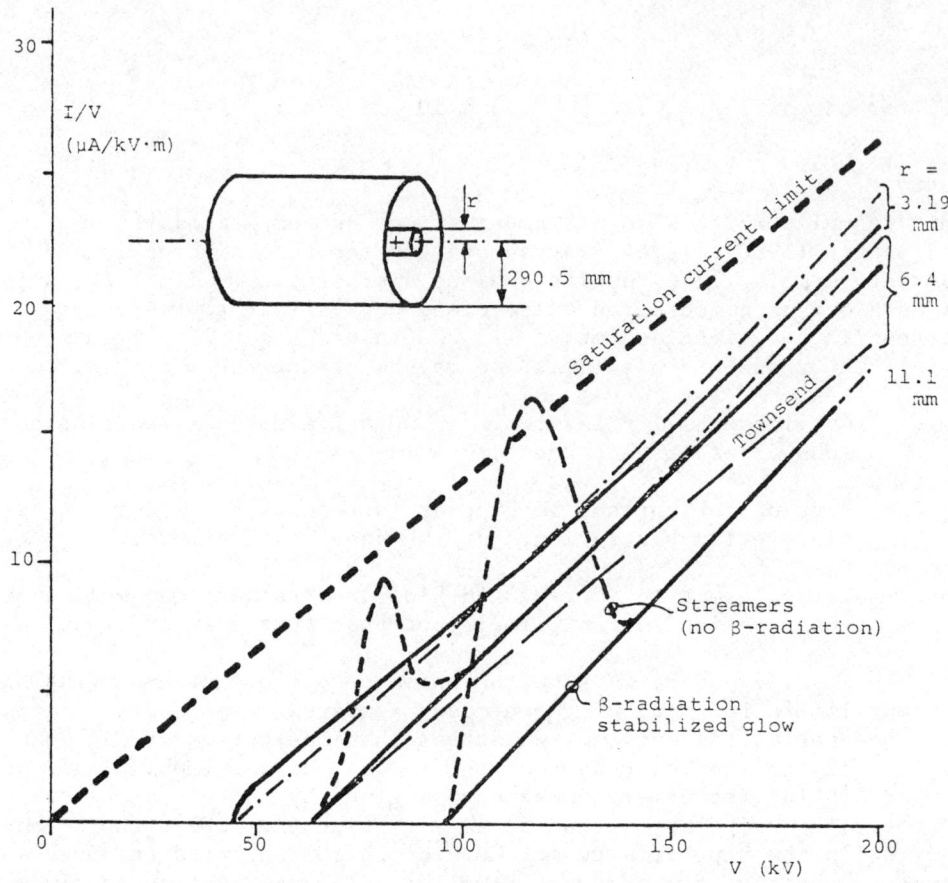

Fig. 7. Positive corona in ambient air between concentric cylinders of 290.5 mm outer diameter and varying inner diameter r (Sigmond, 1981). All calculated curves are based on a mobility value $2 \times 10^{-4} m^2/Vs$.

    ──────── Measurements (Waters and Stark, 1975).
    ──·──·── ≃ exact analytical formula (Waters and Stark, 1975).
    ──────── Townsend low-current approximation (Thomson and Thomson, 1933).
    ──────── Unipolar space charge saturation current limit.

distinguish coronas from other discharges also introduce peculiar diagnostic possibilities and difficulties:

• The corona geometry, widely different from a parallel-plate gap, makes it essential to be able to calculate properly the current

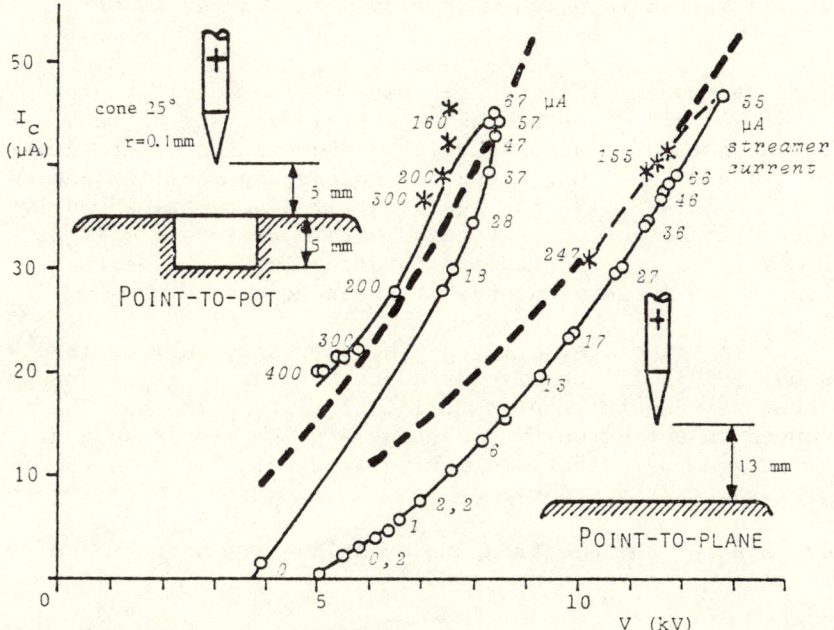

Fig. 8. The continuous ("background") component I_c of the corona current vs. corona voltage, for a 13 mm point-to-plane and a 5 + 5 mm point-to-pot gap in ambient air, compared with the unipolar saturation current curves (Goldman and Sigmond, 1981a; Sigmond, 1981). Numbers along the curves indicate time-averaged streamer currents.
o—o—— Streamer corona
----- Regular, periodic sparking
------- Unipolar space charge saturation current limit.

induced in the corona circuit by a charge moving in the gap, → the Shockley-Ramo theorem.

• The low field in the corona drift region means that the energy of the drifting ions is nearly thermal. This seems to make quantitative ion sampling for mass spectrographic analysis quite problematic, despite optimistic earlier reports (Shahin, 1969). Large, laminar flow extraction orifices sample the corona gas faithfully but may have undercooled jet regions where clusters may form. Molecular flow orifices, tested in the Elion laboratory, show mass discrimination effects which seem due to surface charges or even to surface catalytic activity.

• From a plasma physics point of view, the whole corona is one thick boundary layer with perhaps one little blob or filament of decent plasma inside. Thus, ordinary Langmuir probes and probe theory must be applied with great caution, or as mere potential indicators (Kondo, 1978).

• As mentioned, the electric wind will make difficulties for any retarding field or retarding potential analysis of ions. However, a combination of wind velocity measurement and retarding field analysis has also been exploited for corona ion mobility determination (Goldman et al., 1976). The electric wind carries nearly all the momentum imparted by the total electric field and will thus interfere seriously with attempts to determine the electrode fields by surface force ("field balance") measurements.

Of the above points, only the Shockley-Ramo theorem (Shockley, 1938; Ramo, 1939) will be discussed further, as it is of fundamental importance, very useful, but surprisingly little known. The current $i(t)$ induced in the circuit connecting a given electrode and a reference electrode by a charge q moving with velocity $\vec{v}$ in the interelectrode space, is found by:

(1) Placing the electrode in question at a potential of 1 volt relative to the reference electrode.

(2) Placing all other electrodes in the system at the reference electrode potential (0 volt).

(3) Finding the resulting Laplacian field $\vec{E}_1$ at the location of the moving charge. Fields due to this or other space charges do not contribute to $\vec{E}_1$, which is a purely geometrical parameter.

(4) Inserting:

$$i(t) = q\, \vec{E}_1 \cdot \vec{v} \qquad (25)$$

The theorem is derived under the assumption of infinite light velocity and is valid as long as the dimensions of the apparatus are much smaller than the wavelengths of the Fourier components of the induced current.

The theorem is illustrated by Fig. 9, showing the currents induced in the point and center probe circuits of a 16mm hyperboloid point-to-plane gap in atmospheric air when an O_2^- ion drifts across the gap (Sigmond, 1978). Note how both the point and the probe currents nearly equal the instantaneous charged particle currents leaving/arriving at the surfaces in this special geometry. The center probe arrangement of Fig. 9 was used by Marode (1961a) to investigate positive streamers and is discussed later.

Finally, let us mention the perhaps most widely used (and misused) diagnostic method of all: to study coronas in different gases

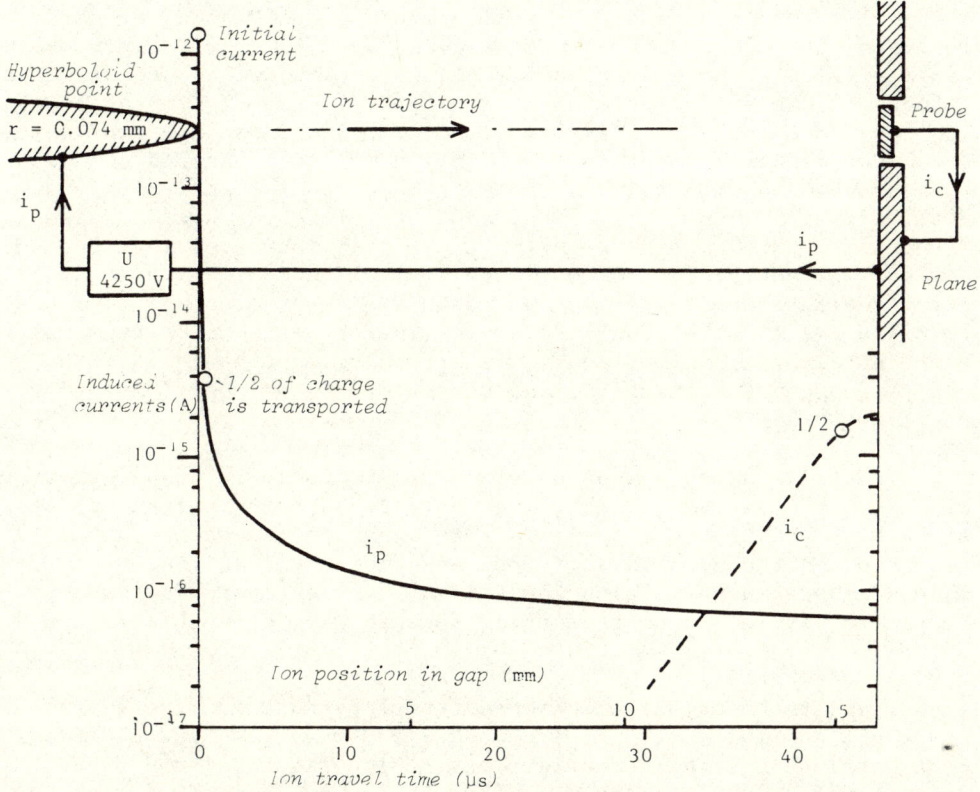

Fig. 9. Illustrating the Shockley-Ramo theorem: Point current i_p and center probe current i_c induced by one O_2^- ion drifting from the hyperboloid point to the plane through the Laplacian field in 87.6 kPa dry air (Sigmond, 1978).

and with different electrode materials in order to see the effects of known changes in specific properties, like γ, electron attachment, ease of photoionization, and the occurrence of metastables. A main uncertainty in these studies has always been the possible effects of unknown impurities in the corona gas. Processes affected are, in order of increased vulnerability:

(a) Drift velocities, diffusion coefficients, photon absorption. The effects of impurities are roughly proportional to the relative impurity contents.

(b) Electron attachment. Strongly attaching admixtures to nonattaching gases works proportionally to their amounts at low levels, but at low E/n values a saturation level is soon reached where the electrons stay attached over most of their drift path.

(c) Ionization and photoionization via the Penning effect. In gases with a metastable energy level U_m, a 10^{-8}-10^{-4} admixture of impurities with ionization potentials $U_i < U_m$ will have dramatic effects on the effective ionization coefficients. The metastables (or resonance excited particles with their imprisoned photons) created by electron collisions with the main gas will diffuse and collide in the gas until an impurity particle is found and promptly ionized. Small admixture contents are the most efficient, as these create a "Penning mixture" where the effective ionization level is among the lowest excited levels, which otherwise would compete with ionization for the electron energy. We may roughly state that it is very easy to clean a gas sufficiently to avoid impurity effects of type (a), quite hard for type (b), and very difficult for type (c). No one has yet done a satisfactory measurement of the primary ionization coefficient of He.

Thus, in the experimental studies cited below, for negative as well as for positive coronas, one must always bear in mind the possible effects of undetected impurities. For instance, a report about the presence of Trichel pulses or positive glow oscillations in an allegedly pure nonattaching gas will always be regarded with some suspicion, while a report of their absence will be much more conclusive.

As to the study of electrode effects by changing the electrode material, one must bear in mind that the corona gas, the electrode bulk metal, and certain constituents or impurities like H_2O, CO, O_2 and oil vapor, join together to form an electrode surface that is characteristic of neither of the constituents. The surface secondary electron coefficient γ, for instance, will be very sensitive to impinging ion energy and to clean-up effects by ions and heat. Firther, for potentials below 1-10 V, the surface will very often be an insulator or a semiconductor.

NEGATIVE CORONAS

Short Survey: Negative Coronas in Different Gases

The most characteristic common feature of all negative coronas is that the cathode borders on the ionization zone, thus securing for this region a prompt supply of secondary cathode electrons, unimpeded even by strongly electron-attaching gases. Positive space charge dominates the ionization region, while the drift region will have a weak or strong negative space charge according to the importance of electron attachment.

Ordinarily, the $\gamma_c = \gamma_p + \gamma_m$ will thus be high, $> 10^{-4}$, making avalanches $< 10^4$ electrons at self-sustainment levels of ionization region voltage. Thus, streamer conditions $N_e > 10^6 - 10^8$

will not be met except in pulsed, highly overvolted gaps. On the other hand, the differential resistance of the ionization region will be markedly negative at pressures above some kPa because of positive space-charge augmentation of the electron multiplication and cathode clean-up augmentation of the γ_c at higher currents. As shown by Fig. 5 and the discussion following Eqs. (6-8), the corona will then be stable, oscillating, or go to spark/arc according to the impedance R_d of the drift region. With coronas in electron-attaching gases, the influence of the external series resistance R_o is often small; if not, the gap + wiring capacitance C_o has to be minimized by careful design.

In nonattaching gases in short corona gaps, the high electron mobility may make R_d small enough to satisfy the inequalities (6) or (7), even at the smallest currents, if R_o is not large enough and C_o is not properly minimized. Such coronas would then seem to have an existence region of zero [see argon in Table 2 below (Weissler, 1943)]. Regrettably, for this case, as well as in many other experimental corona papers, the external circuit parameters are not adequately given.

Corona surveys are preferably still carried out by one experimenter, thus, hopefully ensuring the constancy of unknown parameters while varying the known. The most illustrative surveys of the effect of gas type on low voltage (< 20 kV), short point-to-plane coronas seem still to be those of Weissler (1943) and Das (1961). Both used outgassed Pyrex tubes with baked electrodes, and comparison with more recent specialized investigations suggests that their gases were satisfactory regarding electronegative impurities although not regarding Penning effects. The Pt points were cylindrical with polished hemispherical ends of radii (Weissler/Das) 0.25 mm/0.5 mm; the plates, 40 mm dia. Pt/70 mm dia. Ni; and the gap lengths, 31-46 mm/25 mm. Weissler specifically states that all positive and negative corona measurements in the same gas were done under identical conditions. The results of Weissler concerning both polarities are summarized in Fig. 10 and Tables 2 and 3.

Weissler and Das', most important negative corona result is the appearance of Trichel pulses whenever the gas contains even traces of electron-attaching gases. Weissler's results indicate further that in air, cathode feedback (γ_c) in negative coronas is less efficient than feedback to the gas (γ_g) in the corresponding positive coronas. The slow change from diffuse cathode regions at low currents to concentrated cathode regions at high currents depicted in Fig. 10 is ascribed to cathode surface clean-up. External circuit differences are probably the cause of the only marked deviation between Weissler and Das: Weissler finds direct spark breakdown in pure Ar but steady glow in Ar + 0.3 % N_2, while Das finds up to 10 % N_2 can be added to Ne and Ar without preventing direct spark breakdown. The role of hydrogen in mixtures with oxygen is

Table 2. Essential Data of the Negative Point-to-Plane Corona in Different Gases and Mixtures at Atmospheric Density (Weissler, 1943).

Gas	Localized corona			Trichel pulses	Approx. breakdown potential (kV)	Peculiarities
	Onset potential (kV)	Onset current (μA)	Visual character			
H_2 pure [1]	2·6	140	Unsteady glow	None	—	Alternation of uniform and localized glow
$H_2 + 0.1\% O_2$	4·2	70	Unsteady glow	Yes	—	As above
$H_2 + 0.6\% O_2$	5·8	2	Unsteady glow	Yes	—	As above
N_2 pure [1]	3·7	90	Steady glow	None	14	—
$N_2 + 0.1\% O_2$	4·3	50	Steady glow	Yes	—	—
$N_2 + 1\% O_2$	5·5	0·1	Intermittent [4]	Yes	—	5 to 7 kV intermittent corona
Room air [1]	8 [3]	0·1	Intermittent	Yes	—	8 to 11 kV intermittent corona
A pure [2]	—	—	—	—	3·5	No corona
$A + 0.3\% N_2$	2·5	200	Steady glow	None	4·1	—
$A + 1\% N_2$	3·9	250	Steady glow	None	—	—
$A + 0.1\% H_2$	3·2	400	Unsteady glow	None	4·1	—
$A + 1\% H_2$	3·3	300	Unsteady glow	None	—	—
$A + 0.1\% O_2$	3	1	Steady glow	Yes	—	—
$A + 0.4\% O_2$	3	0·1	Intermittent	Yes	24	3 to 6 kV intermittent corona
$A + 1\% O_2$	3·3	0·1	Steady glow	Yes	—	—

[1] Gap length 3.1 cm
[2] Gap length 4.6 cm
[3] The value may be a misprint, as Weissler in his paper mentioned that "in gases with efficient photoionization we find the same onset potentials for both signs of corona.... Pure argon confirms this even stronger than air." Compare Table 3.
[4] Due to lack of triggering electrons.

strangely ambiguous. Workers measuring primary ionization coefficients add H_2 to O_2 to remove negative ions by rapid associative detachment, while corona investigators, with equal success, add small amounts of O_2 to H_2 to obtain negative ion effects. The negative ion involved, probably OH^- (Elford and Rees, 1974), apparently must be unstable at higher values of E/n.

The "Onset current" column of Table 2 shows the lowest self-sustained current possible in Weissler's set-up. Any doubt about the free-electron character of his coronas in alledgedly nonattaching gases is dispelled by comparison with the unipolar saturation current limit (23), yielding a maximum of about 2 μA for a negative ion corona across 40 mm at 4 kV. A comparison with the "Current"

Table 3. Essential Data of the Positive Point-to-Plane Corona in Different Gases and Mixtures at Atmospheric Density (Weissler, 1943).

Gas	Pre-onset streamers	Onset potential (kV)	Visual character	Burst pulses	Current[3] (μA)	Onset of breakdown streamers (kV)	Approx. breakdown potential (kV)
H_2 pure [1]	None	3·5	Localized[4]	None	4	10	18
$H_2 + 0·1\% O_2$	Yes	5	Intermittent Streamer-corona	None	—	8	—
$H_2 + 0·5\% O_2$	Yes	4·5	Streamer-corona	None	3	9	—
$H_2 + 1\% O_2$	Yes	4	Uniform[5]	Small	3	15	18
N_2 pure [1]	None	4·8	Localized	None	1	11	20
$N_2 + 0·2\% O_2$	Incipient streamers	4·5	Streamer-corona	None	1	9·3	—
$N_2 + 1\% O_2$	Incipient streamers	4·8	Localized; for higher volts uniform glow	Yes	1	12·5	20
Room air [1]	Yes	5·5	Uniform	Yes	—	15	—
A pure [2]	None	None	None	None	—	—	3·5
$A + 0·1\% H_2$	Yes	3·3	Localized	None	0·5	—	—
$A + 0·5\% H_2$	Yes	3·5	Localized	Small	0·5	—	—
$A + 1\% H_2$	Yes	3·45	Localized	Small	0·5	—	—
$A + 0·3\% N_2$	None	None	None	None	—	3·5	3·5
$A + 0·5\% N_2$	Yes	3·6	Streamer-corona	Small	—	4·6	4·6
$A + 1\% N_2$	Incipient streamers	3·5	Localized	Small	0·5	6·5	6·5
$A + 0·1\% O_2$	Yes	3·8	Streamer-corona	None	—	4·1	4·1
$A + 0·4\% O_2$	Yes	4·5	Streamer-corona	Small	0·6	—	7
$A + 1\% O_2$	Yes	4·0	Uniform	Small	0·6	—	15

[1] Gap length 3.1 cm
[2] Gap length 4.6 cm
[3] Current at a potential ≃ 20 % above the corona onset.
[4] Localized glow, due to electron avalanches approaching the point in the highest field region.
[5] Uniform glow, caused by the burst pulse corona.

column of Table 3 (positive coronas) is also most instructive.

The average-current vs. voltage curves found by Weissler and Das were, as usual, very compatible with the Townsend type formula (Rajapandyan and Raju, 1972).

Loeb (1967) reported that negative corona gaps in all free electron gases in the range 0-30 kPa investigated suffered spark breakdown at onset, caused by "an entirely new breakdown sequence, hitherto not reported, and one which is not a streamer process." These observations seem easily explained by our stability considerations above and in an earlier section.

Finally, we mention the recent work by the Tartu group (Korge et al., 1979) on negative point-to-plane corona in pure stagnant

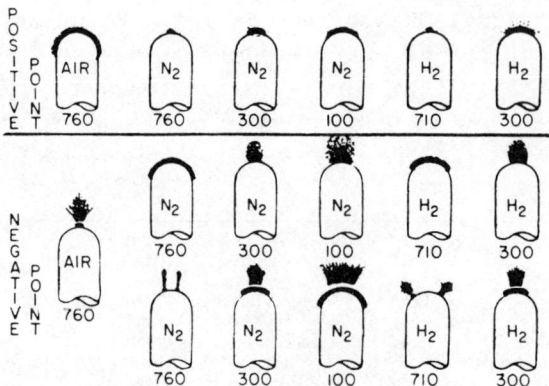

Fig. 10. Visual forms of corona discharges under indicated conditions in the onset region (Weissler, 1943). The forms in the second row show the uniformly spread discharge at low currents that contracts to the concentrated forms of discharge with high currents, seen in the last row. Gas pressures in Torr (1 Torr = 133 Pa).

nitrogen. Figure 11 shows the strange current-voltage curve obtained for this continuous-glow type corona in a 40 mm gap at 103 kPa. Add to this the fact that lower A-E part is slowly drifting in time so that at 6.05 kV one registers a 25 min "formative time lag" from 2 nA at point B to 12 μA near C, and one then appreciates that there is more between cathode and anode than is told in the present simple tale. From point E-F, the current stays at about 50 percent of the saturation current from Eq. (23) for electrons in N_2 (mobility 5.5 x $16^{-2} m^2/Vs$).

Negative Coronas in Air and Other Electron-Attaching Gases

General Remarks. An Air Cornona Computer Model. Air is still economically the most important corona medium, readily available, and with so many natural "impurities" (H_2O, CO_2) that it is fairly insensitive to the addition of more (though corona in a stagnant volume of air introduces changes too drastic (Gosho, 1980)). It is the most widely studied corona gas and, for gaps in the meter range, the only one. This subsection will therefore concentrate on air coronas, with interpersed comments on coronas in other gases of importance, like SF_6.

A glance at Fig. 3 tells us that air is a weak electron attaching gas, compared to, for instance, SF_6. In atmospheric density drift regions, an electron will suffer typically 0.2 - 1 attachments per mm drift path. Also, it is sensitive to density variations below about $E/n = 50 \times 10^{-21}$ Vm^2 and may induce attachment instability

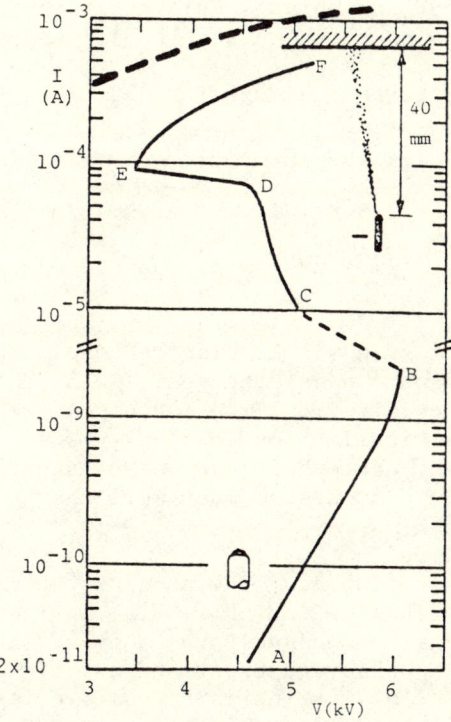

Fig. 11. Current-voltage curve for negative point-to-plane corona in 103 kPa nitrogen (Korge et al., 1979). The curve A - E is time-dependent and badly reproducible. Also drawn (in heavy, dashed line): The electron space charge saturation current limit (Sigmond, 1981).

between 40 and 90 x 10^{-21} Vm^2. A last point of importance is that α' for air is substantially lower than for SF_6 in very high fields, which indicates that air may be less susceptible than SF_6 to breakdown from very fine points. Apart from this, a study by Dupuy and Gibert (1981, personnal communication) indicates that SF_6 has corona properties very similar to air at five times higher density.

To give a more quantitative picture of processes and charge distributions in coronas and to serve as a basis for further discussion, Fig. 12 shows a computed solution (Sigmond, 1978) of the continuity equations along the axial Laplacian field line in the negative point-to-plane corona gap of Fig. 9, in 87.6 kPa dry air. It is based on conversion coefficients and mobilities presented in Sigmond (1978), with a gap voltage about equal to the experimental onset voltage. Note that the distance scale is logarithmic, which makes the relative potential V/V_o nearly a linear function of the abscissa metric scale, but also gives undue visual weight to the charge densities near the point. Relevant transit times from the ionization/drift region boundary to the electrodes are:

Positive ions, to the cathode : $\simeq 0.2$ μs

Free electrons, to the anode : $\simeq 1$ μs

CO_3^- ions, to the anode : $\simeq 500$ μs

with shorter transit times for electrons and negative ions at high currents.

The development of a single electron avalanche s e$^-$, and of the densities s n_+ and s n_- of the positive and negative ions left behind, are also shown in Fig. 12. The dipole character of this initial ion density distribution is clearly seen. However, after a hypothetical steady state has been established, the negative ion density distribution n_- contains much more charge than the positive n_+.

The main negative ion near the anode is seen to be CO_4^-. The curves in the higher E/n region show that, in the increased drift region field at higher corona voltages and currents, CO_3^- and even O_3^- might dominate, as observed experimentally in a 30 mm air gap by Gardiner and Craggs (1977). The main positive ion is O_2^+, as the primarily formed N_2^+ rapidly converts to O_2^+ and NO^+ (not included in the model) by charge exchange and reaction.

The unipolar saturation current limit for this gap is, according to Eq. (23), I_s (μA) = 0.3 V^2 (kV^2), assuming a mobility μ = 2.7 x 10^{-4} m^2/Vs. However, currents appreciably in excess of this are observed at voltages some kV above onset, indicating increasing current transport by free electrons through the drift region. This is readily explained by the weakness of electron attachment in air, as discussed above.

<u>Air Corona Regimes Existence Diagrams.</u> The most extensive general survey of point-to-plane coronas in air is that of Miyoshi's laboratory (Sakai et al., 1958; Miyoshi et al., 1964) which covered the density range 300-27000 x 10^{21} m^{-3} (1.3 - 102 kPa), gap lengths 1 - 30 mm, point radii 0.1 - 5 mm, and voltages <300 kV. The air was dry (filtered through P_2O_5), and the point materials were brass, Ag, Al, Fe, Pt and W. However, no corona dependence on point material was mentioned.

Only two of the many illuminating graphs can be shown here. Figure 13(a) and (b) are existence diagrams, each for several point radii respectively below and above a certain radius r_G (1.2 mm at atmospheric density), while Fig. 14 shows voltage-current character-

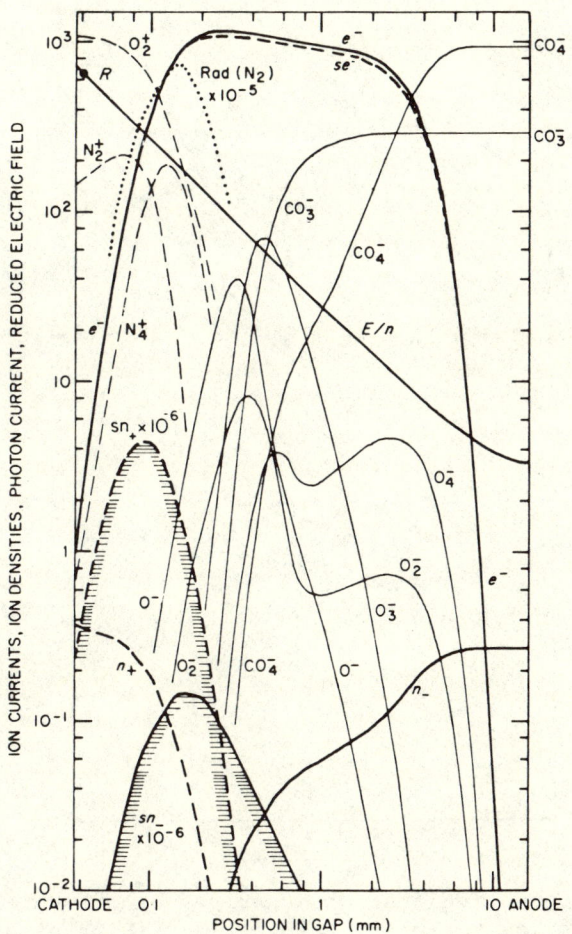

Fig. 12. Computed solutions to the continuity equations along the axial Laplacian field line in a negative point-to-plane corona gap in atmospheric dry model air, at 87.6 kPa 0°C (Sigmond, 1978). Gap dimensions: as in Fig. 9. Voltage: 4250 V (~ corona onset voltage).
se^-, sn_+, sn_- : Electron number in a single avalanche, started from the cathode and summed positive and negative ion densities left in its track, before ion drift, detachment, or conversions set in.
n_+, n_-: Summed positive and negative ion stationary densities, assuming a cathode emission current density of one electron per m^2 per s.
e^-, O^-, N_2^+, etc.: Electron and ion stationary currents, assuming a cathode emission of one electron per s.
R: Rogowski point.

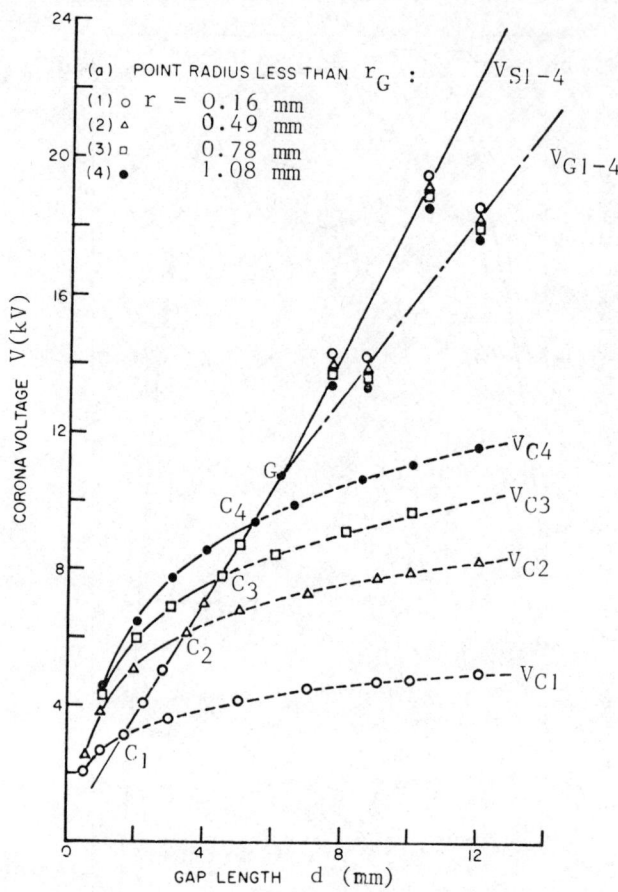

Fig. 13a. Existence regions for negative point-to-plane corona forms in atmospheric density air (Miyoshi et al., 1964). Part B of this figure is on the following page.
V_C : Trichel pulse onset
V_G : Pulseless glow onset
V_S : Spark breakdown

istics for the same coronas. In Fig. 13, solid lines mark spark breakdown voltages V_S, dashed lines mark Trichel pulse onset voltages V_C, and dot-dash lines mark continuous glow onset voltages V_G, all as functions of the gap length d. Note that sparking occurs along the lines C - C' for $d_C < d < d_{C'}$, while for $d_{C'} < d < d_G$, one randomly gets direct sparking at the V_G line C' - G' or first glow at V_G and then sparking at V_S.

The observations of the Miyoshi laboratory show some remarkable and illuminating features:

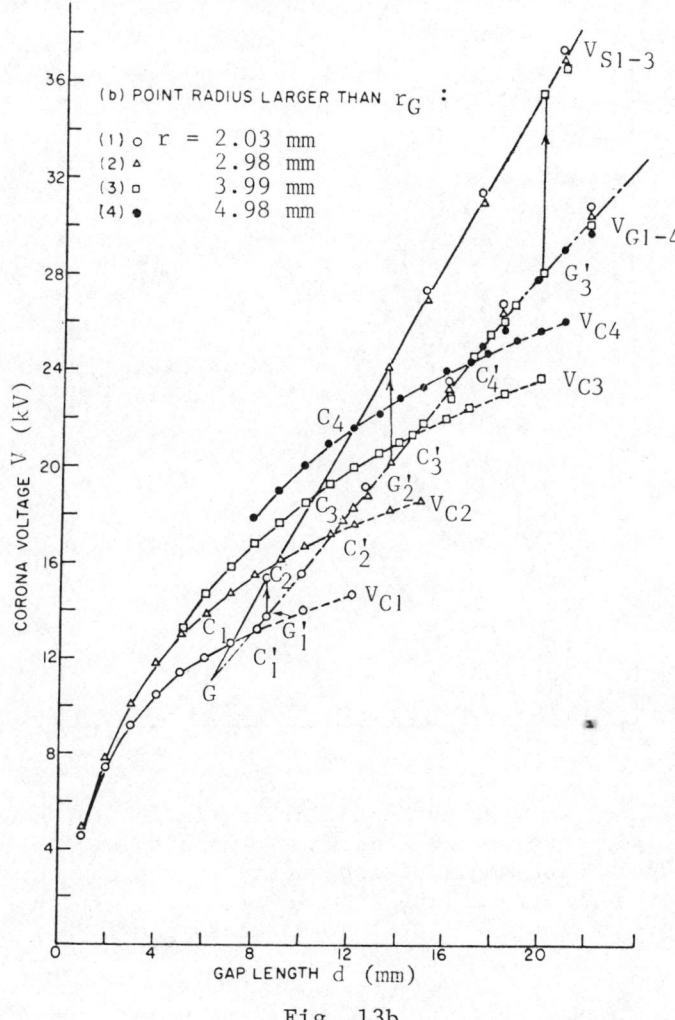

Fig. 13b.

• *The curves for the lowest discharge threshold (onset) voltage vs. gap length, shown as solid lines passing into dashed lines, are markedly dependent on the point radius and join each other without discontinuities or changes of slope.* These lines clearly mark the voltages where $\mu_r = 1$ and the ionization region breaks down, while the resulting discharge form (spark or Trichel) depends on the length (impedance) of the drift region. Direct sparking occurs when the inequalities (6) or (7) are satisfied already at onset, i.e. at small gap lengths where the drift region resistance R_d is lowest.

• *The transition from Trichel coronas into spark or continuous glow occurs along voltage vs. gap length curves that are independent*

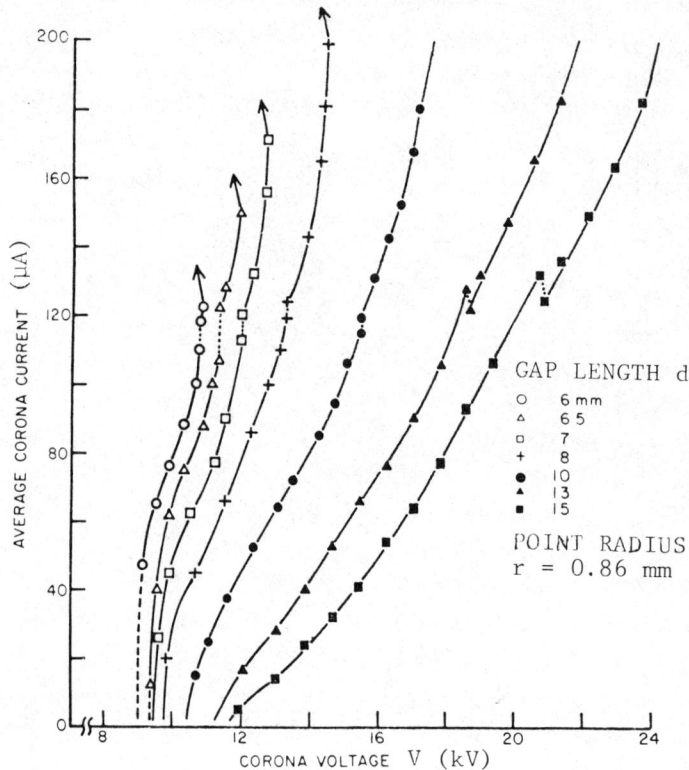

Fig. 14. Typical voltage-current characteristics for negative point-to-plane corona in atmospheric density air (Miyoshi et al., 1964). The small discontinuity in the characteristics marks the transition from lower current Trichel pulse corona to higher current pulseless glow.

on the point radius. The direct Trichel-to-spark transitions are confined to the short gap lengths, while the Trichel-to-glow transitions take place at currents somewhat greater than the unipolar ion saturation current limits. The spark threshold currents (at constant gap lengths) decrease with increasing point radii. This shows that the ionization region discharge, once formed, must have electrical circuit parameters that are fairly independent of the point radius. Not that even the smallest radius employed (0.1 mm) is appreciably larger than the normal cathode fall thickness of a glow discharge in atmospheric density air. Thus, the transition from the Trichel to higher current regimes will depend primarily on the drift region. At the spark breakdown voltage, inequalities (6) or (7) are just satisfied, i.e. $\Delta V/\Delta I \simeq 0$. Thus, a large ΔI causes a small ΔV, and the larger currents necessary to level out the drift region field at smaller point radii can be passed without great change of voltage.

• *The average current vs. voltage curves of Fig. 14 are remarkably unaffected by the Trichel-continuous glow transition.* This shows once again that the Trichel pulses are a phenomena localized around and in the ionization region, without much influence on the corona as a whole.

Negative streamers are not found in these or other existence diagrams for short, nonpulsed corona gaps. The concept originated and still causes discussion for parallel-plate gaps, and they definitely are observed in long rod-to-plane gaps (see other papers in this book). However, they need (in atmospheric air) about 1.8 MV/m field for stable propagation, opposed to about 0.5 MV/m for positive streamers. Thus, positive streamers from the plane cathode are more probable than negative streamers from the point in negative air coronas.

Onset Phenomena and Trichel Pulses. Single avalanches in a negative point-to-plane corona gap in air are too small ($\sim 10^4$ electrons) to be readily seen on an oscilloscope screen. However, near onset, trains of avalanches connected by photon (γ_p) feedback to the cathode merge together into "equivalent" avalanches, which by stochastic fluctuations and positive space charge augmentation can grow into easily detectable, random pulses. The positive space charge rapidly disappears at the cathode, while the accumulated negative ions further out in the gap suppress further pulses while they drift to the plane anode (see transit time information above). The next pulse must also wait for a new seed electron. A little above onset ($\mu_r \gtrsim 1$), but still in the autostabilization regime, the discharge still may die out because of negative ion quenching, but now the first pulse after reignition will be large with all characteristics of a concentrated negative glow discharge, including heavy cathode sputtering (Buchet and Goldman, 1970); see Fig. 15.

By increasing the average corona current in air, O_2 or CO_2, the pulses become smaller and very regular. These so-called Trichel (1938) pulses have pulse shapes approximately independent of the corona average current and consequently, a repetition frequency proportional to the current, except near to the Trichel-continuous glow transition. For atmospheric air coronas between a negative hemispherically capped cylinder of radius r and a plane at distance d, Fieux and Boutteau (1970) find the Trichel pulse amplitude to be in the 0.5 - 5 mA range, and the respective frequency:

$$F \text{ (kHz)} = 2.27 \text{ I } (\mu A)/r(mm) \tag{26}$$

valid for $r \geq 0.125$ mm and $d \geq 10$ mm.

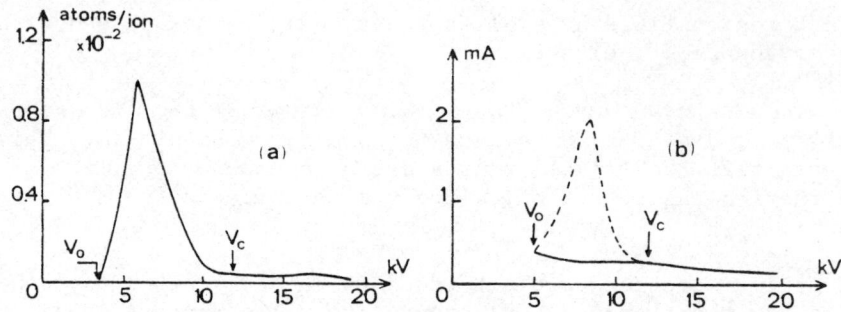

Fig. 15. Properties of the negative point-to-plane corona in atmospheric density air (Buchet and Goldman, 1970). Gap dimensions: d = 20 mm, r = 0.065 mm.
a) Sputtering rate at the point vs. applied voltage
b) Trichel pulse heights (solid line) and large pulse heights typical of the autostabilization regime (dashed line).
The similarity between the sputtering curve and the large curve shapes of pulse heights illustrates the correlation existing between both phenomena.

Typical Trichel pulse shapes for air and O_2 are shown in Fig. 16. The O_2 pulse shapes shown are also similar to these in air at atmospheric density. The Trichel pulse risetimes have been studied most exhaustively by Zentner (1970 a,b) and are ≃ 1.5 ns in atmospheric density air for point radii above 0.02 mm, increasing with decreasing air density.

The spatio-temporal development of the light emission during the Trichel pulse stage has been very thoroughly investigated by Ikuta and collaborators (Ushita et al., 1968; Ikuta et al., 1975), as well as in Sigmond's laboratory (Bugge and Sigmond, 1969; Torsethaugen and Sigmond, 1973), all working in air or N_2 - O_2 mixtures at pressures mainly below 10 kPa, where the processes are conveniently extended in space and time. Figure 17 shows spatio-temporal crossplots of the development of a Japanese Trichel pulse, while Fig. 18 shows image intensifier stroboscopic exposures of a Norwegian one. In all cases, the identified spectral bands belong to the second positive system of N_2 and the first negative system of N_2^+, with the latter radiation only found in the "negative glow" that characterizes the Trichel pulse.

A description of the physical processes occurring in the Trichel pulse cycle in air is best started when, after a pulse, the positive space charge has drifted into the cathode and the remaining free electrons have reached the anode. The breakdown (μ_r = 1) voltage

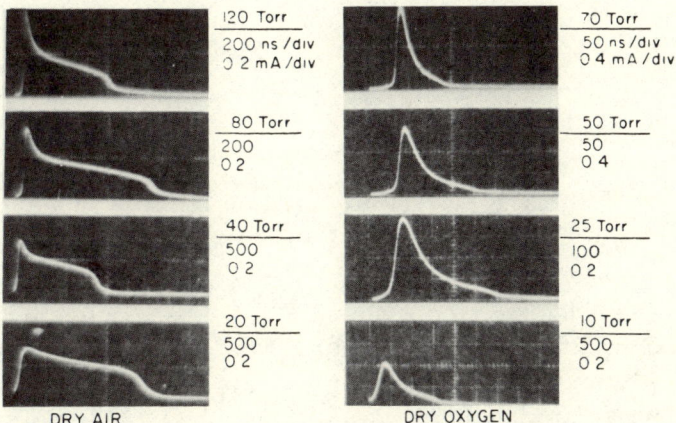

Fig. 16. Trichel current pulses in dry air and oxygen at different low densities (1 Torr = 133 Pa) (Bugge, 1968). Gap dimensions: as in Fig. 9 with Mo cathode point.

of the ionization region has risen to the Laplacian-field value, well above the voltage actually available, which is kept low by the negative space charge accumulated in the drift region during preceding Trichel pulses. However, optical observations (Zenter, 1970 a,b; Bugge and Sigmond, 1969; Torsethaugen and Sigmond, 1973) show that a non-self-sustained discharge is still burning in the ionization region, with a spatial light density distribution in excellent agreement with computations for a Laplacian field, as that in Fig. 12 above. The cathode electron current I_o driving this discharge falls off as $t^{-3/2}$, where t is the time elapsed since the end of the preceding pulse, thus indicating I_o to be caused by diffusing metastables formed by that pulse (Sigmond, 1978).

While I_o is decreasing, the negative space charge in the drift region is transported away, raising the ionization region voltage towards the breakdown ($\mu_r = 1$) value, at a rate dependent on the gap voltage. When μ_r reaches unity, positive space-charge intensification of the cathode region field sets in, raising μ_r above 1 and rapidly increasing the current into the mA range. The rise is checked by the rapid disappearance of positive space charge into the cathode and accumulation of negative ions outside the ionization region. However, the simultaneous contraction of the discharge into something very akin to a classical negative glow spot also seems to lower the burning voltage of the ionization region (see curve B of Fig. 1), so that the discharge persists for some time before disappearing, often abruptly as shown in Fig. 16 and discussed further in Sigmond (1978).

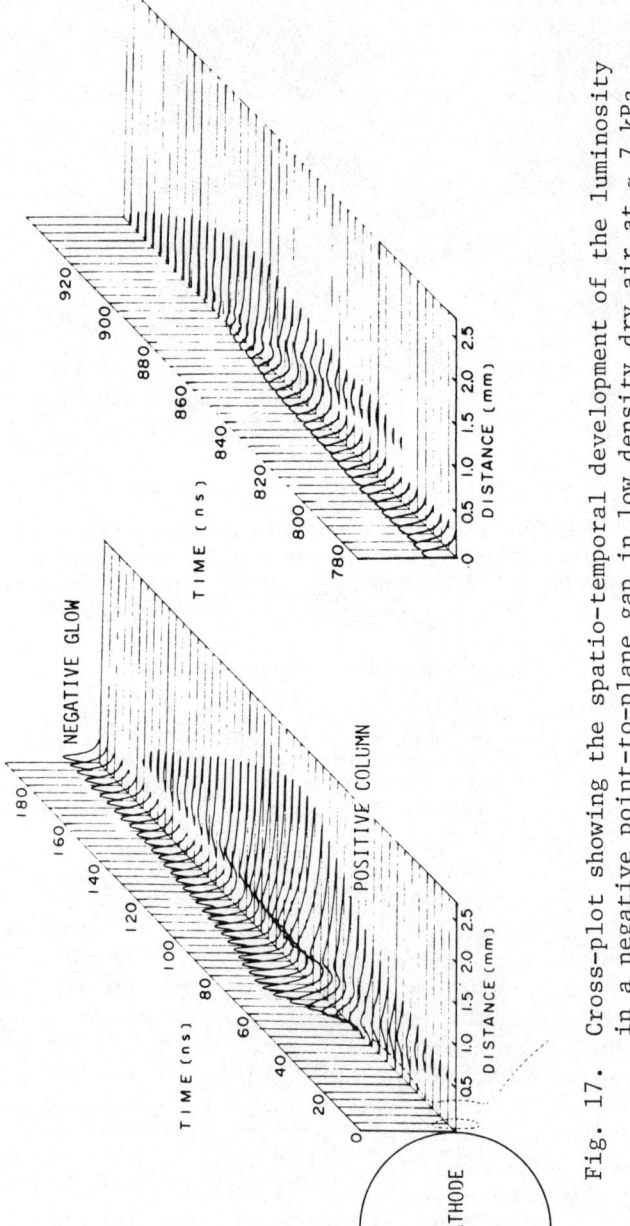

Fig. 17. Cross-plot showing the spatio-temporal development of the luminosity in a negative point-to-plane gap in low density dry air at ~ 7 kPa 0°C (Ikuta et al., 1975). Gap dimensions: d = 30 mm, r = 1 mm. Voltage: 2600 V. Note the disappearance of the positive column ("brush") during the Trichel pulse plateau phase.

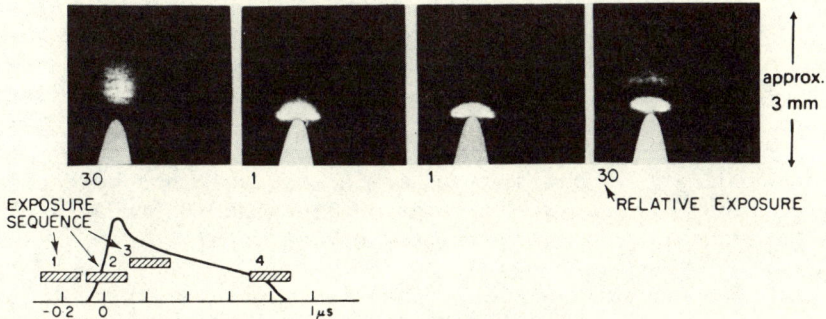

Fig. 18. Image intensifier frames of the Trichel pulse luminosity development in a point-to-plane gap in a low density 50% N_2 - 50% O_2 mixture at ~ 5 kPa 0°C (Torsethaugen and Sigmond, 1973). Gap dimensions: as in Fig. 9, but with Au point r = 0.046 mm. Voltage: 1460 V. The horizontal "striations" are of instrumental origin.

It is important to realize that the repetition period of regular Trichel pulses usually is shorter than the negative ion transit time, so that several negative ion clouds from former pulses may be drifting simultaneously (Buchet et al., 1966; Lama and Gallo, 1974). In fact, the pulses get irregular for longer periods, and the discharge passes into the autostabilization regime.

In O_2 and CO_2 regular repetitive Trichel pulses are found. In SF_6, however, Trichel-like pulses form but are very irregular both in time of occurrence and in amplitude, just as if the mechanism described above was operative without sufficient cathode seed electrons to start a new pulse exactly when the gap recovered and μ_r passed unity.

An example of the current density distribution over the plane in a negative Trichel pulse point-to-plane corona in air is shown in Fig. 6 (Goldman et al., 1978). The reasons for the current density dip in the center are not yet known.

<u>The Negative Continuous Glow and the Transition to Spark</u>. The transition from the Trichel pulse to the negative continuous glow regime is no dramatic event (see above). When raising the current, the Trichel pulses on the oscilloscope screen are seen to approach each other and, at the same time, to lengthen. When they nearly touch, trains of pulses may join in long, steady plateau pulses with separate Trichel pulses in between. A very small further increase in current makes all pulses join in one continuous plateau, the continuous glow.

The negative continuous glow corona has been studied quite thoroughly by Kondo and Miyoshi (1978) and by Kurimoto and Farish (1980). As this is a relatively high-current discharge, thermal gas density effects start to be important, and a 10-12 percent density reduction is observed in the center. As shown in Fig. 6, this does not seem to change the average current density distribution away from the Warburg law, but strong current fluctuations were observed in the center. Figure 19 shows one of Fieux and Boutteau's photographs of a negative continuous glow corona in air.

Apart from existence limits, mentioned previously, not much is known about the transition from Trichel pulse or negative continuous glow coronas into the spark. As noted, negative streamer formation seems hardly probable in nonpulsed short corona gaps. Figure 20 shows how, instead, luminous spots, which should become positive streamers at atmospheric pressure, originate at the plane when a negative point-to-plane gap approaches breakdown in air at pressures lower than atmospheric pressure where the processes are conveniently extended in space and time. This phenomenon occurs only after a conditioning period leading to the growth of microcrystals on the surface (Goldman et al., 1965).

POSITIVE CORONAS

General Features of Positive Coronas

<u>Visual and Electrical Aspects</u>. The most spectacular aspect of positive coronas is the bright streamer filaments. Their luminosity, their crackling noise, and the ozone they give off are known by everybody who handles high voltage equipment. For those who work, for instance, with electrical insulation, in radio-engineering or aircraft, they may also be symptomatic of trouble since, due to their

Fig. 19. Typical aspects of a negative point-to-plane corona in atmospheric density air (Fieux and Boutteau, 1970). Gap dimensions: d = 100 mm, r = 2 mm. Left, in the Trichel regime, and right, in the glow regime, at 55 kV and 195 kV, respectively.

CORONA DISCHARGE PHYSICS

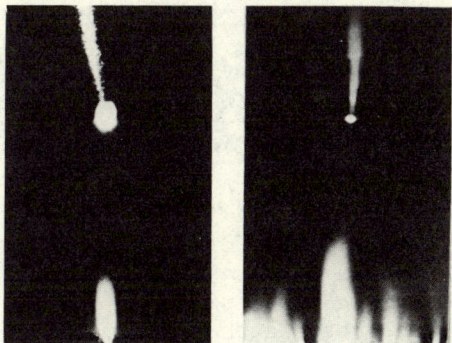

Fig. 20. Anodic phenomena observed in negative point-to-plane corona in air below atmospheric pressure (66 kPa), after a conditioning period (Goldman et al., 1965). Gap length: 20 mm. Point Au cone 26°, plane Cu. Voltage: 13 kV, increased at 13.5 kV at the end of the conditioning period. Conditioning period: 1 hour for the picture at left, 9 hours at right.

high current amplitude (10-200 mA) and their extension outside the conductors (1-100 cm), they are also a source of energy losses, insulation failures, and electromagnetic and acoustic noise (Gary and Moreau, 1976). The other types of corona pulses, Trichel pulses (see earlier section) and positive glow oscillations (see later), cause less damage because of their smaller current amplitudes and spatial extension. Thus, many engineering corona studies have been devoted to the electrical behavior of gaps where positive corona can take place and eventually lead to spark breakdown. Even regarding the average current flowing at spark breakdown, positive coronas are worse than negative from an insulation point of view: positive point-to-plane gaps in air break down under DC voltages at $\simeq$ 100 µA while negative point-to-plane gaps break down at $\simeq$ 200 µA.

Two different aspects of positive point-to-plane streamer coronas in short gaps have been shown in Fig. 4: one with multiple branching typically obtained with pulsed voltage (frame d), the other with a straight single channel produced under DC voltage by repetitive streamers which all take the same path. Even without the help of residual ions, one passes progressively from the branched to the axial structure by using voltage pulses with an increasing repetition rate (Fig. 21). Two processes involving the neutral gaps may play a role in the formation of a preferential path for the successive streamers:

• The creation of a lower-density air channel by the average axial power dissipation, which normally is 0.1-0.5 W/cm (Sigmond and Goldman, 1981);

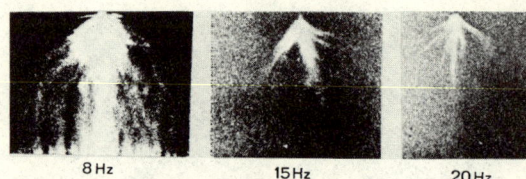

8 Hz   15 Hz   20 Hz

Fig. 21. Photographs showing the influence of the preceding discharges on a positive streamer discharge operated in atmospheric density air with voltage pulses at 8, 15 and 20 Hz, respectively, and with a sweeping voltage in order to eliminate the residual ions between the pulses (Berger, 1974). Gap dimensions: d = 10 mm, r = 0.04 mm. The branching decreases with increasing repetition rate and disappears under DC voltage. When long pulses with very short rise time are used, only the first discharge is branched (Marode, 1975a).

· The persistence between successive streamers of vibrationally excited metastables acting as a reserve of energy for superelastic collisions with the electrons which will participate in the formation of the following streamers (Hartmann and Gallimberti, 1975).

In long gaps, the constriction operates only along a certain length of the gap (Bogdanova and Pevchev, 1971).

Many of the observations cited below have been made on repetitive streamers propagating along a unique path; well defined in space and time, these streamers are also well suited to optical and electrical measurements on the spatio-temporal development of the streamer which, as established by Marode (1975a), does not fundamentally differ for single and branched short streamers. The main difference lies in the stem current, approximately proportional to the number of branches alive at one time.

According to the applied voltage, a positive streamer may stop and disappear midgap or reach the cathode, as shown in Fig. 22, for a positive point-to-plane air gap. Between the onset corona voltage V_o and the spark breakdown voltage V_S, three different regimes are here observed:

· Between V_o and V_G, the autostabilization regime with burst pulses and small-extension streamers ("preonset streamers"[*], frames A-B-C);

---

[*] According to the terminology used in the pioneering studies on coronas. Also applies to "prebreakdown streamers" and "breakdown streamers."

CORONA DISCHARGE PHYSICS 43

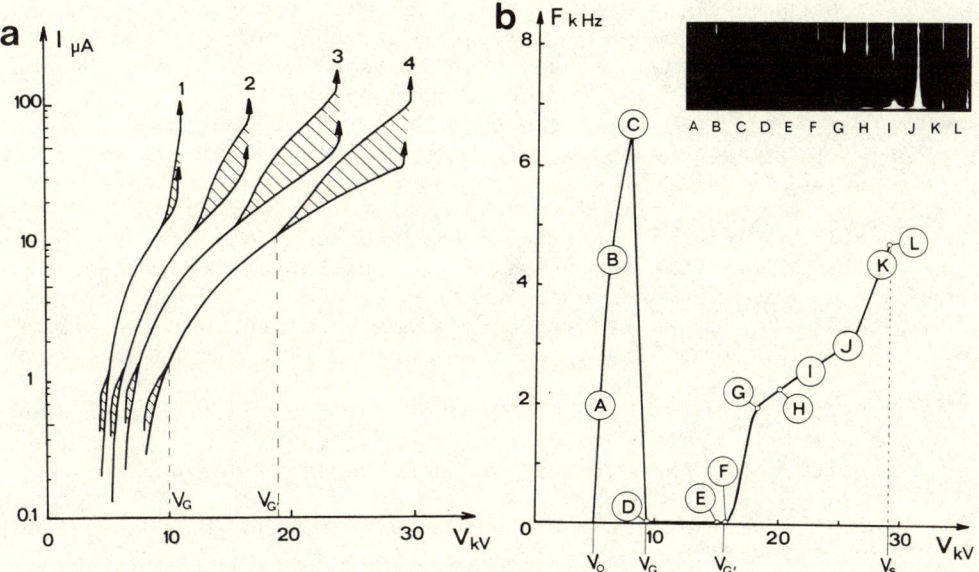

Fig. 22. Typical characteristics for positive point-to-plane coronas in atmospheric density air, illustrating the different regimes
$V_0 - V_G$: autostabilization regime
$V_G - V_{G'}$: continuous glow corona
$V_{G'} - V_S$: streamer corona
a) Voltage-current curves for different gap lengths, d = 10, 15, 20, and 30 mm (corresponding to the curves marked 1, 2, 3, and 4, respectively) and point radius r = 0.2 mm (Buchet et al., 1966). The shaded regions correspond to the streamer average current. The arrows mark the transition to spark breakdown.
b) Streamer pulse frequency as a function of the applied voltage for d = 31 mm and r = 0.17 mm. In the inset at the top is the luminous aspect of the discharge at each of the reference marks indicated on the curve (Hartmann, 1964).

• Between V_G and $V_{G'}$, a perfectly stabilized regime without streamers (frames D-E);

• Between $V_{G'}$ and V_S, a stabilized regime with regularly time-spaced streamers vanishing before hitting the cathode ("prebreakdown streamers", frames F-G) or crossing the whole gap ("breakdown streamers", frames H-L).

The streamers are always associated with current pulses with voltage-dependent frequency (curve of Fig. 22b) and combined with a glow, which is clearly seen in the frames D and E in the absence of streamers and produces the continuous (background) current discussed earlier. Figure 22a gives the relative importance of the average currents corresponding to each of the two components in the case considered.

In AC coronas, necessarily used when insulators are present in the discharge gap, the same type of results is obtained. There is then a glow current synchronous with voltage but, to measure it, it is necessary to compensate the displacement current which may be sufficiently high to hinder the measurements (Crochet et al., 1980).

Existence Domains. We have shown that the positive glow constitutes a stabilizing factor for the discharge. Further, it should be a condition for the existence of a DC positive corona.

The discharge nature differs according to the gas composition, its pressure, and the gap geometry.

In electron-attaching gases and in gaseous mixtures, streamers develop in the gas, but even today their onset conditions are not clearly known. Mysteries remain, for instance, concerning the successive aspects exhibited by corona discharges when small amounts of O_2 are added to N_2. See Table 3 which gives essential data on low voltage positive point-to-plane corona in different gases, pure and mixed with low amounts of other gases, at atmospheric pressure:

N_2 localized glow

$N_2 + 0.2\% \, O_2$ streamers

$N_2 + 1\% \, O_2$ again localized glow.

Mixtures of $N_2 + CO_2$ or $N_2 + Ar$ show equally curious behavior (Buchet and Goldman, 1969).

At atmospheric pressure in SF_6, presently the most widely used insulating gas, positive coronas develop and break down in a manner quite similar to that in air but with a pressure shift: some tens of atmospheres should be necessary to obtain the same behavior with air (Dupuy, J. and Gibert, A., 1981, personal communication). At higher pressures of SF_6, the behavior of the discharge is well described by Fig. 23. Coronas exist as long as a stabilizing glow can be maintained; above a critical pressure, referred to as P_c in Fig. 23, the glow has disappeared and direct spark breakdown occurs (Farish et al., 1979).

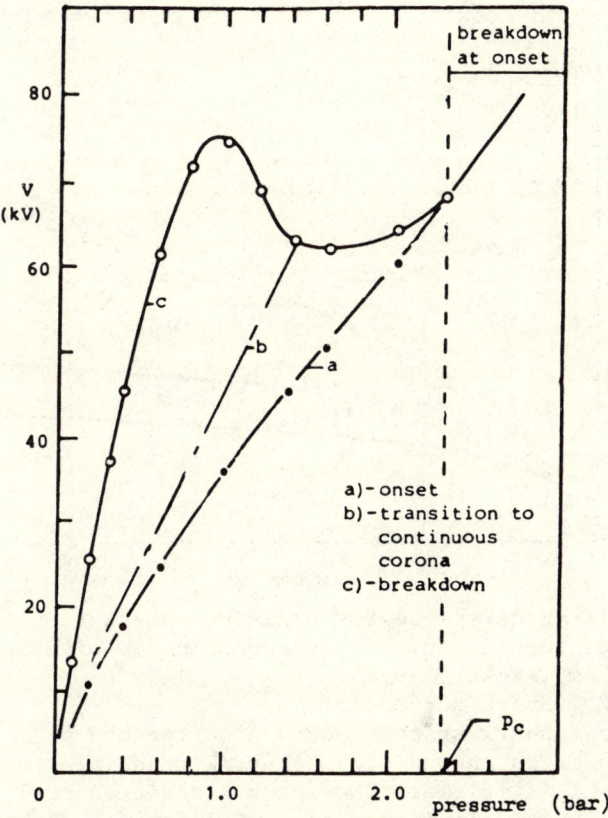

Fig. 23. Existence regions for positive point-to-plane corona forms in SF_6 as a function of pressure (Farish et al., 1979). Gap dimensions: d = 20 mm, r = 2 mm. Gas pressures in bars (1 bar = 100 kPa).

With very fine points in air, one typically obtains the curves in Fig. 22 with less frequent streamers or none at all in the voltage range $V_G - V_{G'}$. With bigger points, pure glow coronas occupy larger voltage intervals in the VI characteristics and, in many cases, spark breakdown directly follows pure glow coronas without intermediate mixed glow-streamer regimes (Fig. 24).

Characteristic Features of the Glow

We stated above that the glow should be a necessary condition for the existence of a DC positive corona. To realize this, we must consider that the burst pulses of the autostabilization regime, formed by trains of small avalanches covering the active electrode,

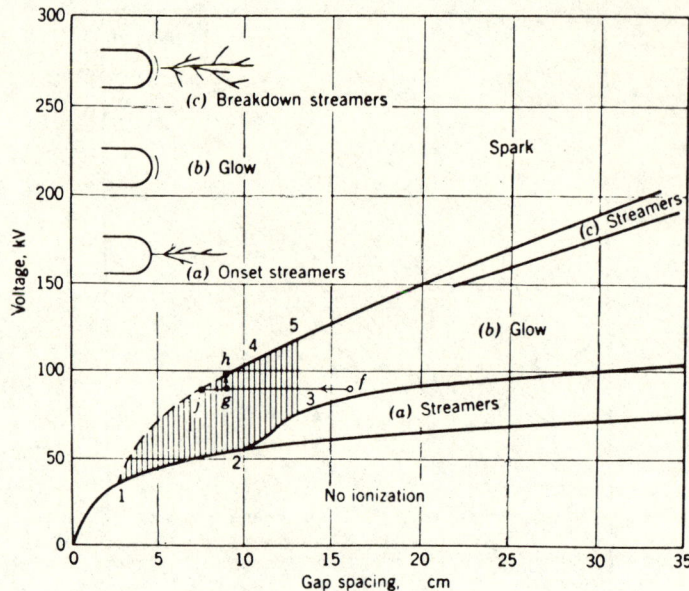

Fig. 24. Existence regions for positive sphere-to-plane high voltage corona forms in atmospheric air with anode radius 20 mm (Nasser, 1971).
1-2 : Spark threshold
2-3 : Spark or continuous glow threshold
4-5 : Continuous glow to spark transition
1-4 : Artificial glow-spark transition realized by establishing a glow at large gap spacings and then entering the shaded region along the path f-g.

are a nonstabilized form of the positive glow before it reaches a steady state when a sufficient supply of electrons is provided to it from the gas or from the cathode. Then, the burst pulses are joined together, and in the case of Fig. 22, we are in the region $V_G - V_{G'}$ where the glow is in its most stable form since it is without any perturbing streamer.

Sometimes, small-amplitude, high-frequency oscillations are seen on the continuous current delivered by the glow in this region. At first sight, the discharge then bears quite a close resemblance to the negative Trichel corona, with a concentrated ionization-region glow and oscillations in the MHz range, but the amplitude of these oscillations is smaller and, contrary to Trichel pulses, proportional to the discharge current, while their frequency hardly varies. Fieux and Boutteau (1970) state for the frequency vs. tip radius:

$$F(\text{MHz}) = 0.5 + 1.25/r \text{ (mm)} \tag{27}$$

for $0.2 \leq r \leq 2.75$ mm, and $2.5 \leq d \leq 200$ mm, and for their highest amplitude reached at voltage $V_{G'}$:

$$\text{Ampl} \cdot (\mu A) = 24 \{1 - \exp[-1.2 \ r \ (\text{mm})]\} \tag{28}$$

for $0.075 \leq r \leq 2.75$ mm and $25 \leq d \leq 200$ mm. For $r > 2.75$ mm, the oscillations become less regular. The physical mechanism for the oscillations is still unclear, but they were observed only in electron-attaching gases, and either photodetachment (Beattie, 1975; Beattie and Cross, 1975) or collisional detachment (Sigmond, 1978) may play a role.

With or without oscillations, this regime is generally known as the positive glow corona, but it would be preferable to call it "positive continuous glow corona" or "positive pure glow corona" since we have seen that the glow is not a specific feature of this regime. The oscillations do not invalidate this proposal because of their low relative amplitude and since, as mentioned earlier, with their high frequencies they do not affect the unipolar current flowing from the glow through the drift region.

Figure 22a shows that the continuous current evolves independently of the streamer current throughout the whole corona voltage range. This is valid not only for its amplitude, but also for its spatial distribution (Hirsh et al., 1979). The average total current distribution shape remains quite unchanged with preonset streamers, but with breakdown streamers a much more constricted current distribution must be associated with it (Goldman and Hirsch, 1980).

The Positive Streamer Corona Discharge

Physical Behavior. The spatio-temporal development of the luminous filaments and the associated current pulses which characterize the positive streamer corona have been thoroughly investigated for point-to-plane short air gaps under DC voltage.

Figure 25 shows the development in total light and current of three individual discharges, in order of increased voltage in the $V_{G'} - V_S$ region of Fig. 22:

• Frame (a) shows the development of a streamer (phase α) at voltage insufficient to permit it to reach the plane.

• In frame (b), the streamer (a term which will here designate both its tip and its trajectory) has crossed the gap. It is followed by the development from the point of a luminous channel (phase δ),

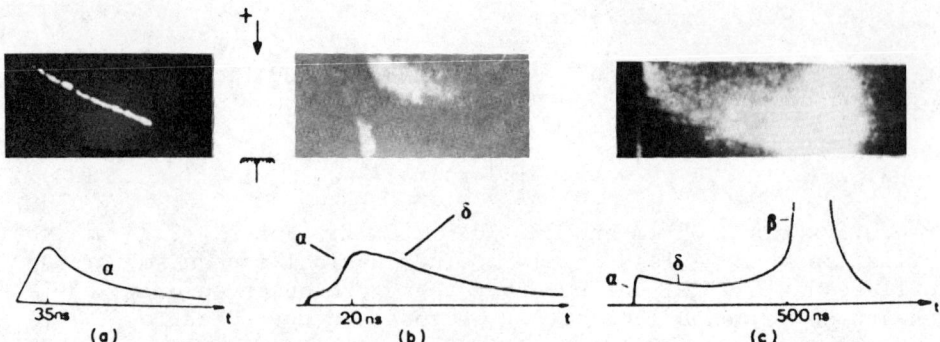

Fig. 25. Time-resolved photographs taken with an image converter camera operating in the streak mode, and associated current pulses illustrating the spatio-temporal development of three individual streamer corona discharges in positive point-to-plane gap in atmospheric density air under DC voltage, in order of increased voltage from the left to the right. Gap dimensions: length in the range 15-31 mm, point radius in the range 0.04 - 0.17 mm.
a) Propagation of the streamer, labelled α, corresponding to the frame F in Fig. 22b; d = 31 mm, r = 0.17 mm.
b) Propagation of the streamer, followed by development of the post-streamer channel, labelled δ (Marode, 1975a); d = 15 mm, r $\simeq$ 0.04 mm
c) Complete development from the streamer initiation to the spark, labelled β (Marode, 1975a); d = 18 mm, r $\simeq$ 0.04 mm.

often referred to in the literature as a secondary streamer as opposed to the streamer itself, called a primary streamer.

• In frame (c), the channel has joined the two electrodes and a spark follows (phase β), after a period of reduced light emission and current which disappears at higher voltages.

Note that, throughout the three phases (α, δ, β), the total light emitted and observed with a photomultiplier, follows a quite similar evolution in time as the current pulses. It is also important to point out that the current flowing in these pulses is essentially electronic. Heights typically increase from the mA range during the streamer propagation and tens of mA during the channel development to the A range or more when the gap breaks down, while frequency increases with voltage as shown in Figure 22b.

After the streamer is initiated, its active head is normally the only visible part with optical radius measured to about 20 µm and electron density 10^{21} - 10^{22} m^{-3} in 40 kPa air (Bastien and Marode,

1977); at 1 atm the radius is probably 10 μm (Marode et al., 1979). The electric field distribution through which the streamer propagates is often determined by the coexistent continuous current flowing in the gap. The streamer can survive in a zero-applied field region, growing by its own space-charge field (Dawson, 1965) and does not always follow the field lines (Fig. 26), but it needs a minimum field, depending on humidity and pressure, for stable propagation (Andersson, 1958; Griffiths and Phelps, 1976). This field, which varies between 4 and 5.5 kV/cm in atmospheric air, has been seen to be current independent in 5-16 mm air gaps (Sigmond and Goldman, 1981). An increase in the electric field ahead of the streamer will increase the electron multiplication in the avalanches developing towards it and thus, increase the speed of the streamer growth. It has been shown that the streamer propagation velocity in small air gaps varies between 2×10^6 and 5×10^7 cm/s, while the mean electron energy in its active tip determined from spectroscopic measurements varies between 12 and 16 eV; this corresponds to a variation of the local field from approximately 150 to 300 kV/cm (Hartmann, 1974).

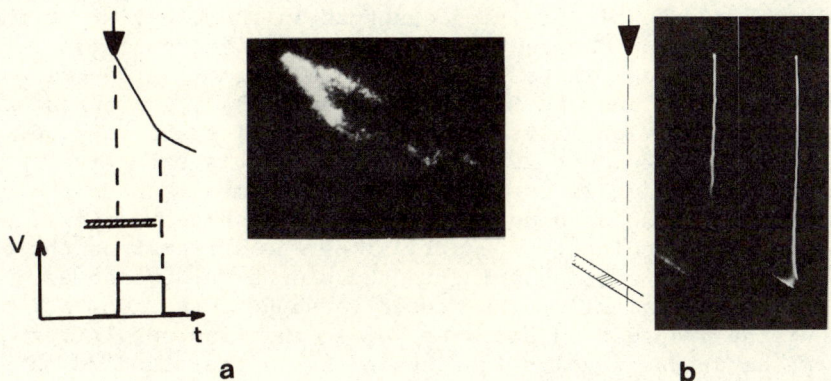

Fig. 26. Frames of positive point-to-plane corona in atmospheric density air, illustrating the concept of the self-propagating streamer (Goldman and Goldman, 1978).
a) Streak photograph of the most vigorous of the streamers of a discharge induced by a voltage pulse. Its reproduction on the trace, correlated with the applied voltage waveform, shows that the streamer continues to propagate towards the plane even after the voltage is no longer applied.
b) Static photographs of a discharge induced by a steady potential. The inclination of the plane with respect to the axis of the point affects the streamer trajectory only near the end of its course; the later it does so, the higher the applied voltage, that is, the more vigorous the streamer.

In air, only nitrogen spectral bands are observed. The absence of emission by the oxygen molecule was explained by the absence of excited oxygen states populated "vertically" from the ground state according to the Franck-Condon principle as happens for the $C^3\Pi_u$ and $B^2\Sigma_u^+$ states of nitrogen. Apart from the result indicated above, other relevant data were obtained by spectroscopic analysis of these last states:

• The production in the streamer of significant amounts of metastables was established by the measurement of high current-dependent vibrational temperatures in the range 1000 - 1500 K, while the gas was observed to remain cold with rotational and translational temperatures that did not exceed 470 K (Hartmann, 1977).

• The light emitted by the channel, which develops from the point after the streamer has crossed the gap, comes essentially from the neutral nitrogen molecule; the mean electron energy is now only about 1.4 eV, corresponding to a local field of about 15 kV/cm for which attachment exceeds ionization (Marode and Hartmann, 1969).

Because of the similarities observed in the behavior of the two kinds of discharges, Marode (1975a) interprets the positive streamer discharge development in terms of glow discharge. He describes the streamer (its head and the filament it builds by its propagation) as a glow discharge positive column with a moving active head, the gas playing the feeding role of a cathode in this formative phase. A true cathode region is formed when the streamer reaches the plane, and the channel then developing from the point appears simply as a luminous positive column. The arrival of the streamer on the cathode marks an important turning point in the discharge development: instead of the several kilovolts needed to support the ionization processes of the propagating streamer before this instant, afterwards only some hundreds of volts (corresponding to the standard cathode fall of a glow discharge) are needed. The difference in potential is then redistributed along the streamer filament through the rapid propagation ($\sim 10^9$ cm/s) of a potential wave from the cathode towards the anode: the return wave (or return stroke), which is easier to see at low pressures (Oshige, 1967; Ikuta et al., 1970; Ikuta and Kondo, 1976). This redistribution proceeds according to the electron density distribution established by the propagating streamer, with a higher potential drop per unit length (thus, a higher electrical field) where the filament resistivity is higher, i.e., on the anode side. Thus, the channel will appear brighter where the streamer was darker. Experimental confirmation of the similarities between positive streamer corona and glow discharge behavior has been provided by the observation of cathode sputtering after the arrival of the streamer at the cathode. The phenomenon has been studied over an extended range of gap lengths: 0.1 - 2 cm by Johnson et al., (1977) and 0.5 - 2 m by Belbel (1976).

Streamer Propagation Modeling. Streamer propagation has been the object of different approaches to establish quantitative models in accordance with the experimental results. Several of them suppose the propagation to occur through a series of steps, one of which is sketched in Fig. 27. At a given time t_n, the streamer tip is represented as a spherical zone of radius r_s containing N_+ positive ions and N_* excited states which are the result of previous ionization and excitation phenomena; a step further, at time t_{n+1}, the head has moved a distance $x_{n+1} - x_n \simeq 2 r_s$ towards the cathode, new ion pairs and excited states have been formed, and the number of positive charges and excited states in the head have respectively changed to $N'_+ \simeq N_+$ and $N'_* \simeq N_*$.

One of the first modeling attempts of a streamer so propagating step by step was made by Dawson and Winn (1965), who assumed that the streamer head is autonomous, bound to the anode by a filament of negligible conductivity. Assuming that photoelectrons are properly distributed ahead of the streamer tip centered at $x = x_n$ and stating that a primary electron triggered by photoionization at a distance a from x_n should create N_+ ion pairs in an avalanche of diffusion radius $r'_s \simeq r_s$ and of length $a - (x_{n+1} - x_n)$ with $x_{n+1} - x_n \geq 2 r_s$ (Fig. 27d), they first determined from calculations on

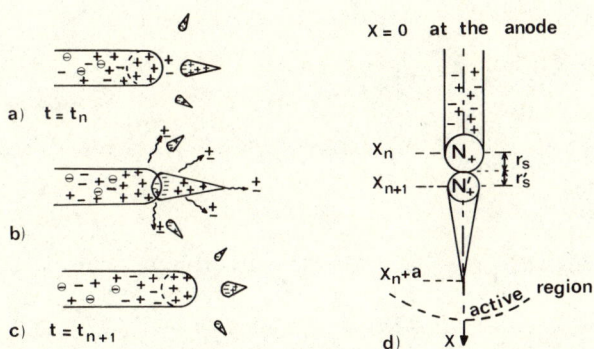

Fig. 27. Schematic representation of the streamer propagation mechanism.
a-b-c) One detailed step of the process showing photon paths and a series of new avalanches which develop around the streamer tip.
d) The streamer active region and the equivalent avalanche simplified model representing a series of new avalanches by a single equivalent avalanche, producing the same space charge as the series of avalanches and used with the assumption that the filament traced out by the streamer tip is of negligible conductivity (Dawson and Winn, 1965; Gallimberti, 1972).

the electron multiplication factor (see Eq. (2)) that $N_+ = 10^8$ ions and $r_s = 2.7 \times 10^{-3}$ cm should provide suitable conditions for self-propagation of a streamer with a = 25.10^{-3} cm and $x_{n+1} - x_n = 6.10^{-3}$ cm. Then, assuming that the energy required to create the new ion pairs and excited states for one step is:

$$W_1 = 2 N_+ eV_i \tag{29}$$

(with eV_i = ionization energy of gas molecule ≈ 15 eV and the assumption that energy lost due to excitation is about equal to that lost due to ionization) and that this energy is taken from the electrostatic energy stored in the positive streamer head:

$$W_p = \frac{N_+^2 e^2}{8 \pi \varepsilon_o r_s}, \tag{30}$$

they found that, with an initial ion population $N_+ = 5.10^8$ ions, the streamer propagating in a zero field region should cross ~ 450 steps over a distance S ~ 2.75 cm before stopping with ten times fewer positive ions, i.e., $N_+ = 5.10^7$ ions. Then, taking into account the influence of an applied field V on a streamer propagating a distance S with an average ion tip population $\bar{N}_+ = 10^8$ ions, they assumed that the energy lost along S due to ionization and excitation phenomena:

$$W_1 = 2 N_+ eV_i S / (x_{n+1} - x_n) \tag{31}$$

should be just compensated by the energy:

$$W_p = N_+ e V \tag{32}$$

gained from the field by the ion population of the initial streamer tip, still supposed to contain 5.10^8 ions. Thus, they found that a potential of 25 kV should give a streamer length of 25 cm, instead of the maximum of 5 cm they actually observed.

Also viewing the streamer as an isolated sphere of space charge at the streamer tip and considering energy balance as a criterion for streamer propagation, Gallimberti (1972) improved the Dawson and Winn's model by developing more rigorous calculations:

(1) To describe the growth of the equivalent avalanche sketched in Fig. 27d, by using the continuity equations:

$$\frac{\partial N_e}{\partial t} + \frac{\partial N_e}{\partial x} v_e - D_e \frac{\partial^2 N_e}{\partial x^2} = (\alpha - \eta) N_e v_e \tag{33}$$

$$\frac{\partial N_-}{\partial t} = \eta N_e v_e \qquad (34)$$

$$\frac{\partial N_+}{\partial t} = \alpha N_e v_e \qquad (35)$$

where N_e, N_- and N_+ are the linear densities, with respect to the direction of avalanche growth, of electrons, negative and positive ions, respectively; the ionization coefficient α was corrected for the self-retarding field of the avalanche.

(2) To solve the energy balance equation:

$$W_1 = W_g + \Delta W_p \qquad (36)$$

which was, for the step from time t_n to time t_{n+1}, computed from the equations:

$$W_1 = \int_{x_{n+a}}^{x_{n+1}} [\alpha V_i + \eta V_a + \Sigma \delta_k V_k + (\alpha + \frac{fc_e}{\lambda v_e}) \frac{mc_e^2}{2}] N_e \, dx \qquad (37)$$

$$W_g = \int_{x_{n+a}}^{x_{n+1}} (N_e e E_g) \, dx \qquad (38)$$

$$W_p = \frac{0.43(N_+ e)^2}{4 \pi \varepsilon_o r_s} + N_+ e \int_o^{x_n} E_g \, dx \qquad (39)$$

The term W_1 takes into account all the losses that the drifting electrons undergo in collisions with the gas molecules; V_i and V_a are the ionization and attachment potentials, δ_k is the excitation coefficient of the energy level V_k, c_e is the square root of the mean square velocity of electrons, λ and f are their mean free path and the fractional energy gained by the electrons from the applied field E_g. As for the term W_p, it takes into account the potential energy due both to the concentration of charges and to their position in the field.

Figure 28 shows that good agreement was obtained by Gallimberti between the computed and the experimental results for the temporal variations of both the length of the streamers and the current pulses associated with them. He also made calculations of the

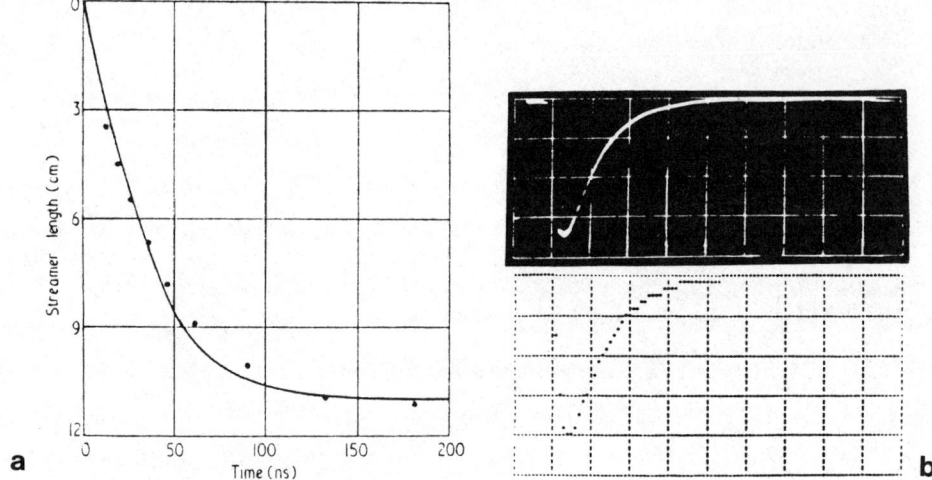

Fig. 28. Computed results on the streamer propagation obtained by Gallimberti (1972) with the equivalent avalanche model for a high voltage positive rod-to-plane corona induced by a voltage pulse in atmospheric air. Gap dimensions: d = 150 cm, rod ϕ = 2 cm with a square-cut end. Inception voltage: 125 kV.
a) Temporal variations of the streamer length. The solid line represents the computed values, the points experimental results.
b) Temporal variations of the streamer current. Scales: 1 A/div and 100 ns/div. for the experimental oscillogram taking into account all the streamers of an individual discharge, 5 mA/div. and 100 ns/div. for the computed curve.

critical field necessary for a sustained streamer growth, but these led him to a value of 7 kV/cm, larger than the value of 4 - 5.5 kV/cm cited above from Andersson (1978) and Griffiths and Phelps (1976). Further, by using a Monte-Carlo method to calculate the probablilty of two equivalent avalanches developing simultaneously, Badaloni and Gallimberti (1973) succeeded in the simulation of streamer corona branching.

In contrast to the models described above, other approaches to simulate the streamer propagation were made (Wright, 1964; Klingbeil et al., 1972) with the assumption that the streamer filament traced out by the streamer tip is conducting, with almost equal densities of positive ions and electrons and with negligible attachment rate. Thus, Wright's model gives an equation which allows calculation of the streamer tip potential V_s as a function of the anode potential V:

$$\frac{\partial^2 V}{\partial x^2} = C R \frac{\partial V}{\partial t} \tag{40}$$

by assuming that the streamer filament can be regarded as a conductor presenting constant capacitance and resistance respectively equal per unit length to:

$$C = \frac{1}{2 \ln(x_k/r_s)} \quad \text{and} \quad R = \frac{4 n_e}{n_e' r_s (v_s + v_{es})} \tag{41}$$

where n_e and n_e' are the electron densities at the streamer tip and along the streamer filament, v_s the streamer propagation velocity, v_{es} the electron drift velocity at r_s, and x_k the distance from the streamer tip to the cathode. Such a model describing the streamer propagation as a continuous process, which takes into account the whole conducting filament, is very attractive; it should be interesting to refine it.

This was partially done by Marode (1975b) in order to obtain precise boundary conditions for the streamer filament which he regarded as an initial stage for the simulation he made on the discharge evolution, considering each point of the gap from the instant it has been crossed by the streamer tip until the end of the streamer current pulse. Combined with a hydrodynamical treatment of the post-streamer channel (phase δ in Fig. 25), this simulation has been later extended until the spark transition (Marode et al., 1979); this will be developed by Marode in his paper on the glow to arc transition.

CORONA APPLICATIONS

If we would have had to wait for a good understanding of the mechanisms involved in the electric arc, no switchgear would be available today. Hopefully, applications will be found to many newly discovered phenomena before extended basic knowledge. The corona is an example of this situation with numerous applied fields, in particular, electrostatics, atmospheric electricity, plasma chemistry and, naturally, gas discharge physics.

It is not our purpose here to review the whole range of corona applications but rather to give an impression about the variety and importance of the old and challenging new ones. So the listing we propose will be rather selective.

Since the different effects produced by coronas provide material for different types of applications, we shall use them as a basis to present the subject. One is tempted to take into account only those which seem a priori beneficial. But this would not be the best way, for different reasons:

(1) Effects regarded as harmful in certain cases may be beneficial in other cases, e.g., surface oxidation.

(2) Both types of effects should sometimes act simultaneously, one dominating the other according to the working conditions; for instance, corona treatment of a few minutes improves the insulating properties of dielectric films while a longer exposure to the discharge acts in the opposite way.

(3) Last, methods and devices adapted to the fight against harmful effects also need to be treated.

Applications Based on the Conduction Parameters

Gas Insulation. The problem is to avoid corona losses, especially from positive streamer coronas, the worst cases for insulation as said before.

Two ways are possible:

(1) One assumes a good understanding of the current and respect for voltage limits. Careful studies of the electrical characteristics in various geometrical, gaseous, and electrical conditions answer this requirement. Much has already be done (Waters, 1978), but much still remains to do.

(2) The other, complementing the preceding one, concerns the improvement of the insulation parameters: the gas, electrodes, insulators, and walls. In this field, much is expected from studies on gas chemistry (see Christophorou's chapter).

Gas Analysis. The ionic mobility in a gaseous medium can be measured by means of corona devices (Goldman et al., 1976), but the most useful applications based on the corona conduction parameters are relevant to the use of current-voltage characteristics. If a new gas component, even in small amounts, is introduced in a vessel, the average ionic mobility will change and produce current changes often sufficient for given measurements. Various applications of this principle can be found in different types of detectors, such as pollution detectors, leak detectors (for instance, for SF_6), and smoke and fire detectors.

Particle Detection in Nuclear Physics. For decades, nuclear particle detectors have taken advantage of corona processes: Geiger

CORONA DISCHARGE PHYSICS

counters (Loeb, 1965), streamer chambers (Tan et al., 1970), spark chambers (Charpak, 1970), and recently, the photoionization proportional scintillation chamber called the PIPS (Charpak, 1980).

Research is still important in this field, especially on the PIPS which gives good energy and spatial resolution over large areas and which can also be used as a photomultiplier or even as an image intensifier.

Applications Based on the Waves Induced by Coronas

The fast current variations produced by streamers induce electromagnetic waves which cause disturbance in electronic devices, in particular, in aircraft, and audio noise, which may be harmful under high voltage transmission lines. These phenomena, generated by coronas (or partial discharges in the case of solid insulating materials containing gaseous bubbles), are used to detect insulation defects in insulators, capacitors, transformers, etc. The detection can be made in situ and the defects localized by triangulation methods.

It is possible to inhibit the streamers and their harmful effects by injecting negative charges from hot wires, radioactive sources, or auxiliary coronas. These charges will help the autostabilization of the discharge by the glow and diminish the noise, but not the current losses (Hermstein, 1960; Popkov, 1962; Waters et al., 1972). In aircraft, substantial reduction of the electromagnetic interference can be obtained by using corona discharge rods made of multiple high-resistivity conducting fibers of a composite material inside an insulating matrix (Brunet and Faure, 1980).

Applications Based on the Corona Light Emission

The diameter of the streamer is, as we have seen, of the order of 20-30 μm, and, at a given point of its trajectory, the light emission lasts about 0.3 ns. Thus, with a slit perpendicular to the streamer path, it is possible to obtain a punctual source of UV light pulses of about 0.3 ns duration, with repetition rates of a few kHz (Hartmann, 1970). Only laser pulses can be shorter and much more powerful, but they will operate with much lower repetition rates and be much more expensive.

Applications Involving Ion Transport

If liquid or solid particles pass through a corona gap, they are charged and can be collected by an electrode. This leads to various corona applications; the most widely industrialized today are often known as electrostics applications (Moore, 1972).

Precipitators for Dust, Ash, etc. The principle was first applied by Cottrell in 1905 to build the first precipitator, commercialized by the Detroit Edison Company in 1923. The fly ash trapped in the USA, at better than 99% efficiency, is estimated at present at more than 20 million tons per year. Cottrell's contribution is hard to overstate!

Electrocoating. Very similar techniques can be used for spraying, for instance, insecticides on plants in such a way that the undersides of the leaves are coated, and there is less insecticide used and less pollution.

The same principle allows us to use electrostatic painting. A variation of the electrocoating approach, called flocking, can be used to give a surface a velvet finish. If the surface is non-conducting, it is first covered with aluminum paint to which an adhesive is applied. The operator holds a hopper filled with short fibers. As the fibers are shaken out of the hopper, they are charged by ions from a set of corona points mounted on the hopper. At this stage, three important effects occur. First, the fibers are impelled along the field lines by the Columomb force. Second, the mutual repulsion of like charges keeps them apart. Third, they align themselves with the field, arriving end-on to stick to the adhesive. By printing the adhesive in patterns on cloth, one can create decorative designs.

Somewhat the same process is used to make coated abrasives. Superior sandpapers and coated cloth abrasives are turned out electrostatically on huge machines. For a "filled" abrasive, the operation continues until all the spaces are filled. For "open" coatings, the rate of feed is lowered, and the electrostatic-spacing effect makes a uniform rather than a patchy coating.

Electrostatic Separation. The corona discharge is also the key for the separation of granular mixtures. For instance, if two kinds of particles to be separated differ from each other in their conductivity, the device represented in Fig. 29 can be used. The mixture comes down from the hopper to form a thin layer on top of the rotating drum which is grounded and passes under a wire that is generating a corona discharge. As a result, ions flood to the drum. The charges of those ions that hit conducting particles are transferred directly to the drum, and accordingly, the particles fall right off the drum. In contrast, the charge of ions that hit insulating particles, coat the outer surfaces of the particles and "pin" them to the drum, from which they later fall or are scraped. In addition to this common type of electrostatic separator used mainly in the mineral industry, a number of quite different designs exist. The world's largest electrostatic-separation plant for the benefication of iron ores, installed in 1965 at the Wabush Mines in Canada, handles six million tons per year. It has been estimated that some

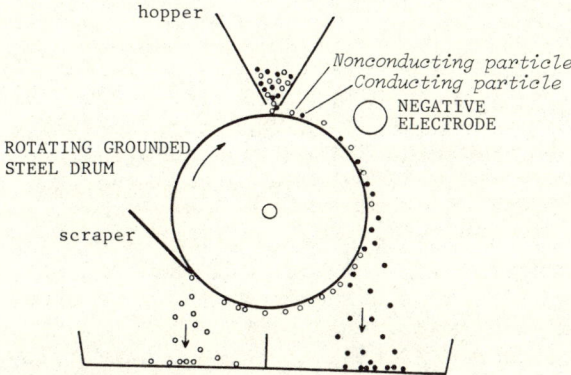

Fig. 29. Schematic representation of a mineral separator employing a corona discharge to separate conducting (•) and nonconducting (o) particles (Moore, 1972).

35 other mineral mixtures are currently being separated electrostatically.

The electrostatic approach has been adopted for a number of others separation tasks. For instance, in the food industry, rodent excreta is removed from rice, garlic seeds extracted from wheat, and nut meats separated from shells.

Elimination of Static Electricity. Electrostatic charges, positive or negative, can be neutralized by corona operating under an alternative voltage; this is the best method to eliminate static electricity.

Reproduction Technics. One of the most important fields of corona applications is xerography and related techniques, which may include telecopying in the near future. The social and economic impact of xerography, which was invented by Chester Carlson in 1930, can be roughly gauged if only by the fact that the worldwide business of the Xerox Corporation reached almost 2 billion dollars in 1970.

Applications Relevant to the Chemical Reactivity Acquired by the Gas in Corona Discharge

These new challenging applications will be seen in A. Goldman's paper devoted to plasma chemistry and in Kristiansen and Guenther's paper devoted to the applications in general.

Acknowledgements. The encouragement and great help given by A. Goldman, the wife of one of us, is gratefully acknowledged.

REFERENCES

Andersson, N. E., 1958, Ark. Fys., 15: 441.
Badaloni, S. and Gallimberti, I., 1973, in: "Proceedings, 11th International Conference on Phenomena in Ionized Gases, Prague," Czech. Acad. Sci., Inst. Phys., Prague, p. 196.
Bastien, F. and Marode, E., 1977, J. Quant. Spectros. & Radiat. Transfer, 17: 453.
Beattie, J., 1975, Ph.D. Thesis, University of Waterloo, Canada.
Beattie, J. and Cross, J. D., 1975, in: "Proceedings, 12th International Conference on Phenomena in Ionized Gases, Eindhoven," North Holland, Amsterdam, p. 170.
Belbel, E., 1976, Docteur-Ingénieur Thesis, University of Paris-Sud, France.
Berger, G., 1974, in: "Proceedings, 3rd International Conference on Gas Discharges, London," IEE Conf. Publ. No. 118, p. 170.
Berger, G., 1981, in: "Proceedings, 15th International Conference on Phenomena in Ionized Gases, Minsk," p. 571.
Bhalla, M. S. and Craggs, J. K., 1962, Proc. Phys. Soc., 80: 151.
Bogdanova, N. B. and Pevchev, B. G., 1971, Akad. Nauk SSSR, Energ. Transport, 6: 75 (in Russian).
Brunet, A. and Faure, F., 1980, in: "Proceedings, 6th International Conference on Gas Discharges, London," IEE Conf. Publ. No. 189, Vol. 1, p. 158.
Buchet, G., Goldman, M., and Goldman, A., 1966, C. R. Acad. Sci., Ser. B, 263: 356.
Buchet, G. and Goldman, M., 1969, in: "Proceedings, 9th International Conference on Phenomena in Ionized Gases, Bucharest," Edit. Acad. Republ. Soc. Romania, Bucharest, p. 291.
Buchet, G. and Goldman, A., 1970, in: "Proceedings, 1st International Conference on Gas Discharges, London," IEE Conf. Publ. No. 70, Vol. 1, p. 459.
Bugge, C., 1968, Technical Report Electron and Ion Physics Research Group, The Norwegian Inst. of Tech., Trondheim, Norway, EIP 68-1.
Bugge, C. and Sigmond, R. S., 1969, in: "Proceedings, 9th International Conference on Phenomana in Ionized Gases, Bucharest," Edit. Acad. Republ. Soc. Romania, Bucharest, p. 289.
Charpak, G., 1970, Ann. Rev. Nucl. Sci., 20: 195.
Charpak, G., Policarpo, A., and Sauli, F., 1980, IEEE Trans. Nucl. Sci., NS-27: 212.
Cribier, F., Goldman, A., and Lécuiller, M., 1976, in: "Proceedings, IEEE International Symposium on Electrical Insulation, Montreal," p. 282.
Crochet, E., Goldman, M., and Haug, R., 1980, in: "Proceedings, 2nd International Symposium on Gaseous Dielectrics, Knoxville," Pergamon, New York, p. 54.
Das, M. K., 1961, Z. Angew. Phys., 13: 410.
Davies, A. J. and Donne, K. E., 1980, in: "Proceedings, 6th International Conference on Gas Discharges," IEE Conf. Publ. No. 189, Vol. 1, p. 138.

Dawson, G. A., 1965, Z. Phys., 183: 172.
Dawson, G. A. and Winn, W. P., 1965, Z. Phys., 183: 159.
Dutton, J., 1975, J. Phys. Chem. Ref. Data, 4: 577.
Elford, M. T. and Rees, J. A., 1974, Aust. J. Phys., 27: 333.
Farish, O., Ibrahim, O. E., and Kurimoto, A., 1979, in: "Proceedings, 3rd International Symposium on High Voltage Engineering, Milan," Paper 31.15.
Fieux, R. and Boutteau, M., 1970, Bull. Dir. Etudes Rech. EdF, France, B2: 55.
Gallimberti, I., 1972, J. Phys. D, 5: 2179.
Gallimberti, I., 1979, J. Phys. (Paris), 40: (C7) 193.
Gardiner, P. S. and Craggs, J. D., 1977, J. Phys. D, 10: 1003.
Gary, C. and Moreau, M., 1976, "L'effect Couronne en Tension Alternative," Eyrolles, Paris.
Goldman, A., Goldman, M., Rautureau, M., and Tchoubar, C., 1965, J. Phys., 26: 486.
Goldman, A., 1974, in: "Proceedings, 3rd International Conference on Gas Discharges, London," IEE Conf. Publ. No. 118, p. 275.
Golamdn, A., Haug, R., and Latham, R. V., 1976, J. Appl. Phys., 47: 2418.
Goldman, A., Selim, E. O., and Waters, R. T., 1978, in: "Proceedings, 5th International Conference on Gas Discharges, London," IEE Conf. Publ. No. 165, p. 88.
Goldman, A. and Hirsh, M., 1980, in: "Proceedings, 2nd International Symposium on Gaseous Dielectrics, Knoxville," Pergamon, New York, p. 43.
Goldman, M. and Goldman, A., 1978, in: "Electrical Discharges," M. N. Hirsh and H. J. Oskam, eds., Academic, New York, Chap. 4.
Goldman, M. and Sigmond, R. S., 1981a, in: "Proceedings, 15th International Conference on Phenomena in Ionized Gases, Minsk," p. 653.
Goldman, M. and Sigmond, R. S., 1981b, in: "Proceedings, Conference on Electrical Insulation and Dielectric Phenomena, Whitehaven, USA," (to be published in IEEE Trans. El. Insul.).
Gosho, Y., 1980, in: "Proceedings, 6th International Conference on Gas Discharges, London," IEE Conf. Publ. No. 189, Vol. I, p. 114.
Graf, D., 1980, in: "Proceedings, 6th International Conference on Gas Discharges, London," IEE Conf. Publ. No. 189, Vol. 1, p. 142.
Griffiths, R. F. and Phelps, C. T., 1976, Quart. J. R. Mat. Soc., 102: 419.
Hartmann, G., 1964, Ing. Thesis, CNAM Paris, France.
Hartmann, G., 1970, in: "Proceedings, 9th International Congress High-speed Photography, Denver," Soc. of Motion Picture and Television Engineers, New York, p. 258.
Hartmann, G., 1974, in: "Proceedings, 3rd International Conference on Gas Discharges, London," IEE Conf. Publ. No. 118, p. 634.
Hartmann, G. and Gallimberti, I., 1975, J. Phys. D., 8: 670.

Hartmann, G., 1977, Doctorat d'Etat Thesis, University of Paris-Sud, France.
Hartmann, G., 1980, in: "Proceedings, IEEE-IAS Conference on Electrostatic Processes, Cincinnati," IEE Conf. Publ. No. 80CH1575-0, p. 1113.
Hermstein, W., 1960, Arch. Electrotech., 45: 279.
Hirsh, M. N., Abbott, R. P., and Benjamin, R. D., 1979, in: "IEEE-IAS Conference on Electrostatic Processes, Cleveland," IEE Conf. Publ. No. 79CH1484-51A, p. 54.
Ikuta, N., Ushita, T., and Ishiguro, Y., 1970, Elec. Eng. Jap., 90: 52.
Ikuta, N., Kondo, K., Ushita, T., and Ishiguro, Y., 1975, in: "Proceedings, Tokyo Symposium."
Ikuta, N. and Kondo, K., 1976, in: "Proceedings, 4th International Conference Symposium on Gas Discharges, London," IEE Conf. Publ. No. 143, p. 227.
Johnson, P. C., Berger, G., and Goldman, M., 1977, J. Phys. D., 10: 2245.
Klingbeil, R. D., Tidman, A., and Fernsler, R. F., 1972, Phys. Fluids, 15: 1969.
Kondo, Y. and Miyoski, Y., 1978, Jap. J. Appl. Phys., 17: 643.
Korge, H., Kudu, K., and Laan, M., 1979, in: "Proceedings, 3rd International Symposium on High Voltage Engineering, Milan," paper 31.04.
Kurimoto, A. and Farish, O., 1980, IEE Proc., 127: (A2)89.
Lama, W. L. and Gallo, C. F., 1974, J. Appl. Phys., 45: 103.
Lécuiller, M., Julien, R., and Pucheault, J., 1972, J. Chim. Phys., 9: 1353.
Loeb, L. B., 1965, "Electrical Coronas," Univ. of Calif. Press, Berkeley.
Loeb, L. B., 1967, in: "Proceedings, 8th International Conference on Phenomena in Ionized Gases, Vienna," Springer, Berlin, p. 73.
Marode, E. and Hartmann, G., 1969, C. R. Acad. Sci., Ser. B, 269: 748.
Marode, E., 1975a, J. Appl. Phys., 46: 2005.
Marode, E., 1975b, J. Appl. Phys., 46: 2016.
Marode, E., Bastien, F., and Bakker, M., 1979, J. Appl. Phys., 50: 140.
Meek, J. M., 1940, Phys. Rev., 57: 722.
Miyoski, Y., Hosokawa, T., and Sakai, O., 1964, Nagoya Institute of Technology Bulletin.
Moore, A. D., 1972, Sci. Am., 226: (3)46.
Morton, P. L., 1946, Phys. Rev., 70: 358.
Nasser, E., 1971, in: "Fundamentals of Gaseous Ionization and Plasma Electronics," Wiley Intersci., New York, Chap. 11.
Oshige, T., 1967, J. Appl. Phys., 38: 2528.
Popkov, V. I., 1962, in: "Proceedings, International Conference on Gas Discharges and Electricity, Supply Industry, Leatherhead, UK," Butterworth, London, p. 225.
Rajapandyan, S. and Raju, G. R. G., 1972, J. Phys. D., 5: L6.

Ramo, S., 1939, Proc. IEE, 27: 584.
Rogowski, W., 1936, Z. Phys., 100: 1.
Sakai, O., Hosokawa, T., and Miyoski, Y., 1958, IEE Jap., 78: 842, 1413 (in Japanese).
Schumann, W. O. 1923, "Elektrische Durchbruchfeldstarke von Gasen," Springer, Berlin.
Selim, E. O. and Waters, R. T., 1980, in: "Proceedings, 6th International Conference on Gas Discharges, London," IEE Conf. Publ. No. 189, Vol. I, p. 146.
Shahin, M. M., 1969, in: "Chemical Reactions in Electrical Discharges," B. D. Blaustein, ed., Adv. in Chem. Series 80, Am. Chem. Soc., Washington, Chap. 4.
Shockley, W., 1938, J. Appl. Phys., 9:635.
Sigmond, R. S., 1961, in: "Proceedings, 5th International Conference on Phenomena in Ionized Gases, Munich," North Holland, Amsterdam, Vol. 1, p. 1359.
Sigmond, R. S., 1978, in: "Electrical Breakdown of Gases," J. M. Meek and J. D. Craggs, eds., Wiley, New York, Chap. 4.
Sigmond, R. S., 1981, J. Appl. Phys. (to be published).
Sigmond, R. S. and Goldman, M., 1981, in: "Proceedings, 15th International Conference on Phenomena in Ionized Gases, Minsk," p. 649.
Tan, B. C., Schmied, H., Rousset, A., Rohrbach, F., Piuz, F., Gross, F., Morgan, C. G., and Cathenoz, M., 1970, CERN Report, Geneva, TC-L/Int. 70.
Teich, T. H., 1981, in: "Proceedings, 2nd International Swarm Seminar, Knoxville," Oak Ridge National Laboratory, Knoxville.
Thomson, J. J. and Thomson, P. G., 1933, in: "Conduction of Electricity through Gases," Vol. II, Cambridge Univ. Press, London, p. 552.
Torsethaugen, K. and Sigmond, R. S., 1973, in: "Proceedings, 11th International Conference on Phenomena in Ionized Gases, Prague," Czech. Acad. Sci., Inst. Phys., Prague, p. 195.
Townsend, J. S., 1914a, "Electricity in Gases," Oxford Univ. Press, London.
Townsend, J. S., 1914b, Phil. Mag., 28: 83.
Trichel, G. W., 1938, Phys. Rev., 54: 1078.
Ushita, T., Ikuta, N., and Yatzuzaka, M., 1968, Elec. Eng. Jap., 88: (1)45.
Van Geel, C., 1939, Physica, 6: 806.
Warburg, E., 1899, Wied. Ann., 67: 69.
Warburg, E., 1927, "Handbuck der Physik," Vol. 14, Springer, Berlin, p. 154.
Waters, R. T., Richard, T. E., and Stark, W. B., 1972, in: "Proceedings, 2nd International Conference on Gas Discharges, London," IEE Conf. Publ. No. 90, p. 188.
Waters, R. T. and Stark, W. B., 1975, J. Phys. D., 8: 416.

Waters, R. T., 1978, in: "Electrical Breakdown of Gases," J. M. Meek and J. D. Craggs, eds., Wiley, New York, Chap. 5.
Waters, R. T., 1981, IEE Proc., 128: (A4)319.
Weissler, G. L., 1943, Phys. Rev., 63: 96.
Wright, J. K., 1964, Proc. R. Soc. London, A, 280: 23.
Zentner, R., 1970a, Z. Angew. Phys., 29: 294.
Zentner, R., 1970b, Elektrotech. Z., A91: 303.

MODELING OF A GLOW DISCHARGE

L. Vriens, A. H. M. Smeets and H. J. Cornelissen

Philips Research Laboratories
Eindhoven, The Netherlands

INTRODUCTION

   The main part of this paper describes the modeling of one particular glow discharge, the low-pressure Na-Ne positive column. The introduction will present a general survey of glow discharges. In the first major section, a phenomenological description is given of what happens when the current density in a gas, subjected to an electric field, is increased from below 10^{-12} A/cm^2 up to 100 A/cm^2. In this way one goes through different discharge regimes. The glow discharge is one of these. In the second major section, the most important kinetic processes occurring in a glow discharge are summarized as well as the resulting deviations from thermodynamic equilibrium. In the next section the various types of glow discharges are subdivided and their characteristic properties described. A later section introduces the low-pressure Na-Ne positive column and the remaining sections of this paper deal with the modeling of this glow discharge. Large parts of the model can be used directly in the modeling of other discharges.

   With the introduction, it will be clear that there are many different glow discharges. For general literature see von Engel (1965), Francis (1956), Hirsch and Oskam (1978), Franklin (1976), Cherrington (1979), and Waymouth (1971). All glow discharges have in common that the deviations from thermodynamic equilibrium are substantial, that the amount of radiation emitted makes them clearly visible, and that they are largely or completely self-sustained. The following definition may therefore be used: <u>Glow discharges are nonequilibrium gas discharges that glow and are largely or completely self-sustained.</u>

A complete model of a glow discharge should include:

(1) A mathematical description of all relevant kinetic processes. This involves the individual atomic processes as well as their mutual dependence due to conservation relations.

(2) A mathematical description of the integrations over the discharge volume, for spatially nonhomogeneous discharges, and/or of the integrations with respect to time, for AC and pulsed discharges. These integrations should account for the appropriate boundary conditions.

(3) A computer program which yields the relevant local and integral characteristics of the discharge as output. Examples are the local particle densities and rates for collision processes, and the integral current-voltage characteristics and radiation efficiencies.

A model gives us insight into the operation of the discharge and can be used for computer optimization. The insight is gained by constructing the model and by comparing its output with experiment. The computer optimization can be used, e.g., for lamps, MHD plasmas, excimer lasers, etc. Quite generally, there are two main problems in constructing a model:

(1) Knowing all relevant processes occurring in the discharge;

(2) Having an accurate mathematical description of these processes, a description also simple enough to be acceptable with respect to computer time.

For the low-pressure Na-Ne positive column, these problems and their solutions are examined in detail. Our solution to the second problem is to use analytical techniques as much as possible. We do not solve the Boltzmann equation for the electron energy distribution. Instead, the computer time is reduced by more than a factor of 100 by using a simple and still accurate analytical approximation, the two-electron-group model (2-EGM) developed by Vriens (1973), Ligthart and Keijser (1980) and Morgan and Vriens (1980). A special study has been made by Vriens and Smeets (1980) to obtain relatively simple analytical formulas to describe excitation, de-excitation and ionization by electrons. Smeets (1982) made another special study to replace numerical integrations in the radiation transport formulas by analytical formulas. These analytical techniques are applicable also for many other discharges.

Before ending these preliminaries, we should say a few words about the presentation of the model. There are two possible ways:

(1) One can start with integral conservation and rate equations

and deal later with all individual terms (processes);

(2) One can start with a local description of all kinetic processes and follow by linking them together using rate and conservation relations, and performing, e.g., integrations over the discharge volume.

In this paper we follow the second procedure.

Phenomenological Survey

Applying an electric field to a gas at low pressure, one observes a number of interesting phenomena when the current-density through the gas is increased. This is illustrated in Fig. 1.

At very low current density, $j < 10^{-14}$ to 10^{-12} A/cm^2, a "non-self-sustained" discharge develops. Electron and ion production are due to external sources such as UV illumination of the cathode and cosmic rays.

When the current density is increased, the Townsend regime: $10^{-12} < j < 10^{-8}$ to 10^{-6} A/cm^2, is reached. Electrons and ions are produced by collisions between electrons and gas atoms in a multiplication process. The external source is still important at the lower j values and becomes negligible at the higher j values.

At still increasing current density, breakdown occurs at about $j = 10^{-4}$ A/cm^2, the voltage decreases, and the discharge becomes visible with dark and light regions arranged in a characteristic manner. Space charges that result from a flow of electrons to the wall, which is initially only partly neutralized by a similar flow of ions, act as an electrostatic lens and tend to make the electric field more uniform. A glow discharge has been established. Positive space charge near the cathode determines the cathode fall. The glow

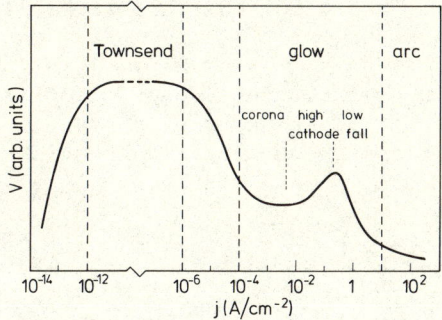

Fig. 1. The dependence of voltage on current density in a wide range of discharge regimes.

discharge derives its name from the luminous zone which develops near the cathode and is separated from it by a dark space. Here the electrons are accelerated to energies high enough to produce a sufficient number of secondary electrons and ions. The electrons entering the glow region form principally two groups: the first consists of the fast ones produced at or near the cathode which have not suffered inelastic collisions in the dark space; the other larger group consists of electrons created in the dark space which have made many inelastic collisions and are relatively slow. Due to their larger number density, the "slow" electrons are more important than the fast ones in the excitation process and produce the negative glow. The electric field is low. In the adjacent Faraday dark space, the electron energies are too small to cause appreciable excitation. The electric field is found to increase with increasing distance from the cathode and becomes constant in the <u>positive</u> <u>column</u>, usually a region of uniform luminosity. Occasionally this region is straited. The positive column fills the rest of the discharge tube almost to the anode. Drift velocities of the charged particles in the field direction are small fractions of their random motion. Because of the small mobility of the ions, the electrons carry practically the whole discharge current in the positive column. When the current density is increased from 10^{-4} to 10^{-1} A/cm^2, the voltage remains practically constant. The area of the cathode surface covered with the negative glow increases in proportion to the current. The voltage increases when the whole cathode is covered. A higher cathode fall is necessary to establish the larger electron emission.

The voltage reaches a maximum and decreases again when effects of cathode heating and gas heating come into play (1-10 A/cm^2). For a cold cathode this maximum is appreciable; for a heated cathode it is only a small maximum. Historically, the maximum was taken as the boundary between glow and arc discharges. Presently, we put the boundary at somewhat higher current values, e.g., for j > 5 to 10 A/cm^2, related to the transition to an equilibrium (LTE) arc (see below and Hirsch and Oskam, 1978). Glow discharges like the fluorescent lamp and the low-pressure sodium lamp have a negative voltage-current characteristic, a low cathode fall (Fig. 1), and operate beyond the voltage maximum.

From the above phenomenological description of the discharge, it is not clear which atomic processes are important. The discussion which follows presents a more quantitative description.

<u>Kinetic Processes, Nonequilibrium</u>

The most important kinetic processes occuring in a glow discharge are:

- excitation and deexcitation of atoms by electrons;

- ionization and recombination;

- relaxation of the electron energy distribution;

- radiation emission and absorption;

- diffusion due to density gradients.

In the model description of the low-pressure Na-Ne positive column all these processes are described in detail. Here we discuss the most characteristic properties pertaining to glow discharges.

We first consider electron-impact excitation and ionization. Until about 1972 it was generally accepted that the ionization was almost exclusively direct (one-step) ionization from the ground state and two-step ionization via the lowest excited states. Because of recent work by Fujimoto et al., (1972, 1973), Biberman et al. (1973, 1979), Vriens (1978), Vriens et al., (1978) and Gallagher et al., (1979, 1980), we have a rather different view of this problem. Direct electron-impact ionization is usually dominant only at low pressure and low electron density, $n_e < 10^{10}$ cm^{-3}. With increasing n_e, up to about 10^{12} cm^{-3}, the lower excited state densities increase linearly with n_e (coronal phase), and the importance of two-step ionization increases proportionally. When n_e increases up to 10^{14} to 10^{16} cm^{-3}, the densities of the lower excited states saturate, while those of the higher excited states still increase considerably. Ladder-like excitation and subsequent ionization via these higher states become more and more important and even dominant (see Fujimoto et al., 1972, 1973). After the saturation phase, $n_e > 10^{16}$ cm^{-3}, a transition takes place to local thermodynamic equilibrium (LTE). In this regime the multistep ionization is balanced by 3-body (2e + ion) recombination, and the excited-state densities increase quadratically with n_e. In this classification, the glow discharges fit into the regime $10^{10} < n_e < 10^{16}$ cm^{-3}. In the above section we took j as a variable whereas n_e is used here. It is to be noted that n_e is not proportional to j due to:

- the nonlinear ionization processes;

- variation of the electron energy distribution.

An essential feature of glow discharges is that the ionization is not balanced by the opposite process, 2e + ion recombination ($\sim n_e^3$). The electron densities are too small. In low-pressure glow discharges, the loss of electrons and ions is mainly due to diffusion to the wall and recombination at the wall. This causes large deviations from LTE and makes it necessary to account for all relevant kinetic processes in the modeling. For high-pressure glow discharges, there is usually another dominant recombination process: molecular ion formation followed by dissociative recombination.

In low-pressure glow discharges, the radiation losses can be substantial. This is the case for example in fluorescent lamps and in the low-pressure sodium lamp. Radiation losses selectively depopulate the excited states and enhance the deviations from equilibrium.

Briefly summarized, the particle densities (e.g., n_e) and rates of kinetic processes in glow discharges are so small that considerable deviations from thermodynamic equilibrium occur. These deviations are caused by:

• The external electric field, which works almost exclusively on the electrons;

• The loss of electrons and ions due to either diffusion to and recombination at the wall, or molecular ion formation and dissociative recombination. These losses usually exceed the 2e + ion recombination by several orders of magnitude; and

• Radiation losses.

Equilibrium can still be achieved, however, with respect to certain degrees of freedom. For example, many of the plasmas studied can be characterized by one electron temperature T_e (Maxwellian distribution) and another, usually much lower, heavy particle temperature T_g. This is possible because of the rapid relaxation to (partial) equilibrium among particles with the same or similar mass: electrons or heavy particles. In contrast, the relaxation to equilibrium among particles of very different mass is slow. The rate of relaxation among the electrons is proportional to n_e^2. At the higher n_e values, e.g., in pulsed high-pressure glow discharges, this leads to a Maxwellian electron energy distribution. At the lower n_e values, as in most low-pressure positive columns, the electron-electron collisions are not frequent enough to reach partial equilibrium resulting in deviations from a Maxwell distribution, in particular at energies above the first excitation threshold (Morgan and Vriens, 1980).

Various Types of Glow Discharges

In Table I a subdivision is made among four types of glow discharges. Characteristic values for electron and gas temperatures, particle densities, and power and current densities are listed. The main processes responsible for loss of electrons and ions and for deviations from equilibrium are listed too. The four types of glow discharges are:

(1) Low-pressure noble-gas positive columns, as studied by Golubovski et al., (1974) and Smits et al. (1974);

(2) Low-pressure metal-doped gas discharges used for lighting

Table I. Some data and processes taking place in different types of discharges. T: Temperature; n_e, n_g and n_m: electron, noble gas and metal vapor density; $f(E_e)$: electron energy distribution. Extreme values of data entries in one row of the Table do not necessarily correspond to extremes in other rows.

	noble-gas low-pressure	metal-doped low-pressure (l-p lamps)	metal-seeded high-pressure (MHD)	metal-doped high-pressure (excimer laser)	arcs noble-gas metal-vapour
T_e (eV)	1 – 3	0.5 – 1	~ 0.5	0.3 – 0.5	0.5
T_e (K)	$10^4 - 3.10^4$	5000 – 10000	3500 – 6000	3000 – 5000	4000 – 6000
T_g (K)	300 – 600	300 – 600	1000 – 2000	300 – 600	4000 – 6000
n_e (cm^{-3})	$10^9 - 10^{14}$	$10^{11} - 10^{12}$	$10^{13} - 10^{15}$	$10^{14} - 10^{16}$	$10^{15} - 10^{17}$
n_g (cm^{-3})	$10^{16} - 10^{18}$	$10^{17} - 10^{18}$	$10^{19} - 10^{20}$	$10^{19} - 10^{20}$	$10^{17} - 10^{18}$
n_m (cm^{-3})		$10^{12} - 10^{13}$	10^{17}	$10^{15} - 10^{16}$	$10^{18} - 10^{19}$
power density (W/cm^3)	0.01 – 0.1	0.01 – 0.1	10 – 1000	10^6	50 – 500
current density (A/cm^2)	0.1 – 10	0.1 – 1	1 – 100	100 – 1000	5 – 50
$T_e = T_g$	no	no	no	no	yes
operation	stationary	stationary	stationary	pulsed	stationary
spatial homogeneous	no	no	no	yes	no
loss of electrons and ions due to	ambipolar diffusion to and recombination at the wall		volume recombination	molecular ion formation and dissociative recombination.	volume recombination
deviation Saha-Boltzmann due to	radiation losses and diffusion of charged particles		radiation losses	pulsed operation, selective dissociative recombination	equilibrium
deviation Maxwellian $f(E_e)$	above first inelastic threshold			equilibrium	equilibrium

purposes, such as the fluorescent lamp (Cayless, 1963; Waymouth, 1971; Polman and Drop, 1972) and the low-pressure sodium lamp (Jack, 1978; Jack and Vrenken, 1980);

(3) High-pressure, metal-seeded plasmas used for MHD generation;

(4) High-pressure, metal-doped, pulsed-excimer-laser plasmas (Rhodes, 1979; Gallagher et al., 1979, 1980).

For comparison purposes, characteristic values and properties are listed also for equilibrium (LTE) arcs, such as used in high-pressure lamps.

It is of interest to make a comparison between the metal-doped low-pressure glow discharge and the pulsed metal-doped high-pressure plasma. From Table I it follows that the values of T_e as well as T_g are very similar in both discharges. Simultaneously, the particle and current densities in the high-pressure discharge exceed those in the low-pressure discharge by three orders of magnitude. This implies that the collision rates in the high-pressure discharge are larger by a factor of 10^6. For normal AC or DC discharges, these larger collision rates would directly lead to thermodynamic equilibrium. It is only due to the pulsed operation of the high-pressure discharge that T_e can be much higher than T_g. In the low-pressure discharge, the loss of electrons and ions is mainly due to ambipolar diffusion and recombination at the wall. The much higher particle densities in the high-pressure discharge make the diffusion less important but enhance the formation of molecular ions. Dissociative recombination is therefore the dominant electron and ion loss process in the high-pressure discharge. The radiation losses are very important in the low-pressure positive column. They are mainly responsible for a loss of fast electrons due to excitation that is not balanced by de-excitation. This is the main cause for the deviations from a Maxwellian electron energy distribution. The effective radiative lifetimes in the high-pressure discharge are very similar to those in the low-pressure discharge. However, the collision rates are larger by a factor of 10^6. This makes the radiative losses negligible. As a consequence, the electron energy distribution is very close to Maxwellian in the pulsed high-pressure discharge.

As far as Table I is concerned we make two further remarks. First, the electron temperature T_e in the noble-gas, low-pressure positive column exceeds T_e in the metal-doped, low-pressure positive column by factors of 2 to 4. This difference can be ascribed directly to the much higher excitation and ionization energies of the noble-gas atoms. A comparable ion production is then possible only at higher T_e and higher electric field strength. A second remark is concerned with the definition of glow discharges where it is stated

that they are largely or completely self-sustained. The reason for
making this restriction is that several excimer-laser plasmas are
partly self-sustained (internal multiplication), and are partly (or
completely) sustained by an electron beam or by UV radiation (Rhodes,
1979). The pulsed-excimer plasmas, for which the internal multiplication is dominant, fall into the category of glow discharges.

The Low-Pressure Sodium-Noble Gas Discharge

Low-pressure sodium lamps were first introduced about 50 years
ago. Their luminous efficacy was about 50 lm/W. At present, due
to many technological improvements, efficacies of about 190 lm/W
are reached (Jack and Vrenken, 1980). At all times it was and still
is the most efficient lamp. A typical lamp is a U-shaped glass tube
with an internal diameter of 1.9 cm, contained in an evacuated outer
bulb provided with an infrared-reflecting coating to maintain the
discharge tube at the desired temperature of about 260°C. The gas
filling is 5.5 torr Ne (with 1% Ar). Under working conditions (260°C),
the Na pressure is a few mtorr. The (AC) current has a RMS value of
0.9 A. The electron temperature is ≤ 0.7 eV, so that only Na is
excited (first excited level 2.1 eV) and ionized (5.1 eV). Neon
serves as a buffer gas (first excited state: 16.6 eV, ion ground state
21.6 eV) to shorten the mean free path of the electrons and ions, to
ignite the lamp, and to heat the wall. The degree of ionization of
the plasma is low ($\bar{n}_e/\bar{n}_g = 10^{-3}$-$10^{-4}$, $\bar{n}$ denoting a radially averaged
density and n_g the gas density).

The first model of the Na-Ne positive column is only six years
old (van Tongeren and Heuvelmans, 1974, 1975). It was a relatively
simple model that correctly predicted general trends but suffered
from large quantitative discrepancies with experiment. For example,
the calculated electric field strengths exceeded the experimental
values by 50 to 100%. In the last six years considerable progress
in our understanding of the discharge has led to a new model in
which many modifications have been included:

(1) The present description uses a 25-level model of the Na
atom, instead of the three-level model (ground state, first excited
state, ion) used earlier (van Tongeren, 1975).

(2) The approximation method used to account for deviations
from a Maxwellian electron energy distribution has been improved by
Ligthart and Keyser (1980), and Morgan and Vriens (1980).

(3) New and more accurate data on several elementary processes
have become available.

(4) A new set of cross section and rate equations has been constructed to describe the electron-induced transitions between excited

states, and to describe ionization from excited states (Vriens and Smeets, 1980).

(5) A new method has been developed to deal with radiation trapping in discharges with, in particular, nonhomogeneous ground-state and excited-state density distributions (Smeets, 1982).

Figure 2 shows which levels of the Na atom are included:

- the 3s ground state and the 4s to 10s excited states (8);
- the 3p to 10p excited states (8);
- the 3d, 4d and 4f, 5dfg up to the 10 dfg ... excited states, with the dfg ... states taken together as one state (8);
- the ion ground state (1).

In the modeling we subsequently included more levels. A rapid convergence was obtained in the series 19 levels (up to p = 8), 22 levels (p = 9) and 25 levels (p = 10). The restriction of the number of levels to 25 leads to inaccuracies of 1 to 2%.

The ladder-like ionization and excitation via the higher excited states of the Na atom ($4 \leq p \leq 10$) are important for the following reasons:

(1) The ionization energy E_{pi} decreases ($\sim p^{-2}$) with increasing principal quantum number p. The cross sections for electron-impact

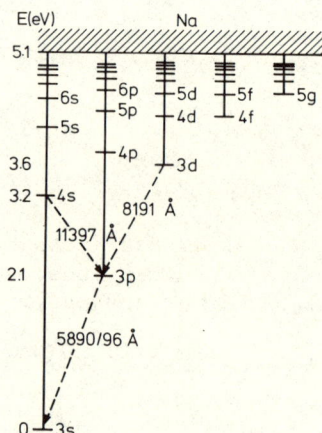

Fig. 2. Energy level scheme of the Na atom with the s,p,d,f and g levels included in the model.

ionization (see next section) become correspondingly larger ($\sim p^4$). The number of electrons that have sufficient energy to cause ionization ($E_e > E_{pi}$) increases because E_{pi} decreases ($\sim p^{-2}$).

(2) The density of excited levels in energy space increases ($\sim p^4$) with increasing p because the level spacing decreases ($\sim p^{-3}$) and because the number of states with one principal quantum number increases ($\sim p$). Moreover, the levels with higher azimuthal quantum number l have larger statistical weights. Their population density is correspondingly larger, due to the complete collisional coupling (Vriens, 1978) between states with $l \geq 2$.

(3) The level spacing decreases ($\sim p^{-3}$), and the cross sections for excitation between neighboring levels increase with increasing p (see next section).

The combined effect of (1), (2) and (3) is to enhance the multistep processes. It overcompensates for the strong reduction of the excited-state densities at higher p.

This reduction is illustrated in a Boltzmann plot in Fig. 3. The radially averaged particle densities $\bar{n}_l$ used for Fig. 3 have been obtained from measured line intensity ratios using known effective radiative lifetimes (Vriens, 1978). In equilibrium, a Boltzmann plot gives a straight line with a slope determined by the electron temperature. The type of plot shown in Fig. 3 is characteristic for a nonequilibrium loss of electrons and ions, possibly combined with radiative losses. In the low-pressure sodium lamp, the nonequilibrium

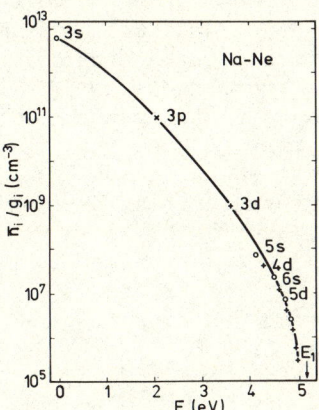

Fig. 3. Boltzmann plot of the reduced ground state and excited-state Na densities $\bar{n}_i/g_i$ versus the level energies for the Na-Ne discharge. g_i is the statistical weight of level i; $\bar{n}_i$ is the linearly averaged population density (Vriens, 1978).

loss is ambipolar diffusion of electrons and ions to the wall and recombination at the wall. This loss exceeds the 2e + ion recombination by several orders of magnitude. It leads therefore to a strong depletion of the higher excited-state and ion densities. Similar bent-shape Boltzmann plots have been obtained for the pulsed high-pressure discharges; see Gallagher et al. (1980). Molecular ion formation and dissociative recombination to excited states in Na, e.g., to states with principal quantum number 5 or 6) are responsible for the nonequilibrium loss of electrons and ions in that case. In order to clarify the bent-shape Boltzmann plot, one can make an analogy with a completely filled bathtub with a continuous stream of water flowing into it (ionization) and an equal amount of water flowing out of it over the rim (2e + ion recombination). In this equilibrium situation the water level is horizontal (straight line in a Boltzmann plot). Next, a large plug is removed from the bottom at one side of the tub (additional loss of electrons and ions due to either diffusion to the wall and recombination at the wall, or molecular ion formation and dissociative recombination). The resulting sizable nonequilibrium loss of water results in a "bent shape" of the water level as a function of the distance to the hole, similar to the bent shape, as a function of the ionization threshold in Fig. 3.

An important feature of the positive column is its quasi-neutrality, i.e., $n_e \sim n_i$. The electrons, which have a much higher mobility than the ions, rapidly drift to the wall, and the negative charge accumulated establishes a radial electric field which accelerates the ions and retards the electrons. As a result, electrons and ions drift with the same speed to the wall, thus maintaining quasi-neutrality. This combined diffusion of charged particles to the wall is called ambipolar diffusion. At the wall, the charged particles recombine, and ground state Na atoms diffuse back into the positive column. Because the ambipolar diffusion coefficient is much larger (10 to 30 times) than the neutral atom diffusion coefficient, the loss of electrons and ions is sizable (large plug in the bathtub), and large gradients in the neutral density are necessary to balance the ion flux. Since the neutral density at the wall has a fixed value (determined by the wall temperature T_w), this can result in a strongly depleted neutral density profile (see later section). In that case, the discharge is very inhomogeneous. In the modeling, it is then necessary first to make a local description of the discharge. Thereafter, radial integrations can be performed to obtain macroscopic properties such as current I, electric field F, and radiant flux P_{rad}.

The model presented below is a DC model. Experimental results are available for commuted DC Na-noble gas (Ne and Ar) low-pressure discharges (van Tongeren and Heuvelmans, 1974, 1975). The results of the model calculation will be compared with these experimental results. From experiment (van Tongeren, 1975), it is further known that commuted DC discharges give already a good impression of the time-averaged behavior of 50 (and 60) Hz AC discharges. The present

MODELING OF GLOW DISCHARGE

model can therefore be used for computer optimization of the AC low-pressure sodium lamp.

In summary, we have presented a general survey of glow discharges, and we have given a special introduction for the low-pressure Na-Ne discharge. Much of what has been said about the latter discharge applies as well for other metal vapor-noble gas positive columns.

In the following, the model of the Na-Ne discharge will be presented. Likewise, large parts of the model can be applied directly in the modeling of other discharges. In order to facilitate the reading of this paper, only the basic formulas are given. Additional formulas are listed in the Appendix.

ELEMENTARY PROCESSES

This section describes the various kinetic processes occurring in the discharge on a local scale. Among these processes, those involving the electrons are the primary and most important ones because the electron mobility μ_e greatly exceeds the mobility μ_i of the heavy Na ions. The electrons are therefore almost solely responsible for the absorption of energy from the electric field. The electrons are also responsible for the excitation, ionization, and heating of the gas due to momentum transfer and the current (electron transport). For these reasons we will first deal with all interactions involving the electrons. Thereafter, we consider the atomic structure of the heavy particles (atoms), the emission of radiation, and the local conservation relations.

The Electrons

Energy Distribution. The acceleration of the electrons in the electric field leads to an anisotropy in the velocity distribution that is responsible for the current. In the noble-gases, at the filling pressures of interest, typically 1 to 10 torr, this anisotropy is very small because of the many elastic collisions with the atoms, in each of which the electron transfers only a minor fraction of its kinetic energy. The interaction between the electrons is much more effective because of their equal mass and the long range Coulomb interaction. The electron densities n_e of interest here are large enough to lead in good approximation to a Maxwellian distribution of the electron energies E_e for the bulk of the energy distribution, i.e., below the first excitation threshold,

$$f(E_e) = 2[E_e/\pi(kT_e)^3]^{1/2} \exp(-E_e/kT_e), \qquad (1)$$

where k is the Boltzmann constant, and T_e is the electron temperature. Above the first inelastic threshold of Na, however, with threshold energy E_1 = 2.10 eV, the 3^2P resonance levels can be excited. Because of the large $3^2S \to 3^2P$ excitation rate, the fast electrons are lost rapidly, and the tail of the electron energy distribution ($E_e \geq E_1$) is depleted. This depletion is accounted for by using the two-electron-group model (2-EGM; Morgan and Vriens, 1980), in which $f(E_e)$ in the tail is represented by a modified Maxwellian distribution,

$$f^t(E_e) = 2S[E_e/\pi(kT_t)^3]^{1/2} \exp(-E_e/kT_t). \qquad (2)$$

The parameter T_t is called the tail temperature; S is a normalization function. For $E_e < E_1$, the Maxwell distribution Eq. (1) is used. For $E_e \geq E_1$, the modified distribution, Eq. (2), is used.

By making extended comparisons between Boltzmann and 2-EGM calculations for Ne, Ar, and Cs-Ar discharges, Morgan and Vriens (1980) obtained an accurate empirical function for S, which only contains T_e and T_t as variables (see Appendix). Equation (2) thus contains only one new parameter, T_t. In order to obtain T_t, the energy conservation relation for the tail group of electrons has to be solved,

$$P_F^t + P_{sel}^t + P_{rec}^t + P^{bt} = P_{el}^t + P_{exc}^t + P_{ion}^t + P_{dif}^t + P^{tb}, \qquad (3)$$

where the subscripts have the following meaning: F—electric field, sel—superelastic collisions, rec—recombination, el—elastic collisions, exc—excitation, ion—ionization, and dif—diffusion. The superscript t indicates that the tail electrons lose or receive the energy, and bt and tb indicate (cross) relaxation from the bulk to the tail and vice versa, respectively. The left-hand side of Eq. (3) represents the energy flow to the tail and the right-hand side, the loss from the tail. P^{bt} and P^{tb} are the largest and most important terms in Eq. (3). They include cross relaxation due to electron-electron collisions, excitation, deexcitation (sel) and ionization, and due to the electric field and elastic losses. All formulas are given in detail in the paper by Morgan and Vriens (1980). What matters is that only one additional parameter, T_t, is introduced, which is uniquely determined by one additional energy balance equation, Eq. (3). Hence, T_t is not a free parameter.

In order to demonstrate how well the 2-EGM works, we reproduce from Morgan and Vriens one figure and a table. In Fig. 4, calculated Boltzmann and 2-EGM electron energy distributions are shown for a 10 torr Ne discharge. E_1 denotes the first inelastic threshold at 16.6 eV; E_2 is the ionization energy of 21.6 eV. In this type of plot, a Maxwell distribution gives a straight line. The deviations

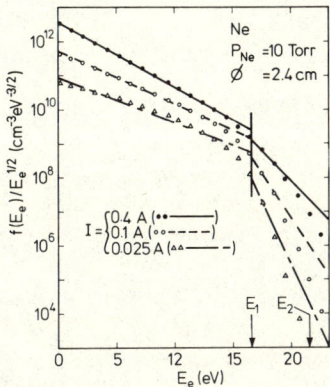

Fig. 4. A. comparison between some of the electron energy distributions calculated from the Boltzmann equation (●●,○○,△△) and, using the two-electron group model (solid, short-dashed, and long-dashed lines), for the 10 torr Ne discharge (axis values) (Morgan and Vriens, 1980).

from Maxwellian are largest at low current (small n_e). The bulk of the distribution ($E_e < E_1$) can be well represented by Eq. (1). In the tail of the distribution there is a large depletion, which is rather well accounted for in the 2-EGM, Eqs. (2) and (3). In Table II, a comparison is made between the rate coefficients for excitation to the first excited state (or group of lowest excited states), as calculated for Cs-Ar, Ar and Ne discharges, using Maxwell distribution, 2-EGM and Boltzmann calculations. The 2-EGM clearly yields very good results, even when the deviations from a Maxwell distribution are very large.

The major advantages of the 2-EGM are its simplicity and the almost negligible computer time and cost required. The 2-EGM is applicable to noble-gas and metal-doped, noble-gas discharges, but not to molecular discharges as these are generally dominated by many inelastic processes each of very small energy loss.

Electron Transport. We first consider transport in the longitudinal (axial) direction, caused by the macroscopic electric field. The current density $\vec{j}$ is related to the drift velocity $\vec{v}_d$, electron mobility μ_e and electric field strength $\vec{F}$ by

$$\vec{j} = -n_e e \vec{v}_d = n_e e \mu_e \vec{F}, \qquad (4)$$

e being the elementary charge. For a stationary distribution function and in the absence of n_e - gradients in the axial direction, as considered here, v_d is constant. The energy gain of the electrons

Table II. Comparison between Boltzmann, 2-EGM and Maxwell excitation rate coefficients K_{01}, indicated by the additional subscripts B, G and M, respectively (Morgan and Vriens, 1980).

Discharge	I (mA)	K_{01B}	K_{01G} (10^{-14} cm^3 s^{-1})	K_{01M}	T_e (10^2 K)	T_t
Cs-Ar (5 torr)	300	287(4)*	295(4)	324(4)	52.1	50.1
	100	376(4)	359(4)	551(4)	61.0	50.5
Ar (5 torr)	400	8.5	10.0	95	158	82.3
	100	9.3	11.9	478	192	62.4
	25	1.6	4.5	1455	225	40.7
Ne (10 torr)	200	38.1	39.3	508	292	120
	50	22.2	22.2	1710	353	77
Ne (30 torr)	400	17.1	16.5	133	245	124
	100	13.2	13.1	595	299	82
	25	7.9	7.7	2010	363	57

*(4) indicates 10^4.

MODELING OF GLOW DISCHARGE

in the electric field is then compensated by the loss due to collisions. The electron distribution in velocity space may be written as the usual Legendre polynomial expansion. For the relatively low electric field strengths of interest here, only the first two terms of this expansion are important, and the drift velocity is given by (Huxley and Crompton, 1974),

$$\vec{v}_d = \frac{4\pi e \vec{F}}{3mN} \int_0^\infty \frac{v^2}{\sigma_m} \frac{df_o}{dv} \, dv, \qquad (5)$$

where m is the electron mass, N is the total particle density and f_o is the first term of the Legendre polynomial expansion. σ_m is the effective momentum transfer cross section,

$$\sigma_m = \sum_j \frac{n_j}{N} \sigma_{mj} + \frac{n_e}{N} \frac{\sigma_{mi}}{\gamma}, \qquad (6)$$

where the summation (j) extends over the various neutral particles, and i stands for the Na-ion. The densities of the excited states of the noble gas atoms and of the higher excited states of Na are too small to contribute appreciably to σ_m. The last term gives the contribution of electron-ion collisions,

$$\sigma_{mi} = \frac{4\pi e^4 z^2}{(mv^2)^2} \ln \Lambda \qquad (7)$$

$$= \frac{\pi e^4 z^2}{E_e^2} \ln \left[\frac{mv^2}{2e^3 z} \left(\frac{kT_e}{2\pi n_e} \right)^{1/2} \right],$$

where z is the effective nuclear charge and $\ln \Lambda$ is the Coulomb logarithm; see Mitchner and Kruger (1973). The momentum transfer cross sections σ_{mj} are taken from Crompton et al. (1970, 1977) for He, Robertson (1972) for Ne, and Milloy et al. (1977) and Frost and Phelps (1964) for Ar. The σ_m for Na (3^2S and 3^2P) are taken from Moores and Norcross (1972), Moores et al. (1974) and Phelps, A.V. and Gallagher, A. (1978), private communication.

When electron-electron interactions are negligible, e.g., at low n_e, Eqs. (5) and (6) apply with $\gamma = 1$. The last term in Eq. (6) is unimportant in this case. When electron-neutral interactions are negligible, as is the case for a fully ionized plasma, only the last term in Eq. (6) is important and $\gamma = 0.582$ for $z = 1$. In that case, the electron-electron interactions are responsible for changing γ from 1 (electron-ion interactions only) to 0.582, see e.g. Mitchner and Kruger (1973). For the more complicated intermediate situation of a weakly ionized plasma, there is no simple function for γ available. Detailed calculations by Keijser et al. (1981), who expanded the electron velocity distribution in Sonine polynomials (see Mitchner and Kruger, 1973), have shown that Eq. (6) with $\gamma = 0.582$ gives values for μ_e that are 10% too low for Na-Ne in the region of interest. To find a simple function for γ is still a subject of study.

Substituting a Maxwell distribution $f_o(v)$ or the 2-EGM distribution in Eqs. (4) and (5) yields

$$\mu_e = \frac{e}{3kT_e} \left\langle \frac{v}{N\sigma_m} \right\rangle, \qquad (8)$$

where $\langle \rangle$ denotes averaging over the electron energy distribution. The (isotropic) coefficient of diffusion, which will be needed in the discussion of radial electron transport, is defined (in the case of a Maxwell distribution) by

$$D_e = \frac{kT_e}{e} \mu_e = \frac{1}{3} \left\langle \frac{v}{N\sigma_m} \right\rangle. \qquad (9)$$

This transport in the radial direction, in which the ions participate too, will be dealt with later.

<u>Gas Heating.</u> The gas heating in the discharge is mainly due to elastic collisions of electrons with atoms and ions. In the energy balance equation, it is a dominant term, and in the actual sodium discharge lamp, it keeps the discharge wall at the desired temperature of $\approx 260°C$. At each electron velocity v, the energy transfer to the heavy particles is proportional to the effective momentum transfer collision frequency ν_m defined by the sum of the respective collision frequencies weighted with the fraction of energy transferred:

$$\frac{2m}{M} \nu_m = \sum_j \frac{2m}{M_j} (n_j v \sigma_{mj}) + \frac{2m}{M_i} (n_e v \sigma_{mi}), \qquad (10)$$

M being the effective mass of the heavy particles,

$$M = \sum_j \frac{n_j}{N} M_j + \frac{n_e}{N} M_i. \qquad (11)$$

The energy loss of the electrons per unit volume and per unit time is (see Shkarofsky et al., 1966),

$$P_{el} = \frac{2kT_e}{\sqrt{\pi}} (1 - \frac{T_g}{T_e}) n_e \int_0^\infty \frac{2m}{M} \nu_m \varepsilon^{3/2} \exp(-\varepsilon) d\varepsilon$$

$$= (1 - \frac{T_g}{T_e}) n_e < \frac{2m}{M} \nu_m E_e > ,$$

(12)

with
$$\varepsilon = E_e / kT_e .$$

<u>Excitation and Ionization.</u> In the modeling of nonequilibrium plasmas it is essential to know the cross sections or, more precisely, the rates for the electron transfer processes. For transitions between low levels of the Na atom, experimental (3S → 3P) and calculated (3S → 4S, 3D; 3P → 4S, 3D) cross sections are available in the literature; for 3S→3P from Enemark and Gallagher (1972), and for the other transitions from Moores et al. (1972, 1974). We have approximated these cross sections with three linear segments, and use a $1/E_e$ dependence for the higher electron energies. Integrating over the electron energy distribution (Maxwell or 2-EGM) then yields simple analytical formulas for the rate coefficients. For ionization from the ground state of Na, the experimental cross sections of Zapesochnyi and Aleksakhin (1969) have been used. These cross sections and the corresponding rate coefficients have also been represented in a simple analytical way.

The approximate analytical cross sections and rate equations of Vriens and Smeets (1980) have been used for all other transitions in the Na atom. These mainly involve excitation, de-excitation and ionization from excited states. The cross section and rate equations of Vriens and Smeets (1980) were obtained by empirical fitting to a wide variety of relevant experimental and theoretical cross sections and rates, available in the literature for many different ground state and excited atoms. They are mainly valid for ground state atoms with one electron in an outer orbital and for excited atoms. The structure of the formulas has been chosen in such a way, that they have the correct limit for large E_e and large kT_e. Because of their validity for many different atoms, we list the most important of these equations and mention their essential properties.

For ionization, the cross section formula is

$$\sigma_{pi} = \frac{4\pi a_o^2 R^2}{E_e + 3.25 E_{pi}} (\frac{5}{3E_{pi}} - \frac{1}{E_e} - \frac{2E_{pi}}{3E_e^2}),$$

where a_o is the Bohr radius, R the Rydberg energy (13.6 eV) E_{pi} the ionization energy and E_e again the electron energy. This formula has the following features:

(1) It reproduces the classical three-body cross sections, obtained to Abrines et al. (1966) using Monte Carlo calculations within their statistical accuracy ($\simeq 10\%$). This classical theory becomes exact for large principal quantum number p.

(2) It gives the correct binary-encounter approximation cross sections for large E_e (this duplicates point 1); see Vriens (1969).

(3) Applied to experiment it gives cross sections that are 10 to 30% higher than measured for H 2S, 15 to 20% lower than for He 2^3S, overlap experiment for Ne*, and are 10 to 15% lower for Ar*. For the alkali atoms, agreement with experiment is found also in a 5 to 30% range. This is illustrated in Table III where experimental and calculated positions and magnitudes of the first maxima of the ionization functions are shown.

Table III. Experimental (EXP) and calculated (BEA, Eq. (13)) positions and magnitudes of the first maxima of the ionization functions (Vriens and Smeets, 1980).

atom	EXP		BEA	
	E_e (eV)	σ_{pi} (10^{-16}cm^2)	$E_e = 2.71 E_{pi}$ (eV)	σ_{pi} (10^{-16}cm^2)
H 2s	13.0±2	9.1±2[a]	9.2	11.4
He 2^3S	12.6±1	7.1±0.4[b]	12.9	5.8
Ne*	14.0±2	5.5±0.4[c]	13.4	5.4
Ar*	12.7±3	8.5±0.4[c]	11.4	7.4
Li	14.1±1	4.3±0.2[d]	14.6	4.5
Na	14.0±1	6.8±0.4[d]	13.9	5.0
K	9.3±0.8	7.8±0.4[d]	11.8	7.0
Rb	10.3±1	9.2±0.4[d]	11.3	7.6
Cs	10.1±1	7.1±0.4[d]	10.6	8.7

a: Dixon et al. (1975)
b: Dixon et al. (1976)
c: Dixon et al. (1973)
d: Zapesochnyi and Aleksakhin (1969), Nygaard (1968, 1975).

The rate coefficient corresponding to Eq. (13) is

$$K_{pi} = \frac{9.56 \times 10^{-6} (kT_e)^{-1.5} \exp(-\varepsilon_{pi})}{\varepsilon_{pi}^{2.33} + 4.38\, \varepsilon_{pi}^{1.72} + 1.32\, \varepsilon_{pi}}, \qquad (14)$$

where $\varepsilon_{pi} = E_{pi}/kT_e$, kT_e is expressed in eV and K_{pi} in $cm^3 s^{-1}$. Equation (14) has the following properties:

(1) For large E_{pi} (or small kT_e), it reproduces the (adiabatic) classical 3-body calculations of Mansbach and Keck (1969).

(2) For small E_{pi} (or large kT_e), the (exact) binary encounter approximation limit is obtained.

(3) In the intermediate range of E_{pi} (kT_e) values, it reproduces the rate coefficients obtained by numerical integration of Eq. (13) over a Maxwell distribution within 1% for $kT_e > 0.2\, E_{pi}$ and within 10% for $kT_e > 0.05\, E_{pi}$.

The cross section formula for $|p\rangle \to |n\rangle$ excitation is

$$\sigma_{pn} = \frac{2\pi a_o^2 R}{E_e + \gamma_{pn}} [A_{pn} \ln(\frac{E_e}{2R} + \delta_{pn}) + B_{pn}], \qquad (15)$$

where

$$A_{pn} = (2R/E_{pn}) f_{pn}, \qquad (16)$$

E_{pn} is the excitation energy, f_{pn} is the absorption oscillator strength. B_{pn} depends on the atomic structure the values of which can be obtained from the experimental or theoretical high energy limit of the cross sections; and γ_{pn} and δ_{pn} are empirical functions, which depend on the ionization energies of the levels $|p\rangle$ and $|n\rangle$, and on A_{pn} and B_{pn}. Analytical functions for B_{pn}, γ_{pn} and δ_{pn} have been given by Vriens and Smeets (1980). Equation (15) has the following properties:

(1) For the first resonance transitions in the alkali atoms and for the H 1S → 2(S+P) transition, it yields cross sections that agree with experiment within typically 5 to 30%. For some of these transitions this is shown in Fig. 5; see also Figs. 3 to 5 of Vriens and Smeets (1980).

(2) For $|p\rangle \to |n\rangle$ transitions in the H-atom, cross sections are obtained which agree with theoretical cross sections of Gee and Percival et al. (1976) within 5 to 20%.

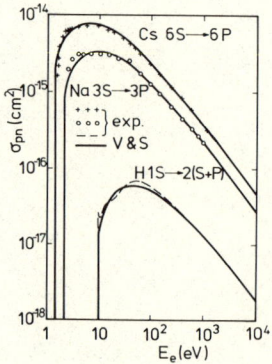

Fig. 5. Experimental and calculated (Eq. (15)) cross sections σ_{pn} for the H1S → 2(S+P), Na 3S → 3P and Cs 6S → 6P transitions induced by electron impact (Vriens and Smeets, 1980).

(3) In the high-energy limit, the functions γ_{pn} and δ_{pn} are negligible, and the Bethe theory high energy formula is obtained (Inokuti, 1971).

The rate coefficients for excitation are well described by the approximate formula

$$K_{pn} = \frac{1.6 \times 10^{-7} (kT_e)^{0.5}}{kT_e + \Gamma_{pn}} \exp(-\varepsilon_{pn}) \times [A_{pn} \ln(\frac{0.3 kT_e}{R} + \Delta_{pn}) + B_{pn}], \quad (17)$$

in $cm^3 s^{-1}$, with kT_e and R in eV. In Eq. (17), $\varepsilon_{pn} = E_{pn}/kT_e$ and Γ_{pn} and Δ_{pn} are empirical functions, which depend on the ionization energies of the levels $|p\rangle$ and $|n\rangle$, and on A_{pn} and B_{pn}. Analytical functions for Γ_{pn} and Δ_{pn} have been given by Vriens and Smeets (1980); see also the Appendix. Equation (17) has the following features:

(1) It gives the correct Bethe-theory limit for large kT_e.

(2) It reproduces the rates obtained by numerical integration of Eq. (15) within 1 to 10% for $kT_e \geq 1$ eV, and within 5 to 20% for $kT_e \geq 0.1$ eV. This is so for a large number of transitions in the H atom, i.e., 1S → 2(S+P) and $|p\rangle \to |n\rangle$ with p varying from 5 to 50 and n from p+1 to p+10. These latter transitions are relevant for the present study because excited alkali atoms are very hydrogenic, in particular for the excited P, D, F... states. The excited

S states contribute little to the total ionization, which make the deviations from a hydrogenic behavior for these states unimportant.

(3) It yields rates for the resonance transitions in the alkali atoms that agree within 10 to 20% with the rates corresponding to the experimental cross sections when integrated over a Maxwellian (or 2-EGM) electron energy distribution.

(4) At low temperatures, it yields rates that agree well with those obtained from the classical adiabatic theory of Mansbach and Keck (1969).

A comparison between the rates obtained with Eq. (17) and the rates obtained from Mansbach and Keck (1969) and Gee et al. (1976) is shown in Fig. 6. At the higher temperatures, good agreement with Gee et al. is obtained; at the lower temperatures there is good agreement with Mansbach and Keck. The rate coefficients of Gee et al. are claimed to be accurate for $T_e > 10^6/p^2$.

Furthermore, we can make an interesting comparison with rates obtained by Delpech et al. (1977, 1979) using time-resolved laser-induced fluorescence spectroscopy. They measured the total depopulation as well as the individual transfer rates in a helium afterglow plasma with temperatures in the range from 300 to 6000°K, for triplet levels with p = 8-17 to triplet levels with n = 5-17. The total depopulation rate (coefficient) is the sum of the ionization rate, the excitation rates, and the de-excitation rates:

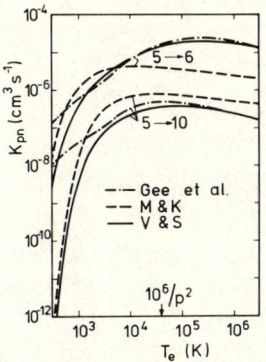

Fig. 6. Rate coefficients K_{pn} for electron-impact excitation of H p = 5 according to Gee et al. (1976), Mansbach and Keck (1969) and Vriens and Smeets (1980). Gee et al. claim their K_{pn} to be accurate above $T_e = 10^6/p^2$ (in °K).

$$K_p = K_{pi} + \sum_{p \neq n} K_{pn} . \tag{18}$$

The de-excitation rate is obtained from Eq. (17) by applying detailed balancing. The de-excitation rate coefficient formula is given in Vriens and Smeets (1980). In Fig. 7, total depopulation rates obtained with Eq. (18) are compared with the K_p measured by Delpech et al. (1977, 1979) for T_e = 400 K and p = 8 to 17. The Vriens and Smeets formulas yield better agreement with experiment than the Mansbach and Keck formulas, whereas the classical theory of Gryzinski (as used by Delpech et al.) is incorrect by one to two orders of magnitude. A comparison with experiment (Delpech) for the depopulation rate of a single level (p = 10) for different T_e is shown in Fig. 8. Equation (18) gives again the best agreement with experiment.

When the threshold for either excitation or ionization exceeds the first excitation threshold E_1, T_e in the rate equations has to be replaced by T_t, and the tail normalization function S of the 2-EGM has to be included.

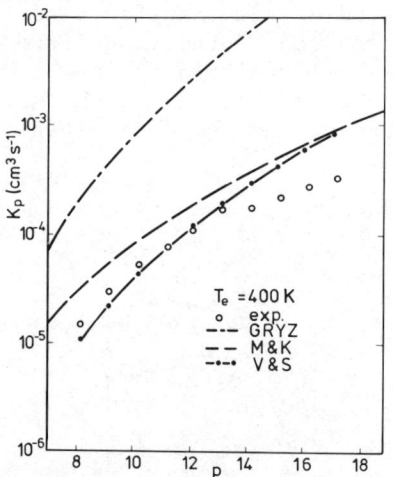

Fig. 7. The total electron-induced depopulation rate coefficients K_p of He 8^3P up to 17^3P according to:
  (1) experiment (Delpech et al. 1977, 1979);
  (2) Mansbach and Keck (1969) with the discrete level structure taken into account (see Vriens and Smeets, 1980);
  (3) the classical binary encounter calculations according to the Gryzinski theory (as reported by Mansbach and Keck, 1969);
  (4) Vriens and Smeets (1980).

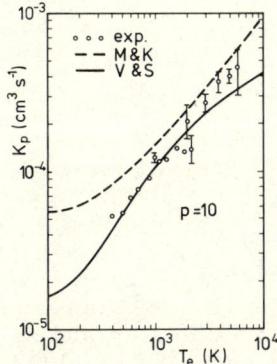

Fig. 8. The total depopulation rates K_p for the initial state He 10^3P as a function of T_e Vriens and Smeets, 1980).

The Heavy Particles

The heavy particles (atoms) interact with the electrons, absorb and emit photons, and interact also with each other. Furthermore, there is transport in the radial direction. Before dealing with the elementary processes, the states involved must be specified. In addition to the notation $|p\rangle$ and $|n\rangle$ used so far for lower and upper states, more specific notations containing a varying amount of information will be used, such as $|p^2D_{5/2}\rangle$, $|p^2D\rangle$, $|pD\rangle$, $|pL\rangle$ and $|pL'\rangle$. Here L and L' are the angular momentum quantum numbers of initial and final atomic states.

Atomic Structure. Since excitation and ionization of the noble gas are neglected, there are only ground-state noble gas atoms in the model. For Na, many atomic states and the ion ground state have to be included. The hyperfine- and fine-structure splitting of the atomic states is, fortunately, extremely small compared to the average electron energy. The resulting rapid collisional coupling between these sublevels leads to relative densities given by the statistical weights. For the electron-atom collisions the sub-levels may thus be taken together. In the radiation transport, it is necessary to account for the subdivision (see later section). For the electron-atom collisions, a further simplification is possible due to the small energy-level separation and complete collisional coupling between the pD, pF ($p \geq 4$), pG and higher L-levels with the same p (Gallagher et al., 1977; Vriens, 1978). These states can be taken together too, and we are left with pS, pP and pQ states with $p \geq 3$, where pQ stands for the pD, pF, pG... states together. The densities are related as follows:

$$n_{pQ} = \frac{g_Q}{g_D} n_{pD} = \frac{p^2-4}{5} n_{pD} \tag{19}$$

and

$$n_{p^2D_{5/2}}^2 = \frac{g_{p^2D_{5/2}}}{g_D} n_{pD} = \frac{3}{5} n_{pD}, \qquad (20)$$

where the g's are the statistical weights. The ionization energies can further be approximated by

$$E_{pi} = R/x^2 = R/(p-\delta_L)^2, \qquad (21)$$

x being the effective quantum number and δ_L the quantum defect. From the well-known ionization energies we obtain for the S, P and Q series of the sodium atom:

$$\begin{aligned}\delta_S &= 1.34 + 0.3/p^2, \\ \delta_p &= 0.86 + 0.2/p^2, \\ \delta &\simeq \delta_{3D} \simeq 0.1/p^2.\end{aligned} \qquad (22)$$

The 25-level model thus contains the atomic pS, pP and pQ levels, with $3 \leq p \leq 10$, and the ion ground state.

Before the electron-induced transition rate Eq. (17) can be applied, the A_{pn} and B_{pn} coefficients must be known. The A_{pn} coefficients are proportional to the oscillator strengths, Eq. (16), which are known from literature (Lindhard and Nielsen, 1977). Analytical equations for B_{pn} are given in the Appendix.

<u>Radiative Lifetimes and Line Profiles.</u> The rate coefficient for photon emission, per excited atom and per unit time is the transition probability. The lifetime is the inverse of the transition probability. For the upper state $|n\rangle$, the total rate coefficient for photon emission is

$$\tau_n^{-1} = \sum_{p<n} \tau_{np}^{-1}, \qquad (23)$$

where τ_n is the natural lifetime of state $|n\rangle$. For optically thin lines, $n_n\tau_n^{-1}$ and $n_n\tau_{np}^{-1}$ directly yield the total and individual rates for photon emission. For optically thick lines, radiation trapping is important, and the line profiles must be known. Since the line profiles may be smaller than or comparable to the hyperfine- and fine-structure splitting, it is necessary to take this splitting into account. It is further necessary to account for Doppler broadening and Lorentz broadening (due to collisions with noble-gas atoms). For the individual transitions, we include the following information about atomic structure (splitting) and the following broadening mechanisms:

3S-3P (fs, hfs, D, L),
3S-4P (fs, hfs, D) ,
3P-3D (fs, D, L) ,
3P-4D (fs, D) ,
3P-4S (fs, hfs, D) ,

where hfs stands for hyperfine structure, fs for fine structure, D for Doppler and L for Lorentz. Characteristic values for the product of absorption coefficient (Mitchell and Zemansky, 1971) and tube radius are:

50 to 100 for the $3S-3P_{1/2}$ transition,

100 to 200 for $3S-2P_{3/2}$,

1 to 3 for 3S-4P,

5 to 8 for 3P-3D,

0.5 to 0.8 for 3P-4D, and

1.5 to 2 for 3P-4S.

The other transitions are optically thin.

On a local scale, we are interested in the difference between the individual rates for photon emission and absorption for the optically thick lines. The rates for photon emission are given by $n_n \tau_{np}^{-1}$. The rates for photon absorption, however, cannot be obtained from only the local properties of the plasma. The photons that are absorbed may have come from any other place in the discharge. Appropriate averaging over all positions is necessary to reduce the problem to a local scale. This will be dealt with in the section on "Macroscopic Properties."

<u>Ambipolar Electron-ion and Neutral Atom Diffusion.</u> On a short time scale, the transport of ions in the axial direction is negligible compared to the transport of electrons. This is due to the fact that the ion mobility μ_i is much smaller than the electron mobility μ_e. On a longer time scale there may be substantial segregation that may be avoided by commuting the DC discharge. The transport of Na in the axial direction is then negligible and need not be considered.

In the radial direction the situation is quite different. The electrons and ions created by electron-atom collisions tend to diffuse out of the positive column towards the wall. The electrons are faster and a positive excess builds up, resulting in a radial space charge field $\vec{F}_r$, which assists the ion and retards the electron current. In the steady state the radial electron drift velocity $\vec{v}_r$ is equal to the radial ion drift velocity. For this drift velocity $\vec{v}_r$, it can be shown, using the Einstein relations $D_e/\mu_e = kT_e/e$ and

$D_i/\mu_i = kT_g/e$, the quasi-neutrality condition $n_i \simeq n_e$, and finally the fact that T_e is almost independent of r and much larger than T_g, that

$$v_r = -D_a \frac{\partial \ln n_e}{\partial r}, \qquad (24)$$

where

$$D_a = (1 + \frac{T_e}{T_g}) D_i . \qquad (25)$$

The flux of charged particles follows from

$$\Gamma_r = n_e v_r = -D_a \frac{\partial n_e}{\partial r} . \qquad (26)$$

For the ambipolar electric field F_r, one finds

$$F_r = \frac{-kT_e}{e} \frac{\partial \ln n_e}{\partial r} . \qquad (27)$$

To estimate the order of magnitude of the ambipolar field we introduce an electrostatic potential $\phi = \phi(r)$ with an axis value $\phi(0) = 0$ and $F_r = -\partial\phi/\partial r$. Equation (27) yields:

$$n_e = n_e(r) = n_e(0) \exp(e\phi/kT_e) . \qquad (28)$$

For characteristic values $T_e = 8000$ K and $n_e(0) = 4 \times 10^{12}$ cm^{-3}, radial field strengths smaller than about 3 V/cm are obtained in a region extending from the axis up to very close to the wall, where the electron density is reduced already by a factor of 100 as compared to the n_e at the axis. The axial electric field strengths are typically smaller than 1 V/cm. At these small field strengths, the ion mobilities μ_i and diffusion coefficients D_i are equal to their zero field values. The values used here are taken from the extended tables by Ellis et al. (1976). In some of the calculations the buffer gas Ne was replaced by a mixture of noble gases. The ion diffusion coefficients are then obtained from Blanc's law,

$$\frac{1}{D_i} = \sum_j \frac{n_j}{N} \frac{1}{D_{ij}} , \qquad (29)$$

where N is the total noble gas density, n_j the density of component j, and D_{ij} is the corresponding ion diffusion coefficient. In the case of weak electric field strength, Blanc's approximation is well justified (see Ellis et al., 1976).

Due to the very small excited-state densities, the neutral atom transport comes almost exclusively from the ground-state atoms. For the neutral (atom) radial flux Γ_{ra}, we have this good approximation

$$\Gamma_{ra} = -D_{3S}\, n_{3S} \left(\frac{\partial \ln n_{3S}}{\partial r} + \frac{\partial \ln T_g}{\partial r} \right). \tag{30}$$

Due to the low 3P density and the smaller diffusion coefficient D_{3P} in the Ne-Na gas, we neglect diffusion of the 3P atoms. The gas temperature gradients are small compared to the Na density gradients in the Na-noble gas positive columns considered. For other positive columns (e.g., Hg-noble gas), T_g gradients may be of greater significance. The diffusion coefficients of Na in the noble gases are taken from literature, e.g., for Ne from Coolen and Hagedoorn (1975). Blanc's law is used again for mixtures of noble gases.

Conservation Relations

So far we have dealt mostly with the elementary processes. On a local scale there are in addition several conservation relations. These link the individual processes together and provide us with the basic equations to be solved in the model.

Particle Balance. In the particle balance, contributions due to diffusion, excitation to and de-excitation from higher levels, ionization and radiative decay have to be included. For the ground state the following holds

$$\frac{\partial n_{3S}}{\partial t} = -\vec{\nabla}\cdot\vec{\Gamma}_{ra} - n_{3S} n_e \sum_n K_{3Sn} + n_e \sum_n n_n K_{n3S}$$
$$- n_{3S}\, n_e\, K_{3Si} + \sum_n n_{nP} (\tau_{nP3S})^{-1}_{eff}, \tag{31}$$

where $\Gamma_{ra} = \Gamma_{r3S}$, K_{3Sn} is the rate coefficient for $|3S\rangle \rightarrow |n\rangle$ excitation, K_{n3S} is the rate coefficient for the inverse process, K_{3Si} represents the ionization, and $(\tau_{nP3S})_{eff}$ denotes the effective radiative lifetime for the $|nP\rangle \rightarrow |3S\rangle$ transition. Note that, e.g., for the $|3P\rangle \rightarrow |3S\rangle$ radiative transition a subdivision is also made with respect to the hyperfine-structure (3S) and fine-structure (3P) states. So, the actual formula used in the modeling is more complicated than Eq. (31). For the excited states, diffusion is neglected and

$$\frac{\partial n_p}{\partial t} = -n_p n_e \sum_{n\neq p} K_{pn} + n_e \sum_{n\neq p} n_n K_{np}$$
$$- n_p n_e K_{pi} - n_p \tau_{peff}^{-1} + \sum_{n>p} n_n (\tau_{np})_{eff}^{-1}. \tag{32}$$

For the electrons and ions,

$$\frac{\partial n_e}{\partial t} = - \vec{\nabla} \cdot \vec{\Gamma}_r + n_e \sum_p n_p K_{pi} \quad . \tag{33}$$

The inclusion of the diffusion terms $\vec{\nabla} \cdot \vec{\Gamma}_{ra}$ and $\vec{\nabla} \cdot \vec{\Gamma}_r$ makes the present model different from, e.g., that of Fujimoto et al. (1972, 1973). They considered a spatially homogeneous plasma and defined a kind of overall ion loss rate coefficient to account for wall losses. Summing Eqs. (31) to (33) leads to

$$\frac{\partial}{\partial t} (\sum_p n_p + n_e) = - \vec{\nabla} \cdot (\vec{\Gamma}_{ra} + \vec{\Gamma}_r) \quad . \tag{34}$$

The left side of Eq. (34) need not in general be equal to zero, due to, e.g., unequal condensation and evaporation at the wall. In the steady state, however, derivatives with respect to time are zero.

Momentum Conservation. One can also write the various momentum conservation relations. These relations have been studied in detail by Mitchner and Kruger (1973). Among the results obtained are the basic formulas used here to describe the ambipolar diffusion. Apart from this, momentum conservation does not lead to results that are of interest for the present analysis.

Energy Conservation. Elastic electron-atom collisions (predominantly with noble-gas atoms) heat the gas, and this energy is transported to the wall by heat conduction. Noble-gas atoms are therefore most important in the energy balance. In contrast, the noble-gas atoms are not important at all in the particle balance.

For the excited Na atoms, it will not be necessary to write separate energy balance equations for each state. Because of the small diffusion terms and the small elastic losses for these states, doing so would only duplicate the particle conservation relations. Instead, only one energy conservation formula will be used for all excited Na states together. Energy and particle conservation are thus also quite different for the excited states.

In the model, energy conservation relations have been used for:

- the electrons and ions together;
- the tail group of electrons;
- the ground-state noble-gas and Na atoms;
- the excited Na atoms and photons together;
- the total energy.

For the low-pressure discharge considered, the ion temperature is equal to the gas temperature, $T_i = T_g$, the radial drift velocities of electrons and ions are equal (ambipolar diffusion), and $M_{Na} \gg m_e$. The total energy per unit volume of the electrons and ions U_{ei} is therefore given by

$$U_{ei} = n_e [\tfrac{3}{2} k (T_e + T_g) + \tfrac{1}{2} M_{Na} v_r^2 + E_{3Si}] . \tag{35}$$

The axial drift term $\tfrac{1}{2} m_e v_d^2$ is negligible because the drift velocity is much smaller than the random velocity of the electrons. The drift term $\tfrac{1}{2} M_{Na} v_r^2$ must be retained since it may become appreciable close to the wall. Energy conservation implies that (Mitchner and Kruger, 1973)

$$\frac{\partial U_{ei}}{\partial t} = - \vec{\nabla} \cdot (U_{ei} \vec{v}_r + P_{ei} \vec{v}_r + \vec{q}_{ei}) + \vec{j} \cdot \vec{F}$$
$$- P_{el} - P_{inel} + P_{sel} + n_e E_{3Si} \sum_p n_p K_{pi} , \tag{36}$$

where $\vec{v}_r$ in general includes drift and global motion of the plasma, P_{ei} is the sum of electron and ion pressure $(p_e + p_i = n_e k T_e + n_i k T_i)$; $\vec{q}_{ei} = \vec{q}_e + \vec{q}_i$ is the heat flux vector; and $\vec{j} \cdot \vec{F}$ is the electric field gain term. The following three terms represent the elastic losses, inelastic losses, and superelastic gain, respectively. The last term is the rate of ion production multiplied with the internal energy per ion. After radial averaging (see later section), Eq. (36) yields the value of the electric field $\vec{F}$, since the energy lost by the electrons and ions has to be supplied by $\vec{F}$.

The energy conservation relation for the tail electrons, Eq. (3), is used to obtain the tail temperature T_t. The tail temperature is important in calculating the excitation and ionization rates.

The energy conservation relation for the atoms is in good approximation given by

$$\frac{\partial U_g}{\partial t} = - \vec{\nabla} \cdot [\tfrac{5}{2} k T_g \vec{\Gamma}_{ra} + \vec{q}_g] + P_{el} , \tag{37}$$

where $U_g = (3/2) N k T_g$, $\vec{q}_g = - \lambda_g \vec{\nabla} T_g$ is the heat flux vector. λ_g is the thermal conductivity of the gas, which for Ne is taken from Ho et al. (1972). Equation (37) implies that elastic losses are carried away mainly by heat conduction. For the present application to a stationary discharge this is correct. The term $(5/2) k T_g \vec{\Gamma}_{ra}$ corresponds to the term $(U_{ei} + P_{ei}) \vec{v}_r = (5/2) k (T_e + T_g) \vec{\Gamma}_r$ in Eq. (36).

For the excited Na atoms and photons, we write

$$\frac{\partial}{\partial t}(U_{rad} + U_{exc}) = P_{inel} - P_{sel}$$
$$- n_e E_{3Si} \sum_P n_p K_{pi} - \sum_n n_n (\tau_{np})_{eff}^{-1} E_{pn} , \quad (38)$$

where U_{rad} is the energy per unit volume due to the photon density, and $U_{exc} = \sum_n n_n E_{3Sn}$ is the energy contained by the excited atoms. Equation (38) implies that the total energy of photons plus excited atoms is increased by inealstic electron-atom collisions P_{inel} and decreases by superelastic collisions P_{sel}, by ionization and by emission of photons.

The total energy per unit volume $U_{tot} = U_{ei} + U_g + U_{rad} + U_{exc}$ does not change in the steady state. The derivative with respect to time is zero, and summing Eqs. (36), (37) and (38) then yields:

$$\frac{\partial U_{tot}}{\partial t} = -\vec{\nabla} \cdot [(\frac{5}{2}kT_e + E_{3Si} + \frac{1}{2}M_{Na} v_r^2) n_e \vec{v}_r$$
$$+ \vec{q}_{ei} + \vec{q}_g] + \vec{j} \cdot \vec{F} - \sum_n n_n (\tau_{np})_{eff}^{-1} E_{np} = 0 . \quad (39)$$

The energy absorbed by the electrons in the electric field, $P_F = \vec{j} \cdot \vec{F}$, is thus lost and balanced by:

- radial transport of electrons, i.e. thermal energy;

- radial transport of ions, i.e. internal energy (E_{3Si}) and drift energy;

- heat conduction of electrons, ions and gas atoms;

- radiation.

MACROSCOPIC PROPERTIES

Boundary Conditions

In discharge modeling it is essential to use sufficiently accurate boundary conditions. It is further far from trivial to deal with the boundary conditions at the wall. We therefore give a detailed analysis of this problem.

Because of the cylindrical symmetry, the derivatives of densities and temperatures with respect to the radial coordinate are

MODELING OF GLOW DISCHARGE

zero at the axis. Hence,

$$\left(\frac{\partial n_p}{\partial r}\right)_{r=0} = \left(\frac{\partial n_e}{\partial r}\right)_{r=0} = \left(\frac{\partial T_g}{\partial r}\right)_{r=0} = \left(\frac{\partial T_e}{\partial r}\right)_{r=0} = 0 . \tag{40}$$

In a (discharge) tube with uniform wall temperature T_w and current I = 0, the Na vapor pressure is uniquely determined by T_w. According to Fairbank et al. (1975), the sodium density is given by

$$\ln n_{Na} = -3.00652 \ln T - 13099.1 \, T^{-1}$$
$$+74.7015 \quad +1.15167 \times 10^{-3} \, T \tag{41}$$
$$-2.12972 \times 10^{-7} \, T^2 + 6.945 \times 10^4 \, T^{-1} \exp(-13600 \, T^{-1}) ,$$

where $T = T_w$ is expressed in °K and n_{Na} in cm^{-3}. when I ≠ 0, n_{Na} at the wall (r = R) may depend on I because of the radial depletion of the Na density distribution caused by the ion and electron flux to the wall. The radial depletion itself leads to a lowering of n_{Na}. On the other hand, each electron-ion pair delivers an energy exceeding 5 eV to the wall, which may cause additional evaporation. More important, however, is that the radial flux $\vec{\Gamma}_{ra}$ (= $-\vec{\Gamma}_r$) is at least two and is usually more than three orders of magnitude smaller than the flux of Na atoms arriving at (and departing from) the wall due to the random thermal motion of the Na atoms, corresponding to the gas temperature T_g. Further on in this section we will present a calculation substantiating this large difference. The fluxes $\vec{\Gamma}_{ra}$ and $\vec{\Gamma}_r$ may therefore be regarded as a small perturbation, and the Na density at the wall will in good approximation follow from Eq. (41).

In the model developed by van Tongeren (1975) and also in some other models of metal-vapor noble-gas or noble-gas positive columns, n_e at the wall was assumed to be equal to zero. However, Eq. (28) then gives a potential difference $\phi(0) - \phi(R) = \infty$. Some approximate calculations, given below, show that a reasonable estimate is

$$3.2 \, V < \phi(0) - \phi(R) < 4.8 \, V , \tag{42}$$

which according to Eq. (28) corresponds to

$$10^{-3} < n_e(R)/n_e(0) < 10^{-2} . \tag{43}$$

Once the upper limit of Eq. (43) is accepted (see below), another useful relation can be obtained, linking the local electron density with the local neutral density. When gas temperature gradients are included (Mitchner and Kruger, 1973),

$$\vec{v}_r = \frac{-\mu_e \mu_i}{\mu_e + \mu_i} \left(\frac{\vec{\nabla} P_e}{e n_e} + \frac{\vec{\nabla} P_i}{e n_i} \right) , \qquad (44)$$

where $\nabla = \partial/\partial r$, $p_e = n_e k T_e$ and $p_i = n_i k T_i$. Equation (44) is quite general; when $T_e \gg T_g$, $\mu_e \gg \mu_i$ and when T_e is almost independent of r, it reduces to Eqs. (24) and (25). Using the fact that $\mu_e \gg \mu_i$ and that the ratios D_i/D_{3S} are almost independent of T_g, we can deduce from Eqs. (26), (30) and (44)

$$\nabla \left(\frac{D_a}{D_{3S}} n_e T_g + n_{3S} T_g \right) = 0 . \qquad (45)$$

This relation implies

$$\left(\frac{D_a}{D_{3S}} n_e(r) + n_{3S}(r) \right) T_g(r) = \left(\frac{D_a}{D_{3S}} n_e(R) + n_{3S}(R) \right) T_w . \qquad (46)$$

We now wish to show that only the n_{3S} term need be included on the right side of Eq. (46). On the left side of Eq. (46) we therefore take $r = 0$ and consider the case of large current and thus of large radial depletion. This implies that only the electron term is important on the left side of Eq. (46). At large depletion, $n_e(0)$ is already much larger than $n_{3S}(0)$. Furthermore, at $T_e = 8000\,°K$ and $T_g(0) = 600\,°K$, $D_a/D_{3S} = 15$ for neon. Equations (43) and (46) then result in

$$\frac{D_a}{D_{3S}} n_e(0) < \left(10^{-2} \times \frac{D_a}{D_{3S}} n_e(0) + n_{3S}(R) \right) \frac{T_w}{T_g(0)}$$
$$< n_{3S}(R) . \qquad (47)$$

Using Eq. (43) again yields

$$\frac{D_a}{D_{3S}} n_e(R) < 10^{-2} n_{3S}(R) . \qquad (48)$$

When $n_e(R)/n_e(0) < 10^{-2}$, Eq. (46) thus becomes in good approximation

$$\left(\frac{D_a}{D_{3S}} n_e + n_{3S} \right) T_g = n_{Na} T_w , \qquad (49)$$

where the left side refers to any $r < R$ (n_{3S} at the wall is understood to be slightly less than n_{Na} to avoid inconsistencies).

Immediately adjacent to the inner tube wall surface, there is a region called the sheath in which there is no charge neutrality

MODELING OF GLOW DISCHARGE

so that the theory of ambipolar diffusion is no longer valid, and in which the radial electric field-strengths may become quite large. The spatial extension of the sheath is of the order of the Debye length

$$\lambda_D = \left(\frac{kT_e}{4\pi n_e e^2}\right)^{1/2} = 10\ g0\ \left(\frac{T_e}{n_e}\right)^{1/2} . \qquad (50)$$

For $T_e = 800\ °K$ and $n_e = 10^{11}\ cm^{-3}$, $\lambda_D = 2 \times 10^{-3}$ cm. This is negligible compared with the tube radius. In order to get more information about the boundary region near the wall, we present some additional calculations and approximations. The Bohm sheath criterion (Bohm, 1949), says that the ions must enter the sheath region with a velocity greater than the acoustic velocity, $v_s = (kT_e/M_{Na})^{1/2}$. We now take the position where $v_r = v_s$ as the boundary of the diffusion controlled positive column plasma and the free fall (of the ions) sheath. This position is somewhat arbitrary, as is the Bohm criterion (Ecker and McClure, 1962). Here we want to mention the difference between the diffusion of the electrons and of the ions to the wall. For the electrons, the velocity v_r is merely a small fraction of their thermal velocity, whereas for the ions the velocity v_r greatly exceeds their thermal velocity. The ionic motion to the wall can be depicted as a beam, a directed flux, whereas the electrons move like a gas. At the plasma sheath boundary we have (Chen, 1965), viewing the wall as a probe at floating potential,

$$\frac{n_e}{4} \left(\frac{8kT_e}{\pi m}\right)^{1/2} \exp\left(\frac{e\phi_s}{kT_e}\right) = n_e \left(\frac{kT_e}{M_{Na}}\right)^{1/2} . \qquad (51)$$

The right-hand side of Eq. (51) is the directed flux of Na ions crossing the boundary surface, the left-hand side gives the electron flux multiplied by the Boltzmann factor where ϕ_S is the potential difference across the sheath. From Eq. (51),

$$\phi_S = \frac{kT_e}{2e} \ln\left(\frac{2\pi m}{M_{Na}}\right) . \qquad (52)$$

At the plasma-sheath boundary, $v_r = v_s$ and Eqs. (24) and (27) give

$$F_r = \frac{(kT_e)^{3/2}}{D_a e M_{Na}^{1/2}} . \qquad (53)$$

We now consider a Na-Ne positive column with $n_{Ne} = 1.65 \times 10^{17}\ cm^{-3}$ (5 torr filling), $T_w = 530°K$ and $T_e = 800°K$. Equation (52) then gives $\phi_S = -3.0$ V, and Eq. (53) gives $F_r = 107$ V/cm. When $n_e(R) = 10^9\ cm^{-3}$, $\lambda_D = 2 \times 10^{-2}$ cm and $\phi_S/\lambda_D = 150$ V/cm. This average electric field in the sheath ϕ_S/λ_D should, however, certainly be significantly larger than the $F_r = 107$ V/cm at the sheath-plasma boundary. Hence $n_e(R)$ must be larger than 10^9 cm^{-3}. On the other hand, when $n_e(R) = 10^{11}$ cm^{-3}, $\phi_S/\lambda_D = 1500$ V/cm, which may be on the high side.

Using these latter numbers a reasonable estimate of $n_e(R)$ is 10^{10} cm^{-3}, yielding $\phi_S/\lambda_D = 500$ V/cm. For $T_e = 8000\,°K$, $T_w = 530\,°K$, $p_{Ne} = 5$ torr, $I = 0.9$ A, and $2R = 1.9$ cm, the axis electron density $n_e(0) \simeq 4 \times 10^{12}$ cm^{-3} (Vriens, 1978). Hence $n_e(R) / n_e(0) \simeq 2.5 \times 10^{-3}$ which substantiates Eq. (43), and $\phi(0)-\phi(R) = 4.1$ V, which is the potential difference between the axis and the plasma-sheath boundary. Further, $(D_a/D_{3S})n_e(R) = 1.5 \times 10^{11}$ cm^{-3}, while $n_{Na}(530\text{ K}) = 5.7 \times 10^{13}$ cm^{-3}. The electron term on the right side of eq. (46) is thus smaller than the n_{3S} term by a factor of 400, which justifies the application of Eq. (49). It is now also possible to compare the flux $\vec{\Gamma}_r = -\vec{\Gamma}_{ra}$ at the boundary:

$$\Gamma_r = n_e(R)\, v_s = n_e(R)\, \left(\frac{kT_e}{M_{Na}}\right)^{1/2}, \tag{54}$$

with the number of Na atoms passing this boundary (from both sides) due to thermal motion:

$$\Gamma_{Na} = \frac{1}{4} n_{Na} \left(\frac{8kT_g}{\pi M_{Na}}\right)^{1/2}. \tag{55}$$

For $T_g = 530\,°K$, $T_e = 8000\,°K$, $n_{Na} = 5.7 \times 10^{13}$ cm^{-3} and $n_e = 10^{10}$ cm^{-3}, Eqs. (54) and (55) give $\Gamma_{Na}/\Gamma_r = 1200$, and the random flux is indeed several orders of magnitude larger than the directed flux due to the ambipolar diffusion. This justifies the use of Eq. (41) for $n_{Na}(R)$.

In summary, this section gives a set of characteristic numbers for potential differences, field strengths, and densities pertaining in particular to the sheath region near the wall. In general, there is an ambiguity in descriptions of the plasma sheath transition and of the sheath region. For the present discharge conditions, we have found that the Debye length λ_D is more than two orders of magnitude smaller than the tube dimaeter, that the electron density n_e at the plasma sheath boundary is a factor of 400 smaller than at the axis and that the directed flux of Na ions to the wall is three orders of magnitude smaller than the random flux due to the thermal motion of the Na atoms. One may dispute whether these numbers are accurate to within a factor of 2, but even with variations of factors of 2 or 3, the conclusions remain valid. Hence, Eq. (41) describes the Na atom density at the wall; Eq. (49) relates the electron and Na atom densities at any place in the discharge to the Na atom density at the wall, and the sheath region need not be considered separately, due to its small spatial extension and small electron density.

MODELING OF GLOW DISCHARGE

Radial Averaging

We will now discuss radial averaging or integrating the local relations, which necessary and useful for the following reasons:

• It yields quantities such as total current and power dissipation, radiation output of individual lines, and total efficacy, quantities that can be directly compared with experiment;

• Some terms in the resulting radially integrated formulas appear to be negligible. This leads to a simplification;

• In the model it is easier, and more efficient with respect to computer time, to work partly with formulas containing only radially integrated quantities and partly with formulas containing local quantities.

Radial averaging will be indicated by

$$\bar{A} = \frac{1}{\pi R^2} \int_0^R A \, 2\pi r \, dr \, , \qquad (56)$$

where A is the local quantity. Averaging over both the electron energy distribution and the radial coordinate is denoted by $<\bar{A}>$. For example, the macroscopic current I is given by

$$I = \pi R^2 \bar{j} = \pi R^2 \, eF \, \overline{n_e \mu_e} = \frac{e^2 F}{3k} \, \overline{\frac{n_e}{T_e} <\frac{v}{N\sigma_m}>} \, . \qquad (57)$$

In the steady state, the energy conservation relations (37) and (38) yield

$$\pi R^2 \overline{P_{el}} = 2\pi R \, [-\frac{5}{2} kT_g \Gamma_r + g_g]_{r=R} \, , \qquad (58)$$

where it has been used that $\Gamma_{ra} = \Gamma_r$ ($\vec{\Gamma}_{ra} = -\vec{\Gamma}_r$), and

$$\overline{P_{inel}} - \overline{P_{sel}} = \overline{n_e E_{3Si} \sum_p n_p K_{pi}} + \overline{\sum_n n_n (\tau_{np})_{eff}^{-1} E_{pn}} \, . \qquad (59)$$

Equation (59) can be simplified, because the internal energy of the ions, produced in the volume, must flow to the wall due to ion transport. There is no volume recombination. Hence

$$\pi R^2 \, \overline{n_e E_{3Si} \sum_p n_p K_{pi}} = 2\pi R [E_{3Si} \Gamma_r]_{r=R} \, . \qquad (60)$$

Recalling that v_r is assumed to satisfy the Bohm sheath criterion, Eqs. (54) and (60) provide us with the electron density at the plasma-sheath boundary,

$$n_e(R) = \frac{R}{2} \left(\frac{M_{Na}}{kT_e}\right)^{1/2} \overline{n_e \sum_p n_p K_{pi}} \quad . \tag{61}$$

Radial averaging of Eq. (39) gives the power dissipation W per unit length of the discharge column:

$$W = IF = \pi R^2 \sum_n \overline{n_n (\tau_{np})_{eff}^{-1} E_{np}} \tag{62}$$

$$+ 2\pi R [(3kT_e + E_{3Si})\Gamma_r + q_g]_{r=R} \quad .$$

Two simplifications have been made in obtaining Eq. (62): first, the Bohm sheath criterium which yields $1/2 M_{Na} v_r^2 = 1/2 kT_e$ at $r = R$; and second, the q_{ei} term is omitted; $q_{ei} \ll q_g$ close to the wall because $n_e \ll n_g$, $\lambda_e/\lambda_g \approx (n_e/n_g)(T_e/T_g)^{1/2}$ and $\lambda_i/\lambda_g \approx n_e/n_g$.

In the free fall sheath, the ions are accelerated and the electrons are retarded. It is assumed that these terms cancel each other in Eq. (62) (see above section). Equations (26), (60), and (61) further yield $[\partial n_e / \partial r]_{r=R}$.

In the formulas given so far, T_g, T_e, $(\tau_{np})_{eff}$, and λ_g (hence) q_g) all depend on position. In the model calculation, we have found that the exact functional dependence of T_g on r is relatively unimportant (a 1% effect). We have therefore approximated this dependence by a parabolic profile, the value of $T_g(0)$ being determined by P_{el}. Two problems are left: the dependence of T_e on r and the values of $(\tau_{np})_{eff}$ for the lines which are optically thick.

For a Cs-Ar discharge, which is very similar to the one studied here, Waszink and Polman (1969) could not detect any T_e gradients. At first sight, this is surprising in view of the small heat conductivity of the electrons. A direct conclusion is that the heat conduction is not responsible for these small gradients. Another much more efficient coupling mechanism is the radiation trapping in combination with the balance between exciting and de-exciting collisions. For the most important lines the radiation trapping is very large, and it couples not only close lying positions in the discharge (as is done in diffusion), but also works over longer distances. In the modeling, we make use of the small T_e gradients by simply taking T_e independent of position.

The effective radiative lifetimes $(\tau_{np})_{eff}$ for the optically thick lines depend in a complicated way on the level splitting and line broadening (see earlier section), and on the radial density distributions of the upper and lower states $|n>$ and $|p>$. In order to save computer time, a separate calculation is made in the model to obtain the radially averaged values of $(\tau_{np})_{eff}$. These radially averaged values (next section) are then taken as position independent and are used in all foregoing equations.

Radiation Transport

A general outline of the method used will be given. Details and final equations will be published separately (Smeets, 1982).

First, the transmission factor of the line is defined by

$$T(k_o \ell) = \int_{-\infty}^{\infty} L(u) \exp[-k_o \ell L(u)] du , \qquad (63)$$

where k_o is the absorption coefficient at the center of a Doppler broadened line; u is defined by

$$u = 2 \frac{(\nu - \nu_o)}{\Delta \nu_D} (\ell n\, 2)^{1/2} , \qquad (64)$$

where $\Delta \nu_D$ is the full width at half maximum (FWHM) of the Doppler line, and $L(u)$ is the normalized line shape, which may have a pure Doppler, Lorentz or in general a Voigt profile. For a Voigt profile, an analytical formula for the transmission factor is constructed, which for $k_o l \to \infty$ reduces to the asymptotic formula given by Walsh (1959) and which reproduces the asymptotic form given by the first two terms of a Taylor expansion in $k_o l$ for $k_o l \to 0$. It reduces to the analytical formula given by Irons (1979) for a Lorentzian line. For a Voigt profile and $0 < k_o l < \infty$, the analytical formula agrees quite well (and is fitted to) the results obtained from a numerical evaluation of Eq. (63).

Second, an escape probability,

$$\theta(r) = \frac{1}{4\pi} \int_S T(k_o |\vec{r} - \vec{s}|) \frac{(\vec{r} - \vec{s}) d^2 \vec{s}}{|\vec{r} - \vec{s}|^3} , \qquad (65)$$

is introduced, which is the probability that a photon emitted at position r will escape from the closed volume, the integral being taken over the closed surface area. For an infinite cylinder,

$$\theta(r) = \frac{1}{\pi} \int_0^\pi \int_0^{\pi/2} T(k_o \zeta \sec\beta) \cos\beta \, d\beta \, d\phi , \qquad (66)$$

where β gives the inclination with respect to the radial plane, and ζ is given by

$$\zeta^2 + 2r\,\zeta\cos\phi - (R^2 - r^2) = 0 \quad , \tag{67}$$

ϕ being the angle between the projection of $\vec{r}-\vec{s}$ in the radial plane and the shortest path to the surface. At the center of the cylinder ($r = 0$, $\zeta = r$), $\theta(0)$ can be rather well described by (Smeets, 1982)

$$\theta(0) = T(\frac{4}{\pi} k_o R) \quad . \tag{68}$$

Next, a rather crucial and new approximation for $\theta(r)$ is made for $r = 0$, analogous to the treatment for an infinite plane parallel-slab (Irons, 1979). The cylinder is approximated first by two half cylinders, one with radius R-r, the other with a radius R+r:

$$\theta(r) = \frac{1}{2} \{T[f_- k_o(R-r)] + T[f_+ k_o(R+r)]\} \quad . \tag{69}$$

The first term on the right side stands for the integration over $|\phi|$ between 0 and $\pi/2$, and the second term must approximate the integral over $|\phi|$ between $\pi/2$ and π. As discussed by Smeets (1982) f_- must tend to 2 for $r \to R$, and f_+ becomes $8/\pi^2$ for $r \to R$. By means of numerical evaluation, it is found that

$$f_- = [(16/\pi^2)(1 - r^2/R^2) + 4\,r^2/R^2]^{1/2} \quad , \tag{70}$$

$$f_+ = (4/\pi)[1 - r^2/R^2 + (4/\pi^2)\,(R^2/R^2)]^{1/2} \tag{71}$$

are good approximations, which account for the deviations from the half-cylinder shapes.

For nonhomogeneous lower-state density distributions, k_o and r are replaced by the effective values

$$\bar{k}_o = R^{-1} \int_0^R k_o(r')dr' \tag{72}$$

and

$$\bar{r} = \bar{k}_o^{-1} \int_0^r k_o(r;)dr' \quad . \tag{73}$$

MODELING OF GLOW DISCHARGE

Thirdly, an excape factor is defined,

$$\theta = \int_V n_y(\vec{r})\theta(\vec{r})\, d^3\vec{r} / \int_V n_y(\vec{r})\, d^3\vec{r} \quad , \tag{74}$$

$n_y(\vec{r})$ being the density of the emitting atoms. For an infinite cylinder,

$$\theta = \int_o^R \{T[f_{-\bar{k}_o}(R-\bar{r})] + T[f_{+\bar{k}_o}(R+\bar{r})]\} n_y(r) r\, dr \\ \times [2\int_o^R n_y(r) r\, dr]^{-1} \quad . \tag{75}$$

Using Eq. (75), the number of numerical integrations required is reduced to a minimum, and nonhomogeneous density distributions of lower and upper state can be dealt with.

Finally, analytical forms for $T(k_o l)$ have been derived applying to lines that are split into several components (Smeets, 1982). These latter formulas have to be used in the model.

THE COMPUTER MODEL

Structure of the Model

The structure of the computer model is schematically illustrated in the block diagram in Fig. 9. The left side shows the iteration steps used to arrive at the radially integrated values, such as total current and radiative output, and the right side gives a magnification of that part of the diagram which involves the calculation of the necessary local properties.

On a local scale, the central differential equation that is solved in the model is the equation for the electron density. It can be derived from Eqs. (26), (33), and/or (34) and is written in the form (see van Tongeren, 1975)

$$\Delta_r n_e(r) + Xf(n_e, T_e, T_t, \tau_{eff}) = 0 \quad , \tag{76}$$

where the parameter X, which must converge to T_e, is introduced. A method for solving the nonlinear eigenvalue problem of Eq. (76) has been given by Bouwkamp (1968). To evaluate the function f in Eq. (76), the contributions to ionization from all levels must be known, which implies that the population densities and the electron tem-

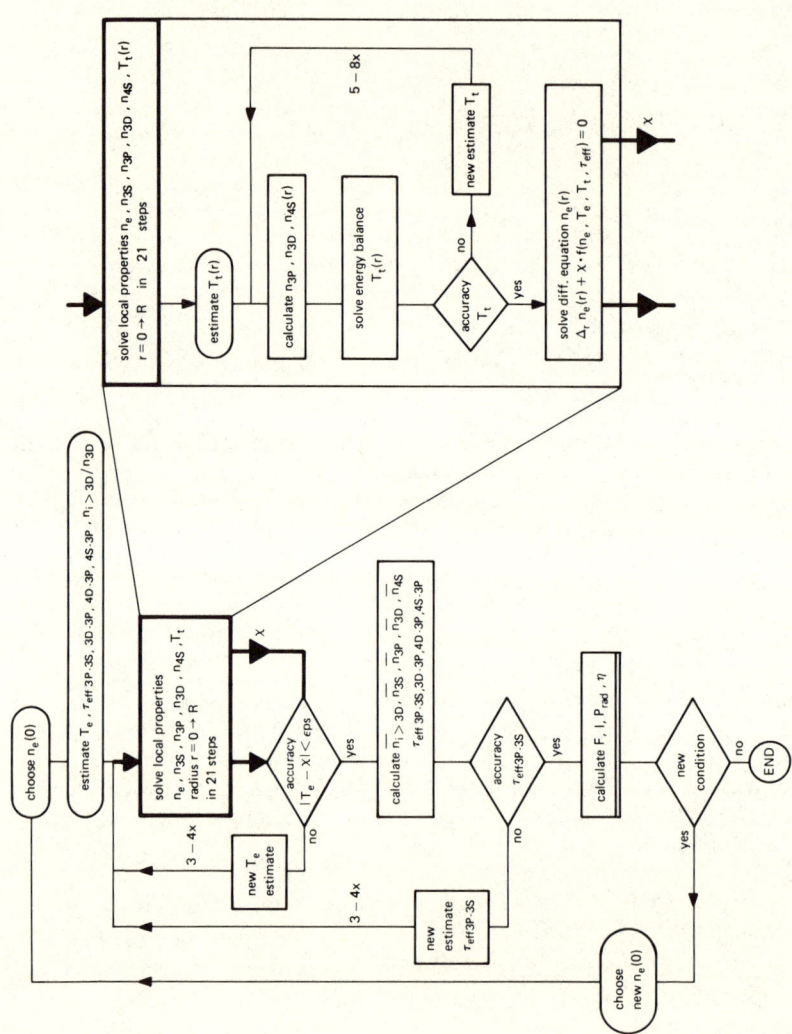

Fig. 9. Block diagram of the numerical procedure for solution of the DC model equations. The oval box indicates an input or output operation; the symbol eps in the decision box (diamond) denotes a small quantity, eps << T_e. The double underlined rectangular box gives the relevant integrated quantities; e.g., $\eta = P_{rad}/FI$ indicates the efficiency for visible light production in the column.

peratures T_e and T_t must be known. This means that the individual particle balance equations (Eqs. 31-33) have to be solved simultaneously. The collision rates in these equations are obtained from the sections entitled "Energy Distribution" and "Excitation and Ionization." After making a first estimate of T_e, of the values τ_{eff} for the optically thick lines, and of the population densities, the program calculates the densities of the most important states in a local iteration in which X converges to T_e. This iteration includes a subroutine in which T_t is determined by solving the energy balance for the tail electrons. The subroutine takes about eight steps.

Next, the differential equation (76) is solved using a 21 point grid in the radial direction. In order to account for possible large gradients near the wall, the grid is wide near the axis and dense near the wall. Changing the 21-step radial subdivision into a 49-step subdivision resulted in only about 1% changes in electric field strength, radiant power, etc. The solution of the differential equation yields a new value of T_e. When T_e differs too much from the preceding estimated value, a new estimate is made, and the local iteration starts again. The T_e iteration converges rapidly, usually in 3 to 4 steps.

The program next solves the energy balance equations cited earlier on an integral scale. New values result for the radially averaged particle densities. The density profiles for the lower and upper states of the optically thick lines yield new values for τ_{eff}. When $\tau_{eff3P-3S}$ differs too much from the earlier estimate, the local iterations start again, using new estimates for τ_{eff} (all optically thick lines) and for the particle densities. This integral iteration converges in typically 3 to 4 steps. The separation of the T_e and τ_{eff} iterations works very well. We have also tried to combine these two iterations. The result was an unstable procedure.

The following simplifications have been used in the program (1, 2 and 3 have been noted already):

(1) T_e is taken independent of position.

(2) The values for τ_{eff} are taken independent of position.

(3) A parabolic $T_g(r)$ profile is used.

(4) For the higher excited states (i > 3D), the radial density profiles $n_i(r)$ are taken proportional to the $n_{3D}(r)$ profile. This simplification is justified by the fact that the 3P and 4S profiles are found to be only slightly different, and that the 4S and 3D profiles are already very similar. In the program it is therefore only necessary to calculate the ratios of the radially averaged densities; i.e., $\overline{n_{i>3D}}/\overline{n_{3D}}$.

At the end, the program yields the values of F, I, W, P_{rad}, τ_{eff}, and a number of local quantities, for the chosen values of the tube diameter, wall temperature, filling gas (e.g., Ne) pressure and electron density at the axis, $n_e(0)$. To determine a voltage-current characteristic, we start with some thirty different values for $n_e(0)$.

Results and Comparison with Experiment

Some of the results obtained are presented in Figs. 10 to 13.

Figure 10 shows calculated radial density distributions for the most important states of Na. The ne filling pressure, the wall temperature $T_{\overline{w}}$ and the tube diameter ϕ are those of a typical Na lamp. The current I = 0.9 A corresponds to the I_{RMS} = 0.9 A in this Na lamp. Figure 10 illustrates that the radial density distributions are very nonhomogeneous due to the fast ambipolar diffusion of electrons and ions and the slow diffusion of Na atoms. This figure shows clearly why a local description of the discharge is required.

Calculated electron-impact ionization rates for the ground state, the first excited state and for the higher excited states are depicted in Fig. 11. The rates have been normalized, i.e., their fraction of the total ionization is plotted. Figure 11 confirms that the (ladder-like) ionization via the D, F, G,.. states is dominant. In Fig. 11, the K's indicate the rates and not the rate coefficients.

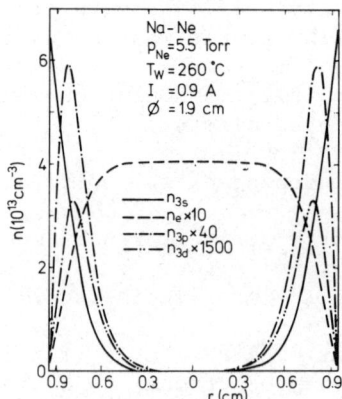

Fig. 10. Calculated radial density profiles of the Na 3S ground state, the Na 3P and 3D excited states, and the electrons in a DC Na-Ne positive column. p_{Ne}, T_w, ϕ and I indicate the filling pressure, the wall temperature, the tube diameter and the current.

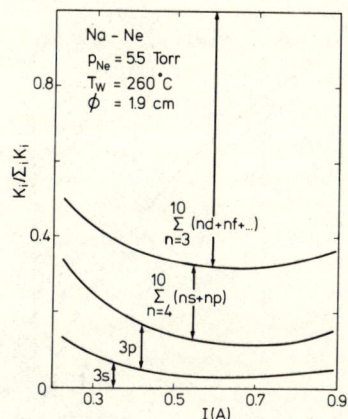

Fig. 11. Electron impact ionization rates K_i from a level i or a group of levels i in the Na-Ne discharge as a function of the discharge current, calculated with the 25-level model. The rates are plotted as a fraction of the total ionization rate $K_{tot} = \sum_i K_i$.

In Fig. 12, a calculated Boltzmann plot is compared with the measured Boltzmann plot of Fig. 3. Very good agreement is obtained. Both plots show the characteristic bent shape of a nonequilibrium glow discharge. Note that the energy scales applying to the two curves have been shifted for clarity in presentation.

Finally, Fig. 13 shows a comparison between calculated and experimental electric field strengths. The experiment was performed in a commuted DC discharge to avoid cataphoresis; the wall temperature was very well controlled using a heat pipe thermostat (van Tongeren and Heuvelmans, 1974 and 1975). Successively more levels have been included in the calculations (2,3,5,9,... up to 25). In the 2-level model there are only ground state atoms and ions. The 3-level model includes the 3P resonance state. Because of the resulting radiative losses, the dissipated power at a certain current increases and the electric field strength increases correspondingly. A large discrepancy with experiment is then found. Including more levels does not increase the radiative losses significantly, but the number of ionization channels is increased. This leads to a much lower field strength at the same current. Good agreement with experiment is obtained in the 25-level model; 19- and 22-level models also give good results already.

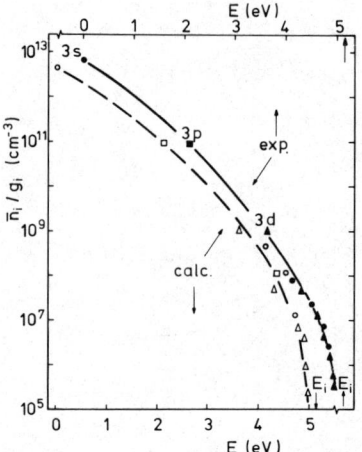

Fig. 12. Comparison of the experimental Boltzmann plot (Fig. 1) with the one calculated using the 25-level model. The energy scale for the experimental Boltzmann plot is shifted 0.5 eV to the right for clarity in presentation.

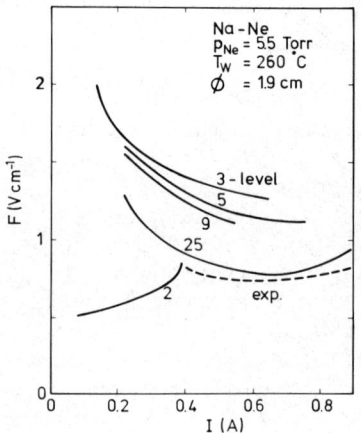

Fig. 13. Comparison of the electric field strengths calculated, using 2, 3, 5, 9 and 25-level models, with experiment (van Tongeren, 1975) as a function of current for the Na-Ne positive column.

CONCLUDING REMARKS

This review paper gives a survey of various glow discharges. It includes low-pressure noble gas and metal-doped noble gas discharges and high-pressure MHD and metal-doped pulsed-excimer plasmas. A comparison with properties of LTE arcs has also been made. Molecular glow discharges are not considered. A detailed model is further presented for the low-pressure Na-Ne discharge. Special emphasis has been given to the examination of kinetic processes and to the application of analytical techniques. The question may now arise: how far can the model be applied for other discharges and discharge conditions? We will attempt to answer this question.

The model is valid for other alkali atoms and other noble gases, provided the appropriate atomic data on momentum transfer cross sections, diffusion coefficients, line broadening and splitting, and the excitation and ionization rates for the lower states are used. The 2-EGM for the electron energy distribution remains valid. The analytical rate equations for the excited states also remain valid. The same method can further be used to describe the radiation transport. We have applied the model already for other noble gases and mixtures thereof. Some restrictions must be made, however.

When the noble-gas density n_g is increased, with metal vapor and electron densities kept constant, gas heating and volume recombination will gradually become significant and ambipolar diffusion will become less important. The electric field strength increases. Volume recombination has to be included, e.g., for $n_g > 10^{19}$ cm^{-3}. When n_g is decreased, the Debye length λ_D increases, the sheath region becomes larger, and $n_e(R)$ increases. For $n_g < 10^{15}$ to 10^{16} cm^{-3}, it will be necessary to account separately for the sheath in the model. The radial depletion is much less in this case.

When the metal-vapor density n_m is increased (higher T_w), the radial depletion will decrease, the radiation trapping will increase, and molecular ion formation and dissociative recombination gradually become important; e.g., above $n_m = 10^{15}$ to 10^{16} cm^{-3}. Other processes like associative ionization may also become significant. When n_m is decreased, the radial depletion increases until there are no longer sufficient alkali atoms available for the ion production. A transition takes place to a higher electric field strength regime where the noble gas is excited and ionized.

When the electron density n_e is increased too much, a similar transition occurs to a regime of large radial depletion with noble gas excitation and ionization (van Tongeren and Heuvelmans, 1974 and 1975). When n_e is decreased, the importance of multi-step processes decreases and the radial depletion becomes less important. Below $n_e \approx 10^{10}$ cm^{-3}, there will be deviations from Maxwellian also in the bulk of the electron energy distribution.

When the alkali atoms in the model are replaced by Hg (fluorescent lamp), it will further be necessary to include other processes like associative ionization (Vriens et al., 1978). This has been done in an unpublished model by Ligthart and Keijser.

Finally, we not that large parts of the model will be applicable also to pure noble gas, low-pressure positive columns. At the lower pressures, these are also ambipolar-diffusion controlled.

APPENDIX

This appendix is used to list additional formulas which are essential for the modeling described.

The tail normalization function S of the 2-EGM (Eq. 2) is given by

$$S = [(T_t/T_e)^{3/2} \exp(-E_1/kT_e + E_1/kT_t)]^r , \qquad (77)$$

where

$$r = 1.09 - 0.17 (1 - 0.5 \, T_t/T_e)^{-1} , \qquad (78)$$

and where r was obtained by empirical fitting to the Boltzmann calculations (Morgan and Vriens, 1980). Since $r < 1$, $f^t(E_1) < f(E_1)$ for $T_t < T_e$, which accounts for the depletion at $E_e = E_1$.

The rate equation for excitation, Eq. (17), contains two empirical functions, Γ_{pn} and Δ_{pn}, and one function B_{pn} which depends on atomic structure. By extended fitting to known cross sections and rate coefficients, Vriens and Smeets (1980) arrived at

$$\Gamma_{pn} = R \, \ln(1 + \frac{X^3 kT_e}{R}) \, [3 + 11(\frac{s}{X})^2]$$
$$\times \, (6 + 1.6ys + \frac{0.3}{s^2} + 0.8 \, \frac{y^{1.5}}{s^{0.5}} \, |s-0.6|)^{-1} \qquad (79)$$

and

$$\Delta_{pn} = \exp(-\frac{B_{pn}}{A_{pn}}) + \frac{0.06 \, s^2}{yX^2} , \qquad (80)$$

where x and y are the effective quantum numbers of states $|p>$ and $|n>$, i.e., $x = (R/E_{pi})^{1/2}$ and $y = (R/E_{ni})^{1/2}$, and $s = |y-x|$.

For the hydrogen atom, we have (Vriens and Smeets, 1980)

$$B_{pn} = \frac{4R^2}{x^3 E_{pn}^2} \left(1 + \frac{4E_{pi}}{3E_{pn}} + b_p \frac{E_{pi}^2}{E_{pn}^2}\right), \qquad (81)$$

where b_p is an empirical parameter, which for the H-atom is very well represented by

$$b_p = \frac{1.4 \ln X}{X} - \frac{0.7}{X} - \frac{0.51}{X^2} + \frac{1.16}{X^3} - \frac{0.55}{X^4}. \qquad (82)$$

When there is no other information available, and when one does not want to make a subdivision between different angular momentum states, Eqs. (17) and (79) to (82) can be used to describe excitation from excited states.

For the Na atom, Kim and Cheng (1978) calculated B_{pn} coefficients for many transitions from the 3S ground State. Values for B_{pn} are not available for excitation from excited states of Na. In order to make a subdivision between the angular momentum levels (this subdivision was not made by Vriens and Smeets, 1980), we use the data of Kim and Cheng for the ground state, and we use a hydrogenic approximation for the excited states. An accurate representation is found to be given by

$$B_{pLnL'} = a_{pLL'} B_{pLn} \qquad (83)$$

where pLnL' indicates the $|pL> \to |nL'>$ transition. B_{pLn} is given by Eq. (81) with b_p replaced by b_{pL}. Further, $\sum_{L'} a_{pLL'} = 1$. Values of $a_{pLL'}$ and b_{pL} are listed in Table IV.

Finally, we note that the diffusion coefficients of Na atoms and ions in the noble gases, and the thermal conductivities of the noble gas, can easily be presented in simple analytic forms, containing the gas density and temperature as parameters.

Table IV. Values of and functions for the parameters $a_{pLL'}$ and b_{pL} in Eqs. (81) and (82); x and y are the effective quantum number of states $|pL\rangle$ and $|nL'\rangle$, $s = |y-x|$; nQ stands for the nD, nF, nG ... states together.

initial state	final state	$a_{pLL'}$		b_{pL}
pS	nS	0.12		$-0.16\ x^{-1} - 0.44\ x^{-3}$
	nP	0.53	(for $p \geq 4$)	
		$0.21 + 0.21 s^{-3/2}$	(for $p = 3$)	
	nQ	0.35	(for $p \geq 4$)	
		$0.50 - 0.05 s^{-3/2}$	(for $p = 3$)	
pP	nS	$0.08 - 0.13\ x^{-1}$		$0.8\ x^{-2}$
	nP	$0.15 - 0.10\ x^{-1}$		
	nQ	$0.77 - 0.23\ x^{-1}$		
pQ	nS	$0.006\ x^{-1}$		$(X^2 - 4)^{-1} [1.4\ x \ln x$
	nP	$0.039\ x^{-1}$		$-0.7X - 0.51 + 1.32\ X^{-1}$
	nQ	$1 - 0.045\ x^{-1}$		$-2.95\ X^{-2} + 0.44\ x^{-3}]$

REFERENCES

Abrines, R., Percival, I. C., and Valentine, N. A., 1966, Proc. Phys. Soc., London, 89:515.
Biberman, L. M., Vorob'ev, V. S., and Yakubov, I. T., 1973, Sov. Phys.-Usp., 13:375.
Biberman, L. M., Vorob'ev, V. S., and Yakubov, I. T., 1979, Sov. Phys.-Usp., 22:411.
Bohm, D., 1949, "Characteristics of Electrical Discharges in Magnetic Fields," A. Guthrie and R. K. Wakerling, eds., McGraw, New York, Chap. III.
Bouwkamp, C. J., 1968, SIAM Rev., 10:114.
Cayless, M. A., 1963, Br. J. Appl. Phys., 14:863.
Chen, F., 1965, "Plasma Diagnostic Techniques," R. Huddlestone and S. L. Leonard, ed., Academic, New York, Chap. 4.
Cherrington, B. E., 1979, "Gaseous Electronics and Gas Lasers," Pergamon, Oxford.
Coolen, F. C. M. and Hagedoorn, H. L., 1975, Physica, C, 79:402.
Crompton, R. W., Elford, M. T., and Robertson, A. G., 1970, Aust. J. Phys., 23:667.
Delpech, J. F., Boulmer, J., and Devos, F., 1977, Phys. Rev. Lett., 39:1400.
Devos, F., Boulmer, J., and Delpech, J. F., 1979, J. Phys., 40:215.
Dixon, A. J., Harrison, M. F. A., and Smith, A. C. H., 1973, "Abstracts, 8th International Conference on the Physics of Electronic and Atomic Collisions," B. C. Cobic and M. V. Kurepa, eds., Inst. of Physics, Belgrade, Vol. 1, p. 405.
Dixon, A. J., Harrison, M. F. A., and Smith, A. C. H., 1976, J. Phys. B, 9:2617.
Dixon, A. J., von Engel, A., and Harrison, M. F. A., 1975, Proc. R. Soc., London, A, 343:333.
Ecker, G. and McClure, J. J., 1962, Z. Naturforsch. Teil A, 17:705.
Ellis, H. W., Pai, R. Y., McDaniel, E. W., Mason, E. A., and Viehland, L. A., 1976, At. Data Nucl. Data Tables, 17:177.
Enemark, E. A. and Gallagher, A., 1972, Phys. Rev. A 6:192.
Engel, A. von, 1965, "Ionized Gases," Oxford.
Fairbank, W. M., Hänsch, T. W., and Schawlow, A. L., 1975, J. Opt. Soc. Am., 65:199.
Francis, G., 1956, "The Glow Discharge at Low Pressure, Handbuch der Physik," S. Flügge, ed., Springer, Berlin.
Franklin, R. N., 1976, "Plasma Phenomena in Gas Discharges," Clarendon Oxford.
Frost, L. S. and Phelps, A. V., 1964, Phys. Rev., 136:A1538.
Fujimoto, T., Ogta, Y., Sugiyama, I., Tachibana, K., and Fukuda, K., 1972, Jap. J. Appl. Phys., 11:718.
Fujimoto, T., 1972, Jap. J. Appl. Phys., 11:1501
Gallagher, A., 1979, "Excimer Lasers," C. K. Rhodes, ed., Springer, Berlin.
Gallagher, T. F., Edelstein, S. A., and Hill, R. M., 1977, Phys. Rev. A, 15:1945.

Gee, C. S., Percival, I. C., Lodge, J. G., and Richards, D., 1976, Mon. Not. R. Astron. Soc., 175:209.
Golubovski, Y. B., Kagan, Y. M., and Lyagushchenko, R. I., 1974, Sov. Phys. Tech. Phys., 19:335.
Hirsch, M. N. and Oskam, H. J., 1978, "Gaseous Electronics," Vol. 1 of "Electrical Discharges," Academic, New York.
Ho, C. Y., Powell, R. W., and Liley, P. E., 1972, J. Phys. Chem. Ref. Data, 1:279.
Huxley, L. G. H. and Crompton, R. W., 1974, "The Diffusion and Drift of Electrons in Gases," Wiley, New York.
Inokuti, M., 1971, Rev. Mod. Phys., 43:297.
Irons, F. E., 1979, J. Quant. Spectro. Radiat. Transfer, 22:1.
Irons, F. E., 1979, J. Quant. Spectro. Radiat. Transfer, 22:37.
Jack, A. G., 1978, Lighting Res. Tech., 10:150.
Jack, A. G. and Vrenken, L. E., 1980, IEE Proc., 127:149.
Keijser, R. A. J., Cornelissen, H. J. and Smeets, A. H. M., 1981, unpublished.
Kim, Y. K. and Cheng, K. T., 1978, Phys. Rev., A, 18:36.
Ligthart, F. A. S. and Keijser, R. A. J., 1980, J. Appl. Phys., 51:5295.
Lindgard, A. and Neilsen, S. E., 1977, At. Data Nucl. Data Tables, 19:533.
Mansbach, P. and Keck, J., 1969, Phys. Rev., 181:275.
Milloy, H. B. and Crompton, R. W., 1977, Phys. Rev., A, 15:1847.
Milloy, H. B., Crompton, R. W., Rees, J. A., and Robertson, A. G., 1977, Aust. J. Phys., 30:61.
Mitchell, A. C. G. and Zemansky, M. W., 1971, "Resonance Radiation and Excited Atoms," Univ. Press, Cambridge.
Mitchner, M. and Kruger, C. H., 1973, "Partially Ionized Gases," Wiley, New York.
Moores, D. L. and Norcross, D. W., 1972, J. Phys., B, 5:1482.
Moores, D. L., Norcross, D. W., and Sheory, V. B., 1974, J. Phys., B, 7:371.
Morgan, W. L. and Vriens, L., 1980, J. Appl. Phys., 51:5300.
Nygaard, K. J., 1968, J. Chem. Phys., 49:1995.
Nygaard, K. S., 1975, Phys. Rev., A, 11:1475.
Ogata, Y. and Fukuda, K., 1973, Mem. Fac. Eng. Kuoto Univ., 35:177.
Polman, J. and Drop, P. C., 1972, J. Phys. D, 5:562.
Rhodes, C. K., 1979, "Excimer Lasers," Springer, Berlin.
Robertson, A. G., 1972, J. Phys., B, 5:648.
Shkarofsky, J. P., Johnston, T. W., and Bachynski, M. P., 1966, "The Particle Kinetics of Plasmas," Addison-Wesley, Reading, Mass.
Smeets, A. H. M., 1982, to be published.
Smits, R. M. M., Lammers, P. H. M., Baghuis, L. C. J., Hagedoorn, H. L., and van der Heide, J. A., 1974, Beitr. Plasmaphys., 14:157.
Shuker, R., Gallagher, A., and Phelps, A. V., 1980, J. Chem. Phys., 72:6081.
Tongeren, H. van and Heuvelmans, J., 1974, J. Appl. Phys, 45:3844.
Tongeren, H. van, 1975, Philips Res. Repts. Suppl., No. 3.
Vriens, L., 1969, "Case STudies in Atomic Collision Physics," Vol. I,

E. W. McDaniel and M. R. C. McDowell, eds., North-Holland, Amsterdam, p. 335.
Vriens, L., 1973, J. Appl. Phys. 44:3980.
Vriens, L., 1978, J. Appl. Phys., 49:3814.
Vriens, L., Keijser, R. A. J., and Ligthart, F. A. S., 1978, J. Appl. Phys., 49:3807.
Vriens, L. and Smeets, A. H. M., 1980, Phys. Rev., A, 22:940.
Walsh, P. J., 1959, Phys. Rev., 116:511.
Waszink, J. H. and Polman, J., 1969, J. Appl. Phys., 40:2403.
Waymouth, J. F., 1971, "Electric Discharge Lamps," MIT, Cambridge, Mass.
Zapesochnyi, I. P. and Aleksakhin, I. S., 1969, Sov. Phys. - JETP, 28:41.

THE GLOW-TO-ARC TRANSITION

E. Marode

Laboratoire de Physique des Décharges (CNRS)
Ecole Supérieure de'Electricité
91190 Gif-sur-Yvette, France

INTRODUCTION

Studies on ionization growth and spark formation in a gas, have developed, in the past, in two complementary ways.

One way was based on the idea that the acquisition of reliable knowledge on discharge processes required the study of situations where the input parameters were, as much as possible, simplified. So, work has been developed in uniform fields and in nonattaching gases like rare gases, nitrogen, etc. In those cases, clear interpretation of the first steps of the ionization growth could be obtained and accurate models could be established (Raether, 1964; Llewellyn-Jones, 1966; Dutton, 1978).

Another research viewpoint led to the extensive study of discharges in nonuniform fields and in attaching gases, cases much more related to practical realities (Loeb, 1965; Goldman et al., 1978; Sigmond, 1978; Waters, 1978; Gallimberti, 1979). Here, the discharge appeared from the beginning as more complex leading to more approximate models, like the streamer model which accounts for the build-up of transient filamentary discharges.

However, this discharge complexity was also encountered in uniform fields as soon as studies on the space-charge controlled phase were carried out (Raether, 1964). In this case, a strong distortion of the applied field appears which removes the prerogative character of a situation where all parameters are known. Here too, approximate models appeared which did not completely clarify the active processes. For example, the respective roles of gas photoionization and the photoelectric effect at the cathode in the self-sustained discharge condition has not completely been established (Davies et al., 1971; Kline, 1974; Yoshida et al., 1976). It follows that the discharge complexity, during the glow-to-arc transition, is more or less the same regardless of the uniformity of the applied field. Perhaps even the nonuniform field case is somewhat clearer since the effect on the gas and the effect on the plane (or large radius) electrode are better decoupled.

This complex situation, i.e., a greater number of unknown processes acting at the same time to give whatever is observed, has, however, not discouraged researchers. A considerable amount of knowledge about the behavior of the glow-to-arc transition has been acquired during the last two decades. The extensive use of high speed techniques — fast oscillography, high speed streak cameras, spectroscopy, schlieren methods, etc. gave us evidence of what remains similar among the different types of transition to spark, and models could be elaborated which predicted, if not precisely, at least with a good likeness, the main features of this transition.

Actually these studies are supported by many practical interests.

The necessity of transporting electrical power with a minimum cost induced and induces studies on the dielectric strength of gases in uniform and nonuniform fields (Christophorou, 1977). The existence of a voltage range where corona glows exist without leading to spark breakdown shows that the breakdown itself is not solely under the control of the self-sustained Townsend condition. Thus, knowledge of the processes acting during the glow-to-arc transition is needed for a complete approach to gas tailoring. Another point is that costly high voltage tests have to be applied to equipment used in transportation of the electrical power, and here again the study of the transition to spark should yield some quantitative criteria which can optimize the test operations (Les Renardières Group, 1974, 1977, 1981). On the other hand, parameters able to accelerate spark formation are sought in fast trigger-gap technology. We must also mention those interested in the use of the glow stage as a reactor for its physico-chemical yields — such as ozone — or for its luminescence property — such as in laser applications. They are very interested in knowing the ways of avoiding glow-to-arc transitions. In the latter case, the formation of a transient discharge constriction leading to spark is catastrophic because the thermal effect destroys the laser effect (Jaeger, Oster, and Phelps, 1976 and references therein).

For the moment it must be recognized that our present stage of knowledge allows us only roughly to predict and thus to avoid or to control the spark formation. But the whole set of models which have emerged during the last few years offers reliable tools which

THE GLOW STAGE

When one refers to the glow phase of a discharge, the first concept which comes to mind is the well-known steady state of the low-pressure, normal glow discharge. This stage is generally described in terms of its spatial structure (Von Engel, 1965; Badareu and Popescu, 1965) (Figure 1), and can roughly be understood as follows.

The few electrons released by secondary effects at the cathode (impact of ions, photons, metastable species, etc.) are multiplied by electron collisions during their acceleration by the applied field. Since the electron mobility is higher than that of the ion, the collection of charged species leaves a net positive charge into

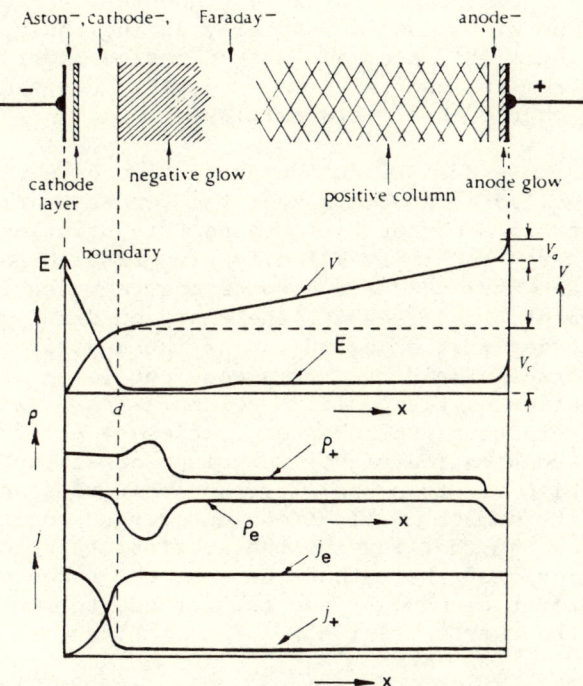

Fig. 1. The glow discharge (after Von Engel, 1965). Typical spatial distribution of
- the potential V and the electrical field E
- the electron and ion densities ρ_e, ρ_+
- the electron and ion current j_e, j_+.

the discharge gap. The drift of positive ions towards the cathode ends by setting the steady state glow regime, characterized by a nearly constant positive space charge located near the cathode. An associated linear decrease of the field appears. This decrease must continue until a very small value of the field is reached. Indeed, if this small value were not reached, the electron collection would remain bigger than that of the ions. The resulting positive space charge would accentuate the field decrease until a zero field region appears which puts an end to the electron sweep. From the cathode, the electron current density increases, and, at the end of the cathode fall (located where the linear decreasing field becomes zero), it approaches the total current density. The positive ion current density shows, in turn, a complementary space variation to allow for the current density conservation. The overall electron avalanche size reaches a value where the self-sustained discharge condition is fulfilled; i.e., each avalanche triggers a following avalanche by producing a secondary electron at the cathode. Further on, the electron enters the negative glow with a high kinetic energy gained in the cathode fall and, thus, is still drifting towards the anode even though the field collapses near zero. However, the collision rate of the electrons slows down drastically during their progression through the neutralized plasma of the negative glow; and since the electron flux remains constant, this velocity decrease results in a growth of the charged species densities.

Furthermore, a decline in the production of excited and ionized species is revealed by a dark space, the Faraday dark space. The zero field situation cannot last beyond this point, since the decrease in electron energy and drift velocity give rise to an accumulation of electrons and form a negative space charge which induces a small field enhancement. As a result, the electron drift velocity increases again avoiding any further extension of the negative space charge so that the electrical field reaches a constant value which characterizes the formation of the positive column. The field takes a value for which the production of charged species by electron impact is balanced by recombination either in volume or at the walls. (In spite of the high electron mobility, no positive space charge is formed since the number of electrons entering a positive column section is the same as that leaving the section; this condition could not be fulfilled near the cathode because the secondary effect could not release enough electrons from the cathode to replace each electron leaving the cathode region.)

In the anode fall, the field reaches a value high enough to produce a number of positive ions which feed the positive column and compensate the ion leak towards the negative glow.

After this quick summary of the usual glow discharge, the question of the use of the word "glow" must be examined since it is used in different cases: positive and negative "glow coronas," dynamical high pressure "glow phase," etc. The question is, are all these glows of the same kind? At first glance, obvious differences appear. For instance, the electron temperature in the positive column will be defined at low pressure by a balance between production and loss at the walls, but no boundary walls are present at high pressure, where a balance between production and recombination plays the dominant role. Thus the rates of physico-chemical yield may vary in a wide range with pressure.

However, a most important feature, found in all these glows, brings them together, and can be used as a definition of a glow. A roughly similar structure prevails with a small active region characterized by a high rate of production of charged species (similar to the cathode fall region), followed by an extended passive (or less active) region where the collection of the charged species happens (the positive column). With the glow thus characterized, a glow-to-arc transition can always be recognized as part of the sequence of a spark formation.

THE SEQUENCES LEADING TO THE SPARK FORMATION

Wave propagations, or states which evolve towards other states, involve spatio-temporal loops of interactions. For instance, a change in a local state induces a change in the surrounding part of the discharge which, in turn, acts on the local situation. But a local change means a change within a volume of the order of some inelastic mean free paths since density or fluorescence variations, for example, are related to the associated collision process. It follows, that a discharge will be more stable at low pressures than at high pressures since at low pressures the mean free paths are often of the same order as the size of the vessel and thus some spatio-temporal coupling of different discharge regions able to trigger unstable discharge evolution are not likely, unless the vessel size becomes very large. (A rough analogy can be suggested: the low pressure discharge acts like a solid body, while the high pressure discharge acts more like an elastic rod.)

Basically this is the reason why the dynamical studies of the glow-to-arc transition are generally carried on at high pressures. In addition, at high pressures the bright sequences can be followed with a more accurate spatial definition and also with a good temporal definition since the lifetimes of the excited states are shortened by quenching pressures, and the emitted light follows closely in time the excitation processes (Mitchell, 1970).

The glow-to-arc transition behavior will now be analyzed by comparing different glow-to-arc transitions at high pressures in small gaps. In Fig. 2, the time-resolved emitted light is schematically reported for various cases. Here the values of the different

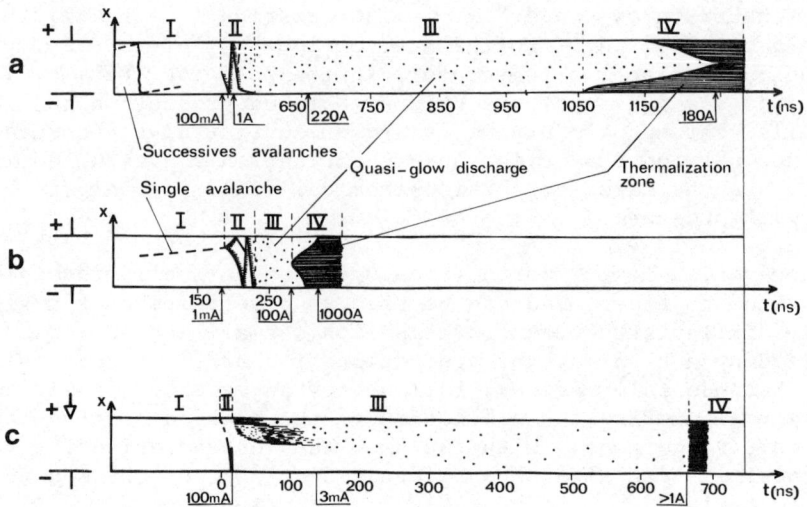

Fig. 2. Comparison between typical streak photographs of the spark formation in different cases.
(a) uniform field gap: Nitrogen, pulsed gap with small overvoltage (7.5%), generation mechanism, p=300 torr, d=2 cm (after Doran, 1968).
(b) uniform field gap: Nitrogen, pulsed gap with high overvoltage (35%), streamer mechanism, p=300 torr, d=2 cm (after Koppitz, 1971 and Chalmers and Duffy, 1971).
(c) nonuniform field gap: Air, DC potential, p=760 torr, d=1 cm, point radius 100 μm (after Marode, 1975).

imposed parameters are quite different:

- gap geometries — parallel-plane electrodes, uniform field (a,b)/point-to-plane gap, nonuniform field (c)

- gas nature — nonattaching gas, N_2, (a,b)/attaching gas, air (c).

- applied potential — impulse: small overvoltage (a), high overvoltage (b)/DC potential (c).

In all these cases, however, the spark formation exhibits similar sequences which can be put into four phases.

During Phase I, one or more avalanches develop under the applied field, but the space-charge field resulting from the produced charged species does not yet change appreciably the applied field. However, since the self-sustained discharge condition (see other papers) is already reached at the beginning of the discharge, each avalanche will statistically be followed at least by another one. In the successive avalanche generation (a), the amplification factor of each avalanche is small compared to that of the streamer mode development (b,c), so that Phase I will last much longer in case (a) before Phase II, characterized by a space-charge controlled discharge evolution, is reached.

Phase II begins when the space-charge field becomes nonnegligible compared to the applied field. In the streamer mode (b,c), an accelerated filamentary discharge expansion is observed when the first avalanche reaches an amplification of the order of 10^8, while this quick expansion is also observed in the generation mode but in a much more diffuse form (see the difference in the current at the beginning of this phase). Axial instabilities, revealed by bright waves, moving up and down through the discharge gap with speeds of the order of 10^7 to 10^9 cm s^{-1} dominate this phase. The ionizing mechanism which acts through these phases is fully described in other papers (see Dutton, Goldman and Sigmond) and some additional remarks will be given about the streamer mechanism further in this paper. But here the following general comment can be made. Under the development initiated by either the generation mode, or by the streamer mode, the successive space-charge rearrangement will tend to push the discharge into a glow stage.

No clear demonstration of this fact is yet available, but a simple approach can illustrate this general behavior.

A Paschen curve (which gives the breakdown potential V as a function of pd, the product of pressure time-gap spacing) for a uniform-field gap is shown in Fig. 3. Along this curve, the self-sustained condition

$$\gamma(e^{\alpha d} - 1) = 1 \qquad (1)$$

is fulfilled (α and γ are the first and second Townsend coefficient), and because γ is a weak function of E/P (the reduced field), the avalanche multiplication factor

$$e^{\alpha d} = M_o$$

may be roughly considered as a constant along the curve.

Let us consider a gap, with electrode spacing d_1, subjected to the breakdown voltage V_1, i.e., whose coordinates (V_1, pd_1), (point A), lie on Paschen curve, to the right of its minimum. Because, without space charge, the potential varies linearly within the gap (curve CJA) the potential V'_2, at an arbitrary position $x_2 = d_2$ (point J) near the anode (x refers to the distance from the cathode) will be less than the breakdown potential V_2 for a gap spacing d_2 (point L). The electron avalanches will build up a positive space charge, near the anode, say, between d_2 and d_1, so that the field

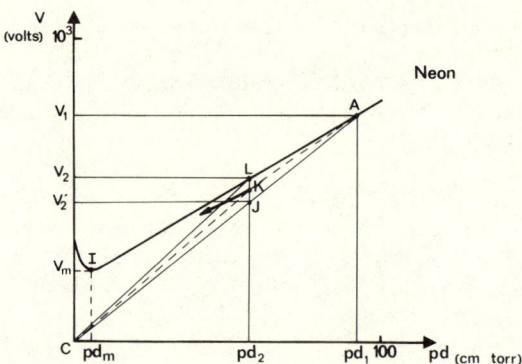

Fig. 3. A typical Paschen curve-sparking potential V versus pd in neon (———). Lines like CJA, CKA and CLA are potential distributions within a gap according to the text.

will rise between the cathode and d_2 and lower between d_2 and d_1 leading to a potential distribution like (CKA).

Now on the one hand, it has been proved (Badareu, 1965; Raether, 1964) that the avalanche size M_o in a uniform field tends generally to be increased by the development of a positive space charge. This is because the increased ionization in the field-enhanced region will be greater than the loss of ionization in the region where the field is lowered. On the other hand, if the distribution of potential is like curve (CLA), then the overall multiplication factor M will be greater than M_o, because L lies on the Paschen curve, and the multiplication factor from cathode to d_2 will already be equal to M_o. These two reasons tend to suggest that under a potential distribution like (CKA), the multiplication factor M will be greater than M_o leading to

$$\gamma(M - 1) > 1 . \tag{2}$$

As a result, the discharge current rises and produces an increased potential drop on the external circuit resistance inducing a gap potential fall which tends to reestablish condition (1). Meanwhile, however, the positive space charge accumulates, causing further field enhancements on the cathode side of the gap so that K tends to shift to the left towards the point $I(V_m, pd_m)$, the Paschen minimum, and the gap potential falls around V_m to fulfill condition (1). No further space change can accumulate, because V_m is the minimum potential able to produce an avalanche size M_o. Additional space charge would thus decrease the avalanche size, decrease the current, and increase the gap potential to restore a Paschen minimum within the gap, so that this situation seems to be stable.

If this description is correct, the discharge would tend to establish a region whose coordinates are close to the Paschen minimum, i.e., close to a region where the ionization rate per unit drop in potential is at its maximum.

This is actually what is observed for a glow discharge. The active cathode region is characterized by its cathode reduced thickness $(pd)_c$ which is very close to the Paschen minimum coordinates, as seen in Table 1. Of course, the difference must probably be because the field is not uniform within the cathode fall.

Table 1. Comparison between the Paschen minimum coordinates (V_m, pd_m) and the cathode fall coordinates (V_c, pd_c).

Gas nature	Cathode nature	V_m (Volt)	V_c	$(pd)_m$ (torr cm)	$(pd)_c$
He	Fe	150	150	45	1, 3
Ne	Fe	244	150	3	0, 72
N_2	Fe	275	215	0, 75	0, 72
O_2	Fe	450	290	0, 70	0, 31
Air	Fe	330	269	0, 57	0, 52
H_2	Pt	295	276	1, 25	1, 00

Thus the unstable runaway situation which is set up when the breakdown potential is applied tends to evolve towards a stable glow discharge.

This optimization mechanism may well be a qualitative explanation to the question of why the space charge phase (Phase II) ends in a glow structure, whatever mode (successive avalanches, streamer mode) controls the space charge rearrangement.

However, whether the gap potential will fall or not with the glow formation will depend on the nature of the applied potential source (capacity voltage generator with low impedance). If the potential does not drop too much, the glow will be in a quasi-stable situation which may evolve into an abnormal glow regime and then towards the spark.

Coming back to Fig. 2, the glow state duration is larger in case (a) than in case (b). In the successive avalanche generation mode (a), a diffuse glow sets in, while in the so-called streamer mode (b) a filamentary glow is formed. Since the spark itself is a filamentary discharge, a constriction of the discharge must occur in case (a) before the spark forms, while the filamentary structure, already present in case (b), implies a much more localized power dissipation and thus a much faster glow evolution.

The filamentary nonuniform case (c) in air is more puzzling. Because of the small value of the mean electric field, the electrons disappear by electron attachment, so that the glow stage tends to be self-quenched, as demonstrated by the reported current decrease during Phase III. Nevertheless, the final spark formation still appears, and this will be explained later in terms of hydrodynamic phenomena. In summary, the filamentary channel is formed and preheated during Phase II and III. In the final spark development, Phase IV, regions of dissociation of the gas molecules and thermal ionization expand through the preheated filamentary channel either from midgap (b) or from electrodes (a,c).

This will now be shown by summarizing the study of Tholl (1970) in hydrogen on streamer to spark transition in a uniform gap, by means of spectroscopy techniques. As long as the radiation is due to molecular hydrogen, the density $n_e(r,x,t)$ is derived from it. When dissociation becomes such that emission of atomic hydrogen dominates, then n_e is determined neither by stark broadening of the atomic lines nor by measurement of the continuum intensity. Fig. 4 gives the n_e (r,x,t) distribution. This figure shows two things:

1) The initial density distribution is very inhomogeneous, with its maximum (at 6.10^{14} cm^{-3}) located in a constructed region, the

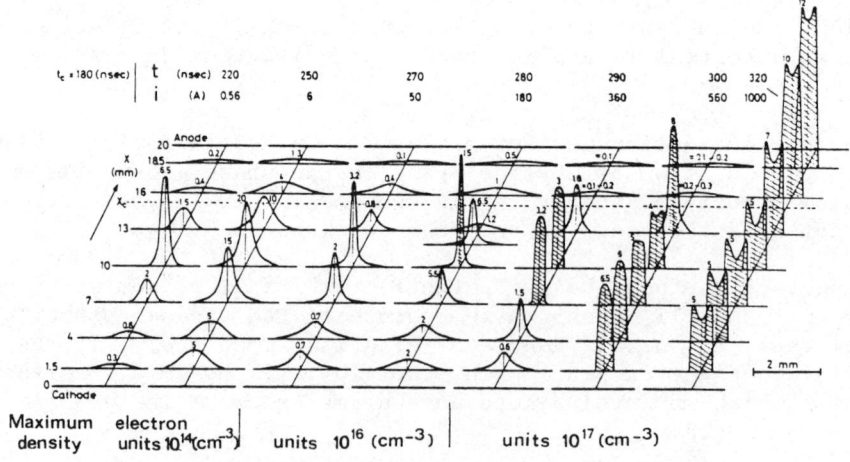

Fig. 4. Radial and axial electron density distribution at various times in a filamentary discharge build in the streamer mode in hydrogen. The zero time is the instant of production of the initial electron released at the cathode (after Tholl, 1970). Uniform field gap, pulsed discharge with overvoltage of 16.8%, p=460 torr, d=2 cm.

"neck", behind x_c (the critical distance of streamer formation from the cathode);

2) In between 260 to 290 ns, the neck constricts more and begins to emit the Balmer lines H_α and H_β, which indicate a substantial dissociation.

After $t \simeq 270$ ns, the density rises above some 10^{17} cm^{-1} in the neck, and the thermalization appears (this will later be commented upon). The molecular hydrogen emission decreases strongly on the channel edge because the channel conductivity increases and leads to an electrical field collapse. But the channel core emission is rising and emits atomic lines and continuum, from "bremsstrahlung" and recombination. The temperature increases in the channel core produce an overpressure of 10 times the initial pressure, and a hydrodynamic expansion occurs as seen on the progressive channel enlargement. This rapid thermalization growth can be followed on the dashed part of Fig. 4. The cathode region, and the anode region (whose density changes from 10^{16} to 10^{18} cm^{-3} in 20 ns) are heated, so that, at $t = 320$ns, a complete spark channel is achieved, with a core temperature of 6.10^4K. Similar behavior is observed in Nitrogen (Fig. 2(b); Chalmers et al., 1972).

THE CONTROL OF THE GLOW-TO-ARC TRANSITION

In the above section a phenomenological approach to the whole spark formation was presented. In this section the two boundary states, the glow and the spark will first be characterized. It then will be shown that for the spark to be formed, it is sufficient to reach an intermediate state beyond which the mechanisms of spark formation are clear. Further, the various modes of glow evolution towards this intermediate state will be described, and this will lead to a distinction between two different levels of control of the glow evolution.

The Problems of the Transition

The positive column of a glow discharge is a cold plasma, i.e. a plasma where the temperature of the heavy species remains in the range of room temperature, with a low degree of ionization (typically $n_i/n_g < 10^{-3}$, where n_i and n_g are the ion and heavy species densities). The electrical energy is mainly transferred to the electrons which gain kinetic energy through the action of the electrical field. The electron velocities are randomized by electron collisions with heavy species so that a thermalization of the electron energy occurs and that confers to the electrons a specific electron temperature T_e. This temperature sets in at a value where losses (function of T_e) balance gain of energy from the field. However, electron energy losses through elastic collisions are negligible, since the fractional loss per elastic collision is of the order of m_e/M (the ratio of the electron to heavy species mass), and this is much smaller than the gain of energy between two collisions. It follows that the heavy species temperatures, T_i for ions, T for neutrals, remain small compared to T_e (of the order of $kT_e \sim 2$ eV).

For the cathode region, the main features, (see second section) are the action of secondary electron emission at the cathode and the existence of a cathode fall which sets in at a value where the self-sustained condition is fulfilled in order to materialize the cathode to positive column interface.

At the other end of the glow-to-arc transition is the transient spark whose spatial structure is very similar to that of a glow discharge. It is also composed of a positive column and a cathode region. However, the channel of the positive column of a spark is a hot plasma, which may reach a transient temperature of the order of several 10^4K (Tholl, 1970). At this temperature, the microreversibility (Von Engel, 1965) of the elementary reactions is satisfied, and local thermodynamic equilibrium (L T E) is approximately reached. Thus, all different temperatures, T_e, T_i, T, become equal for initial pressures above some several tens of torr, thermal ionization prevails, and the ionization degree is given by the Saha equation (Chalmers et al., 1972).

At the cathode, the transition between the electrode and the positive column is established by means of a hot cathode spot, i.e. a region in which the electron emission is no longer obtained by secondary effects, but rather by thermoelectronic or field effect.

Thus, there are two problems arising from the transformation taking place when the glow-to-arc transition sets in. The first is to understand how, in spite of the inefficiency with which the electron energy is transferred to the heavy species, the cold plasma of the positive glow-discharge column can evolve into the hot plasma of the highly conducting positive spark column. The second is to understand how the cathode region which, in the glow phase, yields a limited current density controlled by secondary effects transforms into a hot cathode spot which generates the needed high current density.

No clear answer to the second question is yet available, since the mechanisms by which the cathode surface is transformed and heated are still under examination. Apart from the obvious sputtering because of kinetic impact of positive ions at the cathode, the way in which the excess energy of ion neutralization at the cathode (difference between the ionization potential and the work function potential needed to extract the neutralizing electron) dissipates at the surface is still not understood. The surface action of active chemical compounds like metastable species is also not clear. The cathode heating will be the result of all these different energy channels whose importance depends on the production rates of the various acting species, and this production varies with the discharge regime, normal or abnormal glow. Moreover, to study this complex situation, one must bear in mind, that the charged species within this interface region are not in equilibrium with the electrical field, and proper treatment of this region requires an out-of-equilibrium technique such as solution of the Boltzmann equation or Monte-Carlo simulation.

We will no longer consider the physical processes which determine the cathode region but rather turn to the cathode region behavior, since this region is the boundary condition for the positive column and, as such, affects the state which exists within the glow channel.

The Final Thermalization Phase

In the preceding section we have shown that the glow is a cold plasma because the electrons lose only a small fraction of their energy into translational energy of the heavy species. Now Tholl (1970) remarked that with the rise of the degree of ionization, the number of coulomb elastic collisions between electrons and the ions can strongly increase so that, even with a small fractional loss of the electron energy, the discharge approaches the state where the conditions for LTE in a transient plasma, as given by Griem (1964), can be fulfilled.

The variation $d\varepsilon_i$ of the mean thermal energy of a species i due to elastic collision with species j during a time dt is, when diffusion of species i and thermal conductivity are neglected:

$$d\varepsilon_i = \frac{2 m_i}{m_i + m_j} \nu_{ij} \frac{3}{2} (kT_j - kT_i) dt \tag{3}$$

where the m are the masses, the T are the temperatures and ν_{ij} is the collision frequency for the momentum transfer from species i to species j related to ν_{ji} by:

$$m_i n_i \nu_{ij} = m_j n_j \nu_{ji} \tag{4}$$

where the n are the densities.

Applying (3) and (4) respectively for the thermal energy variations of the ions $d\varepsilon_+$ and the neutral $d\varepsilon_n$, we obtain:

$$d\varepsilon_+ = \frac{2 m_e}{m_+} \frac{n_e}{n_+} \nu_{e+} \frac{3}{2} (kT_e - kT_i) dt \tag{5}$$

$$d\varepsilon_n = \frac{2m_e}{m}\frac{n_e}{n}\nu_{en}\frac{3}{2}(kT_e - kT)dt , \qquad (6)$$

while the mean energy variation of a heavy species will be given by

$$d\varepsilon = \frac{nd\varepsilon_n + n_+ d\varepsilon_+}{n_g} , \qquad (7)$$

where $n_g = n + n_+$ is the overall density of the heavy species. Now, the translational energy exchange between an ion and a neutral species is very efficient because they have the same mass (m_+ is assumed equal to m); thus, if the ions are heated by the electrons, they will very quickly share their thermal energy with the neutral species so that T and T_+ can be assumed to be equal and $d\varepsilon$ becomes

$$d\varepsilon = 2\frac{m_e}{m}\frac{n_e}{n_g}(\nu_{en} + \nu_{e+})\frac{3}{2}(kT_e - kT)dt . \qquad (8)$$

The relaxation time τ for the heavy species to reach a temperature of the order of T_e is the time required for $d\varepsilon$ to be of the order of $3/2(kT_e - kT)$; thus, from (8)

$$\tau = \frac{1}{2}\frac{m}{m_e}\frac{n_g}{n_e}(\nu_{en} + \nu_{e+})^{-1} . \qquad (9)$$

We must now evaluate ν_{e+}, the electron-ion momentum transfer frequency. Introducing the cross sections for momentum transfer σ_{en}, σ_{e+}, and $i = n_i/n_g$, the ionization degree, τ, can also be written as

$$\tau = \frac{m}{2m_e}\frac{1}{n_e}(\overline{\sigma_{en}v}(1-i) + \overline{\sigma_{e+}v}\, i)^{-1} \qquad (10)$$

where ⎯⎯ denotes mean values. This is the form used by Tholl with:

$$\sigma_{e+} = 3.10^{-15}(V_i/kT)^2 . \qquad (11)$$

Here, V_i is the ionization potential of the neutral species as given by Spitzer (1956). Using (11) for the atomic hydrogen case because, as seen on Fig. 4, dissociation is already effective, and taking $n_g = 10^{19}$ cm^{-3}, he found a relaxation time as a function of the electron density n_e as shown in curve (a) of Fig. 5. However, the presence of V_i in (11) comes from the assumption that the Saha equation, which relates n_e and V_i, already holds, so that the state towards which the relaxation is supposed to lead is already assumed to exist before the relaxation. The coulomb collisions can, however, be expressed by the use of the Fokker-Planck coefficients; and with a cut-off of the coulomb impact parameter set at the Debye length, without any assumption about the Saha equation and

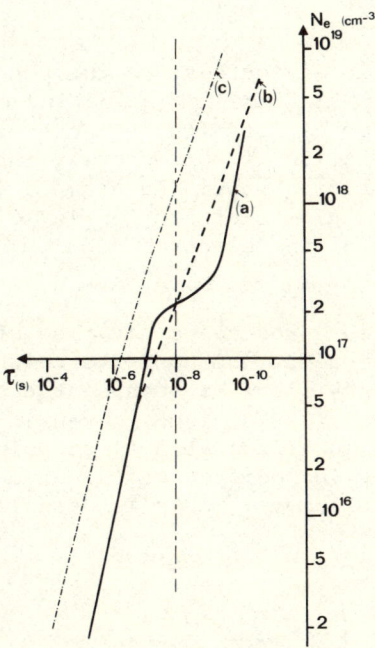

Fig. 5. Thermalization time τ as a function of the electron density.

from Holt and Haskell (1965), (9) can be written as:

$$\tau = \frac{1}{2} \frac{m}{m_e n_e} (6.8 \; 10^5 \; T_e^{1/2} \sigma_{en} (1-i) + \frac{9.42 + 1.5 \ln T_e - 0.5 \ln n_e}{0.38 \; T_e^{3/2}} i)^{-1}$$

(12)

with n_e in cm^{-3}, T_e in °K, σ_{en} in cm^2. If this expression is applied to Tholl's conditions, curve (b) follows, while when applied to atomic nitrogen (whose mass is higher) an increased relaxation time, curve (c), is obtained. In both cases, $T_e = 10^{4}$°K and $n_g = 10^{19}$.

These curves show that a critical relaxation time of, say, lower than 10^{-8}s is reached for densities somewhat greater than that found by Tholl, if LTE is not assumed. Roughly, thermalization by coulomb interaction occurs whenever a degree of ionization greater that 10% occurs.

Therefore, the question becomes what are the processes which allow the discharge channel to reach this critical ionization degree?

Prethermalization Modes of Evolution

One way to obtain an increase in the conductivity of a filamentary channel is via a mechanism which induces a step increase of the charged species density from a background value n_0 to a value n_1. If the location of this density change is moving, it is referred to as an ionizing wave. An example can be shown in the case of the completed glow sketched in Fig. 6(c), with a typical spatial distribution of the electron density, the electrical field, and potential.

The cathode region produces a net rise of the charged species density from the cathode to the positive column. At high pressure, this increase is really stepwise because the cathode region thickness decreases by the order of some microns. However, in that case, the step increase is localized at a fixed point, namely, at the cathode, and in order to modify a whole channel, the location of this step increase must move. This is actually what is observed

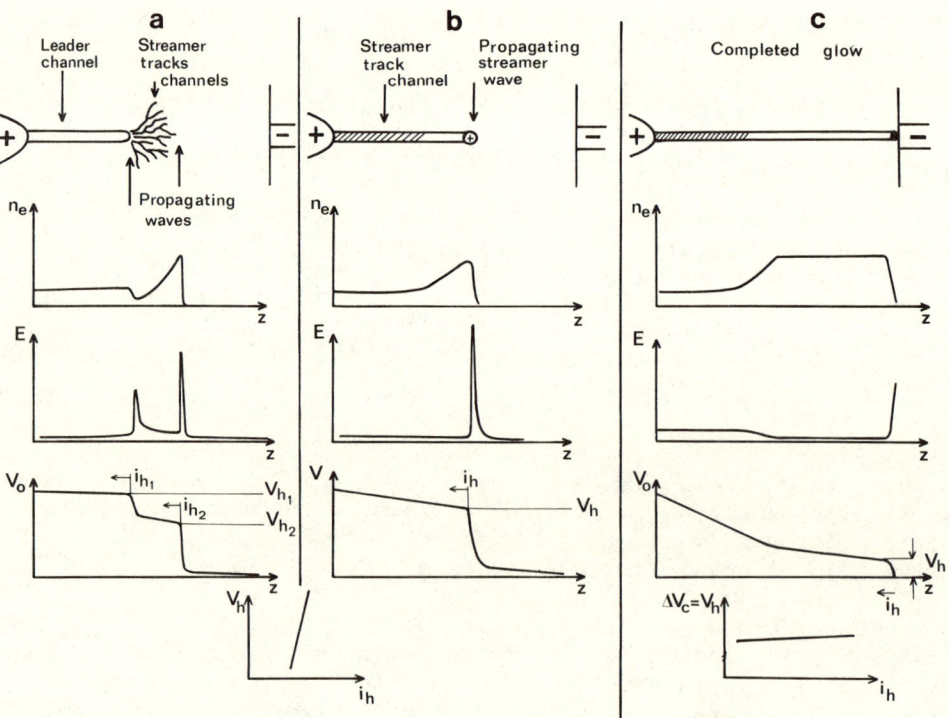

Fig. 6. Various modes of discharge evolution towards spark. V_h, I_h, potential and current at the tip of a filamentary channel, behind the active region.

with a streamer-channel build-up, Fig. 6(b); the streamer located at the end of a filamentary ionized channel produces the required density increase from a background electron or negative ion density to the channel density. As developed in the Sigmond and Goldman paper on corona discharges in these proceedings, the discharge state is here similar to a glow stage, the filamentary channel as the positive column, and the streamer head a moving cathode region using the gas as a cathode. Moving step changes, or ionizing waves, within an already excited channel are observed in larger gaps as sketched in Fig. 6(a). Here, localized field enhancements are first encountered, with streamers building channels as in case (b); all these channels create a background ionized gas within which a second ionizing wave may lead to a second-step increase of the electron density.

This type of behavior is illustrated in Fig. 10 which is a typical streak photograph of a discharge in a 10 m positive rod-to-plane gap in air at atmospheric pressure. The transition between the bright corona region, with its regular advances into the gaps, and the dark leader, which joins it to the positive rod, is clearly seen. Similar transitions are observed in many different cases. In contrast to the example of the long nonuniform positive gap is the case of a 2 cm parallel plate, at 162 torr, in hydrogen (Fig. 7a) taken by Kekez and Savic (1974). Here also, the advance of the tip of a filamentary discharge into an ionized background region is observed. Fig. 7b shows one of the aspects of a 1m gliding discharge in air above an insulating material, as recorded by Larigaldie at ONERA (1980, private communication), another example where a primary discharge builds an ionized region in front of the ionizing wave.

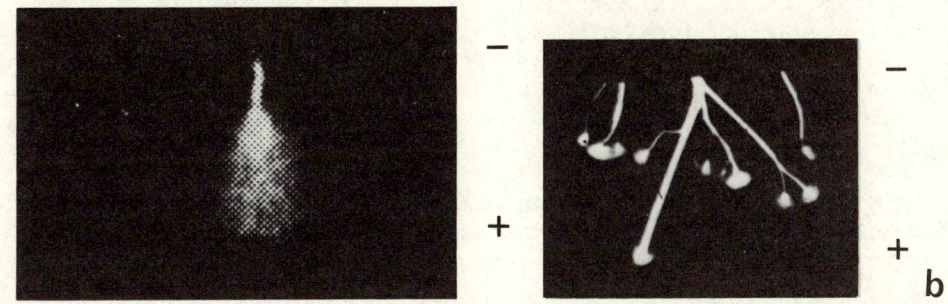

Fig. 7. Still photographs of filamentary discharges developing in an ionized background: (a) parallel plate, H_2, p = 162 torr, d = 1.9 cm (Kekez and Savic, 1974) (b) gliding discharge on an insulator, Air, p = 760 torr, d $\simeq$ 1m (Larigaldie, 1980, private communication).

The other way to increase a channel conductivity is through a process which involves a continuous change in the ionization rate in the whole volume of a channel section. However, the electron continuity equation, for example, (written in one dimension because of the filamentary structure with s, the channel section, and e, the electron charge)

$$es \frac{\partial n_e}{\partial t} + \frac{\partial i_e}{\partial x} = S \qquad (13)$$

shows that the rate of change in n_e depends on two terms: a source term S (which includes ionization terms as well as loss terms like attachment or recombination) and a "collection" or "injection" term, depending on the sign of $\partial i_e/\partial x$ (i_e being the electron current), which corresponds to the rate of electrons leaving or entering the discharge section under consideration.

On one hand, as far as ionization rates are concerned, the source term S depends on the field to neutral density ratio E/N, so that basically the ionization rates may be increased either by an increase of E, or a decrease of N. Because the integral of the field through the gaps is limited by the available gap potential, it has been suggested that the main process is a decrease in neutral species density. This point will be fully developed in the next section.

On the other hand, the $\partial i_e/\partial x$ term will be strongly related to the current i_h supplied by the mechanism of the active tip (or head — subscript h), ionizing wave or cathode region. The current supply i_h, as well as the magnitude of the density step Δ_n produced by the active head, will itself be dependent on the potential V_h existing at the channel-head boundary (as illustrated in Fig. 6). Actually the ionizing wave, or the cathode region, defines the boundary conditions for the channel, and vice versa, so that the magnitudes of the processes acting within the channel and within the active tip region are strongly interdependent. This leads us to distinguish two possible modes which control the evolution of the discharge. These modes will be derived from the dependence of the active tip against a change in V_h.

If the active region is a cathode region, then its behavior is described by the V(i) characteristic of a glow discharge, Fig. 8, which for a small positive column is roughly that of a cathode fall. This figure shows that a large change in the current can be driven by a small change in the cathode fall $V_h - V_c$, V_c being the cathode potential. The head reaction would be, in that case, roughly as plotted on the right curve of Fig. 6.

On the other hand, if the active region is an ionizing wave, or a streamer, then no imposed potential like the cathode potential

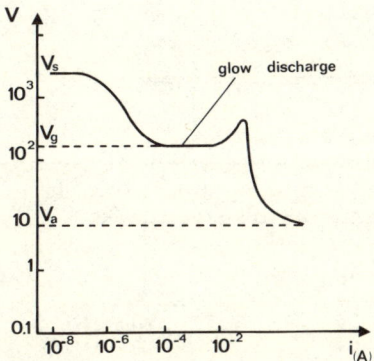

Fig. 8. A typical characteristic of a discharge cell in a uniform field gap, V_s = threshold of the Townsend dark discharge, V_g, glow discharge potential, V_a, spark potential.

V_c prevails near the channel head region, and a change in V_h will be followed by a change in potential of the whole region. Therefore, the electrical field and i_h remain unchanged in the active head. The variation of V_h and i_h, would then be like the left curve on Fig. 6, the opposite of case (c). It follows that the discharge evolves either under a channel-controlled situation or under a channel head-controlled situation as can be explained with the help of Fig. 9.

The channel of length l and the channel-head region are here characterized by their conductivity σ_c and σ_h. If some change in the ionization rate S occurs which tends to increase the electron density within the channel, then σ_c increases, and the potential V_h, determined by the conductivities σ_c and σ_h, also increases. However, because i_h is mainly a function of the head space charge and not of V_h, σ_h will not change, V_h will rise, and the field E_c in the channel will tend to collapse. The ionization frequency ν_i (E/N) decreases, and the source terms $\nu_i n_e - a_e n_e^2$, (where a_e is the recombination coefficient) becomes negative until an electron density decrease compensates the initial n_e increase. A stabilization occurs which ends with $\nu_i n_e$, the ionization rate, balanced by $a_e n_e^2$, the recombination rate. The electron density is then:

$$n_e = \nu_i/a_e . \tag{14}$$

Thus, if S is the channel cross section, μ_e the electron mobility,

$$i_h \simeq e \, S \, \frac{\nu_i}{a_e} \, \mu_e N \, \frac{E_c}{N} , \tag{15}$$

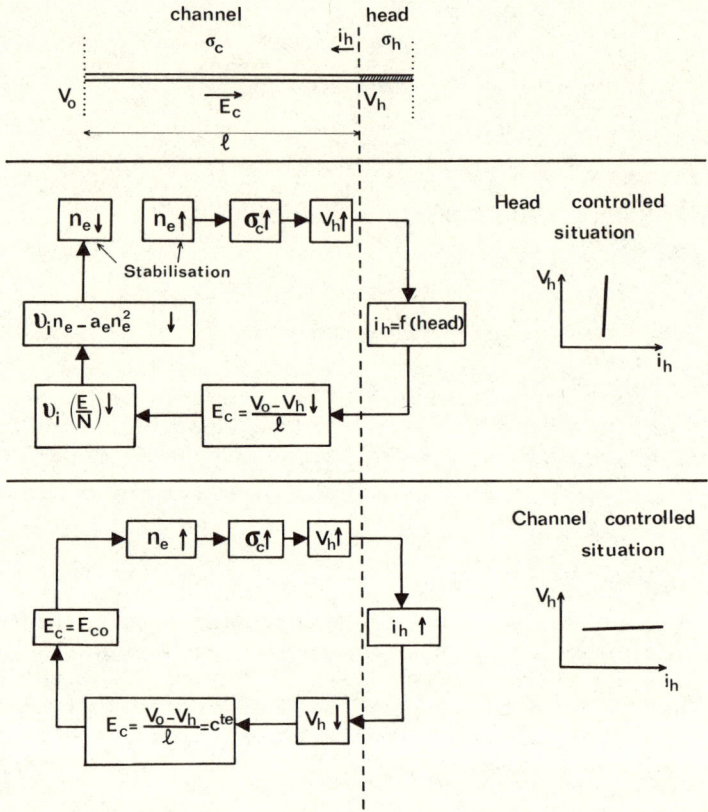

Fig. 9. Diagrams of interactions between a discharge channel and its active tip.

so that ν_i/a_e is determined by i_h; i.e., the channel is head-controlled.

For a channel-controlled situation, the initial increase of V_h leads to an associated increase of the current i_h at the channel head, σ_n increases, the relative change between σ_n and σ_c will not allow V_o to change, and E_c will remain constant. If for this specific value of E_c, the ionization rate increases, this increase will not be perturbed by a head reaction, and thus, we are in a channel-controlled situation.

These two modes of channel development are illustrated by the streak photograph of Fig. 10. In the first discharge sequences, the leader which joins the positive rod to the corona glow could increase in temperature if the corona glow did not limit the current which feeds the leader. However, as shown by Gallimberti (1979), the leader reaches a stage where ionization is balanced by

Fig. 10. Streak photograph of a long gap discharge in air between a positive rod and a plane electrode. Air, p = 760 torr, d ≃ 10 m.

recombination, and this is (see also Brambilla, 1978) a typical head-controlled situation. However, as soon as the corona glow reaches the cathode, the current supplied by the cathode is no longer limited, and the channel is able to quickly reach the spark stage, a typical channel-controlled situation (E = Cte).

But, if in a head-controlled situation, the channel evolution can be computed using a measured or assumed variation of the driving current i_h, the physical basis of the process sustaining the i_h production within the active region still remains a problem.

PROCESSES INVOLVED DURING THE GLOW-TO-ARC TRANSITION

Ionization Growth During the Prethermalization Phase

We have mentioned that the increase in channel conductivity may result either from the propagation of ionization waves, or from some general increase in the ionization rate within the channel, and we have described the coupling between these mechanisms in the preceeding section. However, we have only given some indications about the physical processes which sustain these mechanisms, and the aim of this section is to report in more detail on work dealing with these processes. A brief review will first be given to show how these studies are interrelated.

Within the filamentary channel a dominant factor which may raise the ionization rate is, as already stated, a decrease in neutral species density through hydrodynamic phenomena. Of course, a decrease in the density of heavy species occurs during the final thermalization phase, as reported by Tholl. This process must, however, not be confused with the one suggested for the cold plasma of the prethermalized discharge phase, where density changes of that type have been previously neglected at medium and high pressures.

In uniform fields, Rogoff (1972) analyzed such a process for the final spark formation in hydrogen, while Jaeger, Oster and Phelps (1976) studied constrictions in helium. In nonuniform fields in air and large gaps, Gallimberti (1979) has shown that the leader stage is also associated with neutral density variations. For small air gaps, Marode et al. (1979) suggested that this effect already dominates at an earlier stage of the discharge in order to answer how, in attaching mixtures, the spark can be formed in spite of the quenching effect due to electron attachment.

We now turn to wave phenomena. If a wave is a discontinuity which propagates, various light waves are experimentally recorded in the streak photograph of the spark formation (Fig. 2). However, the physical interpretation of these waves requires careful examination. For example, Davies et al. (1975) integrated the continuity equations describing the ionization growth in a uniform field gap and, in particular, made comparison with the measurements in nitrogen of Chalmers (1972) and Doran (1968), Fig. 2(a), for the low overvoltage case. Figure 11 shows that good agreement is obtained between the computed and experimental values of light fronts and total discharge currents. As seen, however, the current growth does not display a sharp increase associated with the light fronts. So even if these waves are associated with the local field enhancements (and thus ionization rates also), their existence is more related to the need for a potential redistribution after the discharge reaches an electrode. The ionization wave, however, carries with it a very sharp step in spatial electron density distribution. It involves a local stored energy consumption. Either this stored energy is carried with the wave — this

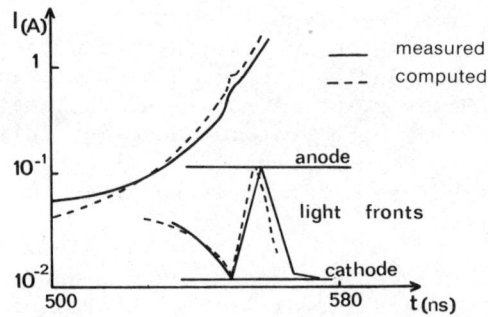

Fig. 11. Comparison between computed and measured light emission and current growth in the case (a) of Fig. 2 (after Davies et al., 1971).

is the case for streamers — or it is already present, as for the electron fluid waves.

In the case of a wave which propagates inside an already ionized region, Kekez and Savic (1974) proposed a mechanism of wave formation by a detonation process. The local concentration of kinetic energy due to the localized charge collection at the channel tip would trigger an explosive process.

Another mechanism suggested by Bareto et al. (1977) which corresponds neither to a decrease in neutral density nor to an ionization wave, consists of a fast energy transmission mechanism through collision-less waves because of collective electron interactions. In this model, the ionization is a further consequence of the rise of the electron temperature. This process is more likely to be important in small gaps.

We shall not develop all these topics because some of them, Gallimberti's work on leaders, and Kekez and Savic's analysis of leader formation and streamer mechanisms, are treated in other papers.

In the following sections, the hydrodynamic phenomena within the channel and the electron fluid ionizing waves will be presented. However, because both of these mechanisms are based on the treatment of hydrodynamic equations, these will first be written.

General Fluid Equations

To describe accurately the dynamic behavior of a gas species "s", the velocity distribution must be obtained; i.e. the Boltzmann equation must be solved (s will be set to e, +, - or n for electron, positive and negative ions, and neutral species, respectively). However, apart from transition regions, like the cathode region, the velocity distribution function is in equilibrium with the electric field, and the source terms as well as the transport coefficients are functions of E/N only. In that case the dynamic behavior of the "s" species can generally be described with enough accuracy by n_s, $\overline{v}_s$, T_s: their density, mean velocity and temperature. The knowledge of the detailed structure of the velocity distribution function is then not necessary. The mean value of $\phi(x,v,t)$, an arbitrary function of particle velocity, position, and time, is given by

$$\overline{\phi}(x,t) \int f(x,v,t)dv = \int \phi(x,v,t) f(x,v,t)dv \qquad (16)$$

where $f(x,v,t)$ is the velocity distribution function, and the integration is over all velocities.

Thus, an equation governing $\overline{\phi}$ is obtained by multiplying the Boltzmann equation by ϕ and integrating it over all velocities. If

ϕ is taken equal, respectively, to 1, $m_s v_s$, and $1/2\, m_s v_s^2$, m_s being the mass of species "s", the so-called moment equations are obtained which give the required equation for n_s, $\bar{v}_s$, T_s.

$\phi = 1$ leads to the continuity equation (referred to as Cs). $\phi = m_s v_s$ leads to the momentum equation (M_s) and $\phi = 1/2(m_s v_s v_s)$ leads to the energy equation (E_s). The following set of equations is obtained:

$$(C_s) \quad \frac{\partial}{\partial t}(n_s) + \nabla \cdot (n_s \bar{v}_s) = \sum_k S_k \tag{17}$$

$$(M_s) \quad \frac{\partial}{\partial t}(m_s n_s \bar{v}_s) + \nabla(n_s m_s \bar{v}_s \cdot \bar{v}_s) + \nabla \cdot \bar{\bar{p}}_s =$$

$$n_s q_s E + \sum_r \bar{A}_{sr}(m\bar{v}_s) \tag{18}$$

$$(E_s) \quad \frac{\partial}{\partial t}(\varepsilon_s + \frac{1}{2} n_s m_s \bar{v}_s \cdot \bar{v}_s) + \nabla \cdot [(\varepsilon_s + \frac{1}{2} n_s m_s \bar{v}_s \cdot \bar{v}_s)\bar{v}_s]$$

$$+ \nabla \cdot (\bar{v}_s \cdot \bar{\bar{p}}_s) + \nabla Q_s = n_s q_s \bar{E} \cdot \bar{v}_s + \sum_r M_{sr}(\frac{1}{2} m_s v_s^2) \tag{19}$$

In these equations $\bar{\bar{p}}_s$ is the pressure tensor, which often reduces to the unit tensor multiplied by a mean hydrostatic pressure

$$\bar{\bar{p}}_s = n_s k T_s \bar{\bar{I}}$$

when viscosity is neglected (k is the Boltzmann factor). ε_s is the thermal energy

$$\varepsilon_s = \frac{3}{2} nkT_s;$$

S_k is the source term (production or loss) due to process k. A_{sr} and M_{sr} are operators which give, respectively, the mean momentum and thermal energy exchange per collision of species "s" with species "r," q_s is the electrical charge of "s" ($q_s = 0$ for neutral species) and Q_s is the heat flux vector.

(C_S) is an equation for the density n_s, which could be solved if $\bar{v}_s$, the mean velocity did not appear. Although (M_s) is an equation for $\bar{v}_s$, it introduces the unknown T_s. (E_s) is an equation for T_s, but introduces the heat flux vector Q_s. An equation for Q_s could be obtained if the next moment equation with $\phi = v_s^4$ is written, but then a new unknown variable would appear, etc... An unlimited number of moment equations are needed for an exact solution. But this would be equivalent to solving the Boltzmann equation, so that usually the set of equations are truncated at the level of the energy equation (E_s) by using the assumption that

$$Q_s = -\lambda_s \nabla T_s, \qquad (20)$$

where λ_s is the coefficient of thermal conductivity; and thus, with the Poisson equation,

$$\nabla E = \frac{1}{\varepsilon_o} \sum_s q_s n_s \qquad (21)$$

the above equations written for all species are a closed set of equations. A_{sr} and M_{sr} are the coupling terms, which for elastic collisions may be written as:

$$A^e_{sr} = m_s n_s \nu_{sr} (\overline{v}_r - \overline{v}_s) \qquad (22)$$

$$M^e_{sr} = 2 \frac{m_s n_s}{m_s + m_r} \nu_{sr} (\frac{1}{2} \overline{mv}_r^2 - \frac{1}{2} \overline{mv}_s^2) \qquad (23)$$

where ν_{sr} is the collision frequency for momentum transfer from species "s" to species "r" (for the hydrodynamic equations see Holt and Haskell, 1965 or Mitchner and Kruger, 1973).

Hydrodynamic Evolution of the Channel

In order to compare the various hydrodynamic approaches to the channel evolution, their main experimental basis and assumptions have been gathered in Table 2.

The light emission, in the sketched streak photograph of discharges in air, corresponds to black lines and dashed regions; in (c), the line which joins the point and the plane is a streamer, while in (d), the black filament region is the corona leader, a region which consists of numerous streamers. In (e), a magnified portion β of the case (d) is drawn, in order to make apparent the time during which a given position in space is exposed to the corona leader phase.

All numerical simulations assume an axial symmetry and compute changes within a small cylindrical section of the filamentary discharge as indicated on Fig. 13.

Rogoff (1972) analyzed the transition from a diffuse glow to a constricted channel in hydrogen (a) which involves a time scale of 10^{-7} s (Cavenor and Meyer, 1969). He applied the hydrodynamic equations to neutral species for a prethermalized channel. Within

Table 2. Major features of the different hydrodynamic approach of the discharge-channel evolution.

	Time scale (s)	Energy channels	Radius pressure	Main features	Main assumptions	Input parameters	Results
(a) H_2 2mm, 60ns, 10A	10^{-7}		75 mm, 500 torr	$E_z = c^{te}$	- Use of fluid equations for neutral species only - Electron energy given to neutral thermal energy - ionization = recombination - $j = f(E_z, N)$ Rogoff (1972)	$I_0 = 6$ A $j = 400$ A cm^{-2}	$T \simeq 2.3 \cdot 10^3$ K $E/N \uparrow \rightarrow$ spark formation
(b) Glow discharge in helium	10^{-4}		1 cm, 1 – 1000 torr	Constrictions Field collapse	- Electron energy given to neutral thermal energy Jaeger, Oster Phelps (1976)	$2 < \frac{E}{p} < 4$ $10^{-2} < \frac{\Omega}{\ell} < 1$	$T \simeq 4 \cdot 10^3$ K $\frac{E}{N} \uparrow \rightarrow$ constriction
(c) Air, 1cm, 25ns	10^{-7}	- Negative ions - No time to release the vibrational energy	10-50 μm, 760 torr	$E_z = c^{te}$	- Gaussian profile of the electron temperature Marode, Bastien (1979)	$I_0 = 100$ mA $j = 3000$ Acm^{-2} $T_0 = 300$ K	$T \simeq 2.3 \cdot 10^2$ K $E/N \uparrow \rightarrow$ spark formation
(d) Air, 10m, 500μs	10^{-4}	- Sustained by vibrational energy - No negative ions formation	1 mm, 760 torr	Field collapses $E(t) = \frac{I}{j}$	- Constante pressure across the channel Gallimberti (1979)	$I_0 = .3$ A $I(t)$ $T_0 \simeq 2 \cdot 10^3$ K	$T \simeq 3 \cdot 10^3$ E/N stabilize where ionization = recombination
(e) Air, 50μs	10^{-5}	- Relaxation time to vibrational energy release	>> mm	- Field collapses $\eta \neq 0$ $n = 0$ $T > T_c$	Gallimberti (1979)		$T \simeq 10^3$ K

the discharge, the ionization rate was assumed to be always balanced by recombination loss. This condition allowed him to write the current density j, as a function of the field E and the density N, and thus to avoid any equations on the charged species. The axial field E_z was assumed to remain constant (channel-controlled situation). A basic assumption was that all the electron energy is transferred into neutral thermal energy. It followed that in a time as short as 60 ns, and for a current density of 400 A/cm^2, a huge increase of the neutral species temperature of 2.10^3K was computed.

The transition from a diffuse glow stage to a constructed discharge channel has also been studied in helium by Jaeger et al. (1976) for a wide pressure range (1 to 10^3 torr). The purpose of this study was to define conditions which avoided spark formation within a laser discharge state. The assumption of an electron energy entirely transferred into thermal energy is also made, but the calculation begins with smaller current densities at an earlier discharge stage than in the case (a). It follows that initially the ionization rate is larger than the rate of loss of charged species, and hydrodynamic equations for charged species must here be treated. For pressures above 10^2 torr, it was found that a phase of charged species growth is followed by a phase where ionization is approximately balanced by recombination. During this second phase, thermal hydrodynamic behavior becomes apparent. The resulting density decrease, more pronounced near the discharge axis, leads to an increased ionization rate near the center, which causes the constriction to develop. In turn, the charged species' density rise induces more and more charged species' losses by ambipolar diffusion. A steady constricted third phase is thus reached, where losses by ambipolar diffusion and recombination are balanced by ionization production. However, this final stage is reached only because the axial field adjusts itself to fulfill this balance situation, and this adjustment results from the potential drop on the external discharge resistance associated with the current increase and not by some "head control" exerted by the active region of the discharge tip. Here, the discharge is assumed to fill the whole discharge gap, and the existence of cathode and anode transition regions, although neglected, is implicitly assumed. The specific functioning of the discharge is based on gap potential adjustment which was assumed to remain constant in the channel-controlled or channel head-controlled, discharge evolution. Thus a basic parameter comes out of this study which is the ratio Ω/l, the external resistance per unit discharge length, and it is shown that the stability of the constriction increases with increasing values of Ω/l, a conclusion which somewhat contradicts the need of large current values for large laser outputs.

Additional and extensive work, unfortunately still unpublished, has been done by Jaeger and Phelps (1978, private communication) in laser mixtures (CO_2, N_2, He) and air which includes the treatment of negative ions and release of vibrational energies.

In the nonuniform case in air (c), hydrodynamic phenomena are assumed to act already at the very early discharge state, just after the channel has been built up by the streamer (Fig. 13). Attachment and recombination dominate so that the hydrodynamic equations for electrons, positive and negative ions, and neutral species had to be explicitly written, as in the former case on helium. In the case of H_2 and He, it is possible to assume that the electron's energy is entirely transferred into the thermal energy of heavy species. This assumption, however, cannot hold in nitrogen or air, since almost all electron energy is stored as vibrational excitation of the nitrogen molecule, and the relaxation time from vibrational energy to translational energy of heavy species is of the order 10^{-5} s (see Gallimberti, 1979). It follows that in contrast with case (a), for a time of the order of some 10^{-7} s and a current density as high as 3.10^3 Acm^{-2}, an increase of only $200°K$ is predicted. Consequences of this small increase will be commented upon later.

The three main features which differentiate the leader stage (d) from the other case are that 1) vibrational energy release dominates, 2) a constant pressure exists inside the core (as in case (b)) and 3) a head-controlled situation sets up, so that the field collapses until ionization is balanced by recombination. In case (e), it is suggested that the formation of the leader is associated with the relaxation time of vibrational energy. Cases (d) and (e) are mentioned here for comparison but will be discussed in detail in the paper by Gallimberti in these proceedings.

Two examples of channel development, channel-controlled and channel-head-controlled, will now illustrate the hydrodynamic channel expansion.

The first example refers to case (c) of Table 2, where the streamer propagation establishes a complete channel from point to plane, as sketched in Fig. 6(c), so that the axial field remains constant in time as long as the gap potential does not drop. The experimental behavior of the glow-to-arc transition (Marode, 1975) is reported in Fig. 12, where five streak photographs are correlated with their associated current measurements. As already indicated on Fig. 2(c), the current may decrease after the initial filament build-up by the streamer (which on the streak photograph appears roughly like an initial straight axial luminosity, due to the slow streak sweep). In (d), the filament luminosity (the transient positive column) decreases and increases again 300 ns after the beginning of the discharge. But the smaller the initial current peak, the larger will be the spark formative time. If the initial peak is below a critical value, the decrease in current will no longer be followed by a spark.

From the applied potential V, the mean value of $E/N = V/Nd$, assuming N equal to the normal density at atmospheric pressure,

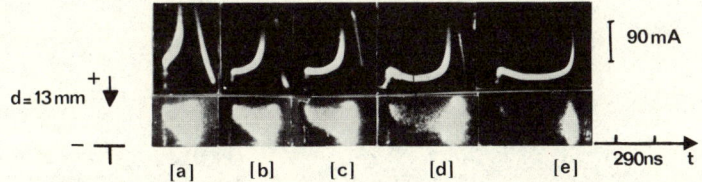

Fig. 12. Streak photographs and associated current pulses in a DC, positive point-to-plane gap. Air, p = 760 torr, d = 1 cm (Marode, 1975a).

gives an attachment rate higher than that of ionization; and since the field within the discharge is made uniform by the filamentary positive column, the attachment rate must dominate, as confirmed by the decrease of the current.

The further increase in current is then explained as follows. In spite of the small heating of heavy species by electrons, the weak increase of temperature of ~ 200°K, raises the pressure inside the channel. The resulting pressure difference between the inside and the outside of the channel will induce a radial leak of heavy species which tends to reequilibrate the pressure. A decrease in neutral density species will result which induces the E/N increase. If E/N reaches the critical value where the ionization coefficient α rises beyond the attachment coefficient η, a new final growth of ionization will occur and lead to the spark formation.

This basic fluid dynamic process is that which is thought to happen in all the cases mentioned in Table 2. The sets of equations (7) to (21) with s = e, +, -, n have been written for cylindrical symmetry, and the expressions obtained can be found in each appropriate paper and for the case commented on now in Marode et al. (1979). We shall concentrate here on the main assumptions and the boundary conditions which often determine the parameters.

The energy equations for charged species are avoided by assuming

$$T_+ = T_- = T \qquad (24)$$

Then T_e is supposed to be independent of time and given by

$$T_e = T_{eo} \exp[-(r/r_o)^2 + T_o] \qquad (25)$$

Where r_o is the discharge radius, $K\,T_{eo}$ = 1.4 eV (Marode, 1975) and T_o is the NTP temperature (Unless vibrational energy plays a significant role, the results are not strongly dependent on T_e). The radial field component is eliminated, leading to radial ambipolar diffusion.

The energy-exchange collision's term is

$$\Sigma_i M_{ri} = (f_e j_e + f_- j_- + f_+ j_+) E_z \qquad i = e, +, - \qquad (26)$$

where the j are the current densities, E_z is the axial field and f_e, f_+, f_- are the fractional energy release of charged species into neutral thermal energy. The parameter f_e is the most difficult to be fixed accurately. Table 3 summarizes the various channels of possible energy dissipation of the charged species.

Table 3. Various possible channels of dissipation of the charged species kinetic energy.

		Type of collision	Chemical energy	Relaxation time (s)	Thermal energy	Radiations
Positive ions		recombination	———————————————————————————— •
		elastic collisions			——————————•	
Negative ions		recombination	———————————————————————————— •
		elastic collisions			——————————•	
Electrons	~ .5 %	elastic collisions			——————————•	
	~ .5 %	rotational excitation		$T_r \uparrow$ —— 10^{-9}		——————•
	~ 80 %	vibrational excitation		$T_v \uparrow$ —— ~ 10^{-4}		——————•
		electronic states excitation	————• ———— ? ————			——————•
		dissociation of molecules	————• ————————————			——————•
		negative ion formation				
		ionization				
		recombination	———————————————————————— •

Since the ion temperature is of the order of the thermal temperature, the ions lose their energy by radiative recombination or by elastic collisions. If recombination is neglected, then $f_+ = f_- = 1$.

Various channels of electron energy dissipation are possible. Rotational and vibrational excitation of the nitrogen molecules raises the rotational and vibrational temperature. Then, the relaxation time of the stored rotational energy falls in the range of 10^{-9} s, while that of vibrational energy may be in the range of 10^{-3} s.

Dissociation of molecules may also be a nonnegligible channel for thermal energy increase because the fragments of the dissociation can keep up to some eV of kinetic energy. Little is known about the method of dissipation of the deexcitation energy of electronic states of molecules. For the time scale involved in this computation, of the order of some 10^{-7} s, only electron loss due to elastic and rotational collisions has been taken into account, and a rough estimation gives $f_e = 5\%$. However, the modification in the computed current growth due to variation of f_e up to 5% (Marode and Bastien, 1978) showed that we need better knowledge of the elementary process mentioned in Table 3 for accurate predictions. The discharge parameters must also be intorduced with a careful estimation. Since the power-dissipation effects depend on the volume of dissipation, the results will depend on the square of the channel radius; thus, a good estimation of this radius must be available. The radius was determined from measurements of electron current density by the Stark broadening method (Bastien and Marode, 1979) and led to a discharge radius of 9 μm.

The constant axial field has been computed from experimental measurements on streamer propagation ($E = 22$ kV/cm).

As seen on Fig. 13, before $t = 0$, no current flows through the analyzed section, and a current i_o appears suddenly with the arrival of the streamer at that section; the value of i_o is an imput parameter. Initially N.T.P. is assumed everywhere.

Figure 14 shows the computed currents for various values of the initial current. The main experimental features reported in Fig. 12 are here confirmed. After an initial decrease, the current increases again to spark breakdown if the initial current is above a critical value, and the formative breakdown time increases for decreasing initial value of the current. The details of the dynamic process can be followed by analyzing a typical case ($i_o = 300$ mA). The radial distributions of various parameters at different times are considered. Three stages can be distinguished. For $t < 34$ ns, $(\alpha - \eta)$ is everywhere negative (Fig. 15) and thus, the electron current decreases (Fig. 14); however, during that decrease the

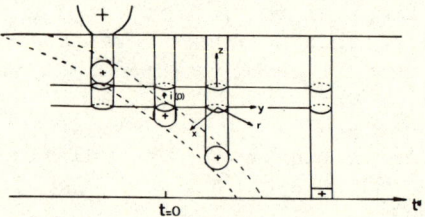

Fig. 13. Sketch of a cylindrical discharge growth. At time t = 0, the discharge reaches the cylindrical space region where the computations follow the discharge evolution. The radius of this cylinder is such that the gas remains unperturbed at the cylinder boundary.

temperature of heavy species increases (Fig. 16). The dynamics of neutral species then comes into action in order to bring the pressure back to the normal external pressure, and the resulting radial leak of heavy species induces a decrease in the neutral density near the axis (Fig. 17). For $34 < t < 50.10^{-9}$ s, the increased ratio E_z/N leads to $(\alpha - \eta)$ positive in the region of the discharge axis, and the electron density rises in the discharge core. However, attachment dominates at the discharge periphery so that in

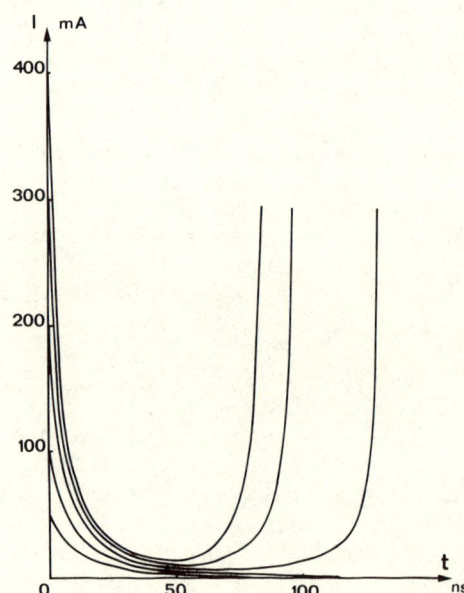

Fig. 14. Computed current evolution for various initial currents. Conditions of Fig. 12 (Marode and Bastien, 1979).

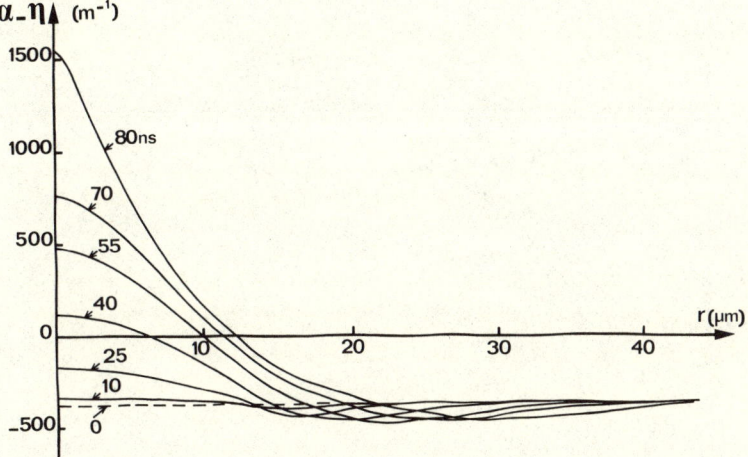

Fig. 15. Computed radial distribution of ($\alpha - \eta$) at various times. Conditions of Fig. 12 (Marode and Bastien, 1979).

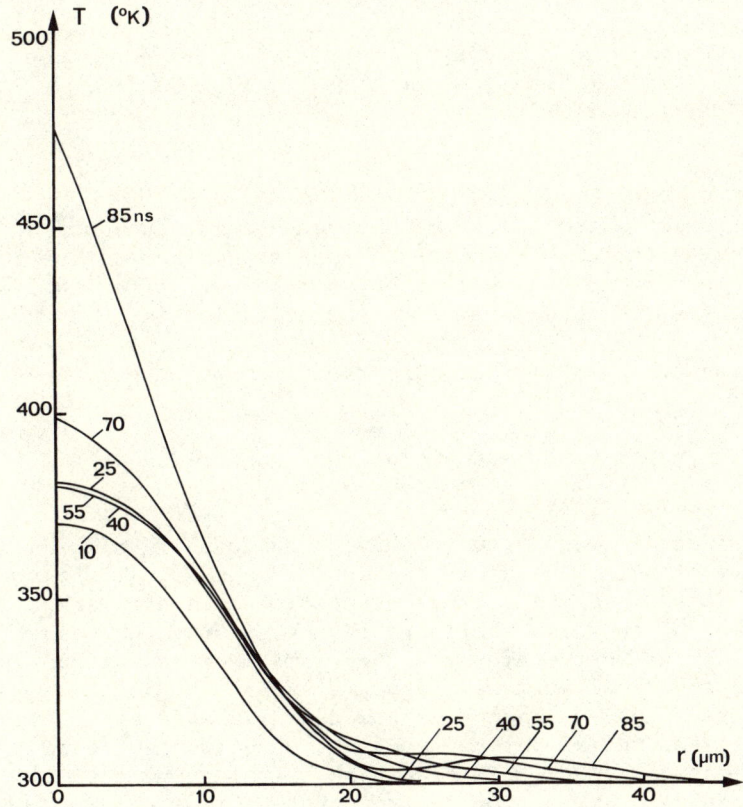

Fig. 16. Computed radial distribution of the heavy species temperatures at various times. Conditions of Fig. 12 (Marode and Bastien, 1979).

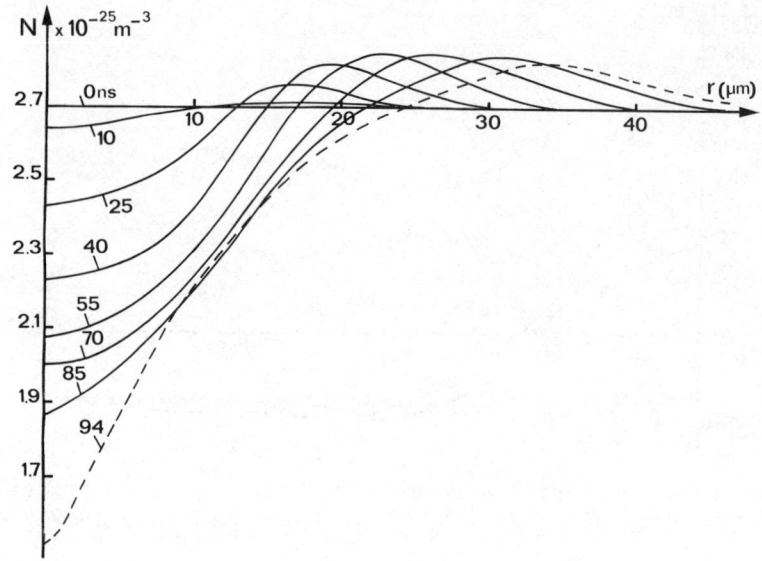

Fig. 17. Computed radial distribution of the heavy species density at various times. Conditions of Fig. 12 (Marode and Bastien, 1979).

spite of the ionization near the discharge axis, the current still decreases. An increase of current appears only beyond $t = 50.10^{-9}$s and leads to a rapid growth at about $t = 80$ ns. After this stage, the critical electron density for thermalization is reached.

The channel constriction, observed in experiment, is here also confirmed.

We will now give an example of head-controlled, channel development.

Experimental hydrodynamic behavior on leader evolution in negative rod-to-plane gap is shown in Figs. 18 and 19. The current evolution is given in Fig. 19, and on the time axis are marked the exposure times of the four schlieren photographs of the leader section viewed by a high speed camera in the frame mode (each exposure lasts 10^{-6}s). A spherical stem wave appears around the negative point, and cylindrical waves, launched from the discharge core, are recorded. If the radial positions of the wave fronts are plotted against time, Fig. 19, it is seen by extrapolation that each wave is triggered by a current pulse, and that their speed is about the speed of sound.

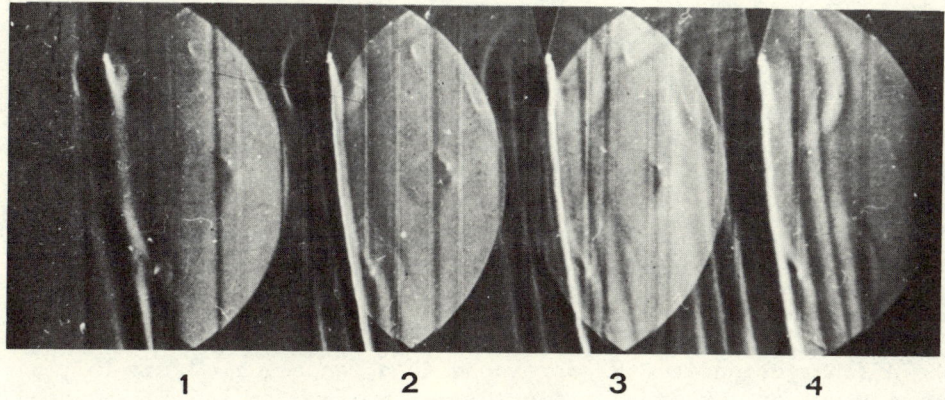

Fig. 18. Still photographs of Schlieren images of a leader channel in negative rod-to-plane gap. Diameter of viewed section equal to 45 mm. Air, atmospheric pressure, d = 2m, exposure times 1 μs. The times of photographs 1, 2, 3, 4 are reported on Fig. 19 (from Dupuy et al., 1981).

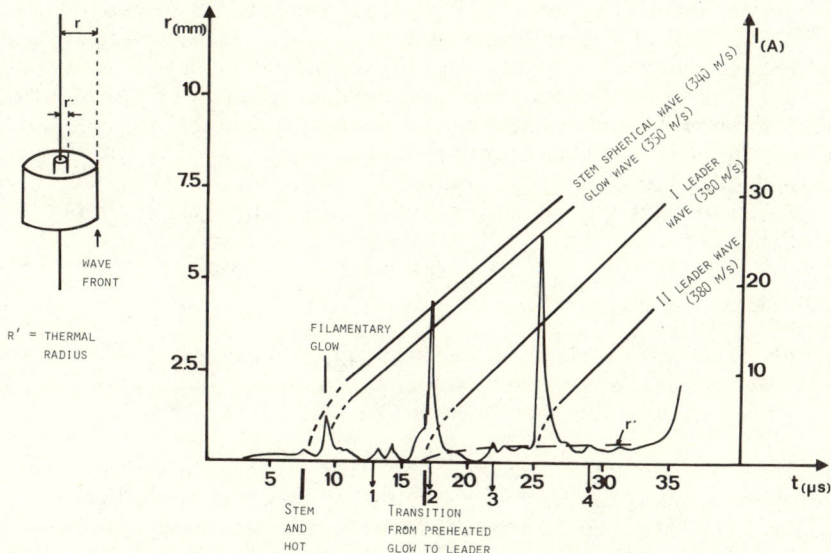

Fig. 19. Current evolution in the discharge of Fig. 18. Radial evolution of the various successive cylindrical waves launched by the discharges, which are seen on the photographs of Fig. 18 (from Dupuy et al., 1981).

Taking an idealized current pulse, Fig. 20, similar to one of the pulses of Fig. 18, we can compute a channel head-controlled situation, using the same hydrodynamic set of equations as that used for the channel-controlled case. In that case the current variations in time are imposed i(t), and the electrical field is computed as:

$$E_2 = \frac{i(t)}{\int_0^\infty en_e e^2 \, rdr} \quad . \tag{27}$$

The discharge conditions correspond to a leader, i.e., an initial temperature of 10^3 °K, a radius of ~ 1 mm, and an initial axial field $E_z \sim 2.10^6$ V/m, (Dupuy et al., 1981). Figure 20 shows the computed results. The radial distributions of pressure and neutral density are plotted for the different indicated times. A peak of overpressure is found to be launched from the discharge axis. The initial density has a gaussian profile because of the initial gaussian profile of the temperature. At other times, the density distribution changes, as seen by comparison with the dotted lines, which corresponds to the initial density distribution.

It is seen that the decrease in neutral species' density near the axis, corresponding to their radial spread, is revealed by the excess density at the channel boundary. This excess density, as well as the overpressure peak, propagates at the speed of sound, so that the observed cylindrical sound waves are well predicted by the model. In this example, the increase in ionization efficiency within the channel will be rapidly balanced by a field drop because the imposed current, at the filamentary discharge head, will impose the field value through relation (27). The head thus controls the channel situation.

Electron Fluid Waves

As previously mentioned, various approaches to ionizing waves exist, and some are developed in other papers. Here aspects of the electron, fluid-wave approach will be given.

Basically, the hydrodynamic behavior of the heavy species during the channel heating may occur in the electron population. We will begin with a brief historical review. Paxton and Fowler (1962) showed that if a stable propagating electron front exists, the electrical energy $1/2 \, \varepsilon_0 \, E^2$ stored ahead of the wave may be dissipated mainly as electron kinetic energy so that the resulting electron-pressure increase triggers a wave close to an electron acoustic wave. Shelton and Fowler (1968) then took into account the heavy species velocity and the electron energy loss due to ionization and

GLOW-TO-ARC TRANSITION

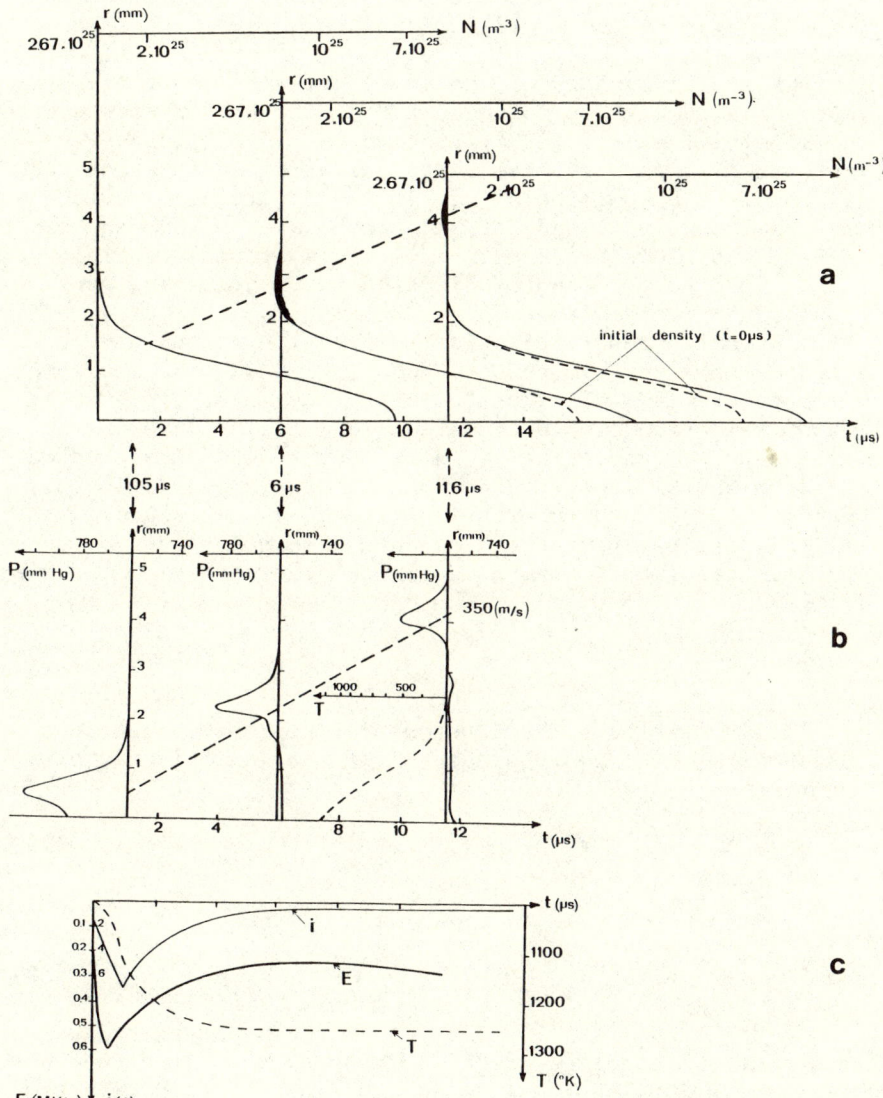

Fig. 20. Computed result due to an initial imposed current i(t) as shown in c. Field E(t) and heavy species temperatures T(t) evolution on the discharges axis (c). Radial pressure distribution (b) and neutral density (a) for the different indicated times.

neglected in the preceeding treatment. They found conditions for proforce waves (whose direction of propagation is in the field-directed electron velocity) and antiforce waves (whose direction of propagation is against the field-directed velocity).

Albright and Tidman (1972), however, presented a stable proforce wave solution without the need for an electron pressure gradient, and Klingbell et al. (1972) found pro- and antiforce wave solutions with allowance for gas photoionization in front of the waves. A complete wave solution was given for proforce waves, by Fowler and Shelton (1974) and for antiforce waves by Sanmann and Fowler (1975) who showed that antiforce waves are in principle possible without the help of photoionization in front of the wave.

A more detailed description follows. All electron wave treatments are done in a one-dimensional form (plane wave). Two complementary points of view, allow us to relate the wave velocities to the electron temperatures and the electrical field. The "0" and "1" indexes denote, respectively, values before and after the wave front. Paxton and Shelton (1962) wrote the fluid equations in a frame moving with an assumed existing front (Fig. 21(a)). In that frame all time derivative terms drop out. Two points can be outlined in their treatment. The momentum transfer equation for electrons is:

$$(M_e) = A_{en} + A_{e+} , \tag{28}$$

where (M_e) designates here the first member of the M_e equations (Eq. (18)). If M_+ and M_n are added (+ for ions, n for neutrals), one obtains:

$$(M_+) + (M_n) = A_{ne} + A_{+e} = - (A_{en} + A_{e+}) \tag{29}$$

for which

$$(M_e) = - (M_+) - (M_n) , \tag{30}$$

It follows that the electron momentum changes, can either be obtained from Eq. (28) with the knowledge of the electron momentum transfer collision ν_{en} and ν_{e+}, or derived from the heavy species properties (M_+), (M_n), (Eq. (30)) in which the wave velocity $V_w = V_o$ appears (V_o being the neutral species speed in the wave front frame). The addition $(M_e) + (M_+)$ in Eq. (30) involves the force term:

$$e(n_+ - n_e)E = \varepsilon_o \frac{\partial E}{\partial x} \cdot E = \frac{\partial}{\partial x} (1/2\, \varepsilon_o E^2) , \tag{31}$$

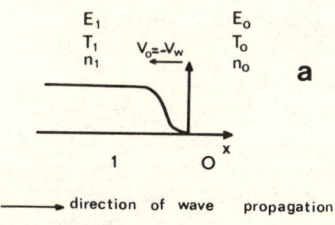

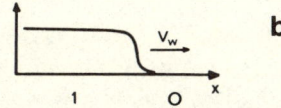

Fig. 21. Sketch of a wave front V_w wave velocity, index "0" refers to values in front of the wave; index "1" refers to values behind the wave. (a) reference frame moving with the wave and in which all time derivatives are zero; the neutral species velocity is $V_o = V_w$. (b) laboratory reference frame in which the heavy species remains at rest, and the wave velocity is V_w.

which is the derivative of the local stored energy, $1/2(\varepsilon_o E^2)$. If the approximation $n_e kT_e \gg n_i kT_i$ or $n_n kT_n$ is made, the integration in space of (30) allows us to obtain a relation between the wave velocity, the electron temperature behind the wave, and the electrical energy consumption. The relation obtained is

$$m_e f n_o V_o^2 = m_e n_1 V_1^2 + n_1 kT_1 + 1/2\, \varepsilon_o (E_o^2 - E_1^2) \;, \qquad (32)$$

V_1 and T_1 being the electron velocity and temperature behind the wave, and f the degree of ionization.

On the other hand, from Maxwell's equation, one gets

$$\frac{\partial}{\partial x}\left(\varepsilon_o \frac{\partial E}{\partial t} + j\right) = 0 \;. \qquad (33)$$

In this frame $\partial E/\partial t = 0$ and since in front of the wave $j = 0$, it follows that $j = 0$ everywhere. Thus, in the energy equations, obtained by adding $E_e + E_+ + E_n$, the gain terms from the field jE are zero, and if it is assumed that the energy dissipates mainly in heating the electrons rather than in ionizing collision losses, a time integration of E_e leads to:

$$m_e f n_o V_o^2 = m_e n_1 V_1^3 + 5 n_1 kT_1 \;. \qquad (34)$$

If V_1 is eliminated between (32) and (34), one gets:

$$V_o^2 = h[T_1, f, 1/2 \, \varepsilon_o(E_o^2 - E_1^2)] \, , \qquad (35)$$

applicable for pro- and antiforce waves. Because the velocity of antiforce waves is only slightly smaller than that for proforce wave, the term $1/2[\varepsilon_o(E_o^2 - E_1^2)]$ is neglected in (35) which becomes:

$$V_o^2 \simeq \frac{16}{3} \frac{kT_1}{n_e} \, . \qquad (36)$$

Of course, this relation is obtained after many assumptions. But it shows that, if the assumption $p_e \gg p_i$, p_n (p being the pressures) is made, the electron pressure gradient leads to a wave whose velocity is something like the electron acoustic velocity; V_o is found to be in the range $1.5 \times 10^8 < V_o < 1.7 \times 10^9$ cm/s.

In the Albright and Tidman (1972) treatment, the assumption of a stable wave front propagation is not made, but, rather, it is demonstrated that, starting with an arbitrary shape of a spatial electrical field front, a stable shape builds up with elapsing time in the proforce case. Here the laboratory frame is used Fig. 21(b), so that heavy species velocities are neglected and $\partial E/\partial t \neq 0$. Far before the front, $j = 0$, and the advancing wave is assumed not to be felt; i.e., $\partial E/\partial t = 0$, so that the integration constant of (33) is zero and thus:

$$\varepsilon_o \frac{\partial E}{\partial t} + j = 0 \qquad (37)$$

everywhere. Now the term $j \cdot E$ in the energy equation is no longer zero, so that from (37)

$$j \cdot E = -E \varepsilon_o \frac{\partial E}{\partial t} = \frac{\partial}{\partial t}(1/2 \, \varepsilon_o E^2) \, . \qquad (38)$$

The local energy $1/2(\varepsilon_o E^2)$ appears again, but this time in the energy equation E_e. The integration in time of E_e, with the assumption of a zero electron speed behind the wave then gives:

$$n_1 = \frac{1/2 \, \varepsilon_o E_o^2}{3/2 \, kT_e + e\phi_i} \, , \qquad (39)$$

which means that the available electrical energy in front of the wave $1/2(\varepsilon_o E_o^2)$ is dissipated into ionization of heavy species ($e\phi_i$ is the ionization potential) and thermal energy of the produced electrons.

A major difference appears then with the treatment of Paxton and Fowler (1962). Here, in the proforce case, since the electrons are driven by the field in the correct wave direction, no additional help from the pressure gradient $\partial p_e/\partial x$ for the electron velocity is needed, so that a quite different expression from (35) or (36) is obtained:

$$V_o = \frac{e}{m\nu} E_o \left(1 + \frac{T_1}{e\phi_i}\right), \tag{40}$$

Since $e/m\nu$ is the electron mobility, the wave velocity is here collision-dominated, the opposite of the preceeding case. A numerical example for leader velocities in air gave, for $E_o = 4 \; 10^5$ V/cm at $p - 760$ torr, $V_o = 2 \; 10^8$ cm/s, $T_1 = 2 \; 10^4$ K, $n_e = 2.3 \; 10^{15}$ cm^{-3} and a front thickness of only $\Delta x = 1$ μm.

In the above treatment, the pressure gradient $\partial p_e/\partial x$ was dropped and the energy equation taken into account; on the other hand in the treatment of Abbas and Bayle, the pressure term was accounted for by solving the momentum equation M_e and M_+, the energy equation being replaced by an assumed constant electron temperature (no E_e equation). They found that a proforce steep electron density front forms starting from an arbitrarily shaped electron cloud released from a cathode in a uniform-field gap. Figure 22 shows the electron density distribution (in full line) and the field distribution (in dashed line) at various time, with $E_o/p = 53$ V/cm, $p = 200$ torr in nitrogen. They claim that if a wave front is defined as an electron density front rather than an electrical field front, a wave is more likely to be recognized in small gaps where the electrical field integration is limited by the applied gap potential.

Further on, a more complete treatment, including all sets of equations (C, M, E) and using fewer assumptions, was made by Shelton and Fowler (1968). They showed that it was not possible to neglect either the velocity or the temperature change of heavy species through the wave front. They found that relations (32) and (34) become, respectively:

$$MN_o V_o (V - V_o) + mnv(v - V) + N_o K(T_i - T_o)$$
$$+ nkT_e = \frac{\varepsilon_o}{2}(E^2 - E_o^2) \tag{41}$$

$$MN_o V_o (V^2 - V_o^2) + mnv(v^2 - V^2) + 5N_o V_o k(T_i - T_o)$$
$$+ 5nvkT_e + nv(2e\phi_i) = 0 \tag{42}$$

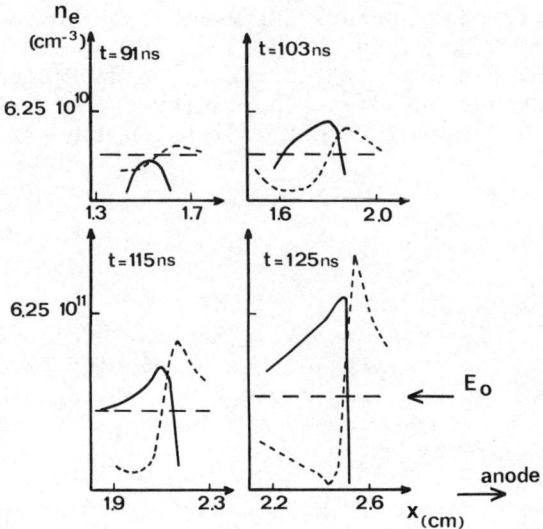

Fig. 22. Axial distribution of the electron density and the electrical field in an uniform field gap in nitrogen at various times. (after Abbas and Bayle, 1980).

Here capital letters refer to heavy species; index "0" refers to values far before the wave front and other values at an arbitrary space position. Shelton and Fowler gave a criterion defining a wave front position x_1. At x_1 the velocity V_1 of the heavy species does not yet have time to change, the same being true for the field E, so that:

$$E_1 = E_0, \text{ and} \tag{43}$$

$$V_1 = V_0. \tag{44}$$

However, nothing is known about the electron density n_1 and speed v_1, so that (41) and (42) become at x_1:

$$n_1[v_1(v_1 - v_0) + \frac{kT_{e_1}}{me}] = 0, \text{ and} \tag{45}$$

$$n_1 v_1(v_1^2 - v_0^2 + \frac{5kT_{e_1}}{me} + \frac{2e\phi_i}{me}) = 0 \tag{46}$$

Condition (45) and (46) can be fulfilled either if the expressions in brackets are zero, or if $n_{e1} = 0$. Shelton and Fowler showed that $n_{e1} = 0$ corresponds to antiforce waves since it will be associated with a smooth increase of all parameters, while the first case will correspond to a step increase in Δn and Δv of the density and velocity of electrons, which corresponds to proforce waves. They analyzed the proforce case and found relations similar to (39) and (40). Then, Fig. 23(a) and (b) shows an example of a more complete analytical solution of the electron-fluid equation (C_e, M_e, and E_e) obtained in the proforce case (a) by Fowler and Shelton (1974) and for the antiforce case (b) by Sanmann and Fowler (1975).

The wave progressions are from right to left, with a speed of 10^9 cm/s. Examples are for helium at 1 torr and $E_o/p = 1.1 \; 10^3$ V · cm^{-1} $torr^{-1}$. In the proforce case, the field decreases from E_o to zero, behind the wave. The electron density rises suddenly to $5 \; 10^{11}$ cm^{-3} and rises slowly, while the electron temperature (given in eV) rises to 6 eV before decreasing behind the wave. The wave front occupies a thickness of 0.5 cm, in contrast with the 50 cm of transition in the antiforce case. In the antiforce case, the electron pressure passes through a sharp maximum of some $3 \; 10^3$ eV, an exceedingly high value. Because of this large maximum in electron pressure, the electron diffusion process allows the electron to move against the field and to reach at the beginning of the wave a zero velocity, i.e., since we are in a frame moving with the wave, a velocity equal to that of the wave in the laboratory frame. At the maximum pressure, the electron can no longer move against the field direction and reverse direction so that $V_e \simeq V_o \simeq V_+$ and, since in

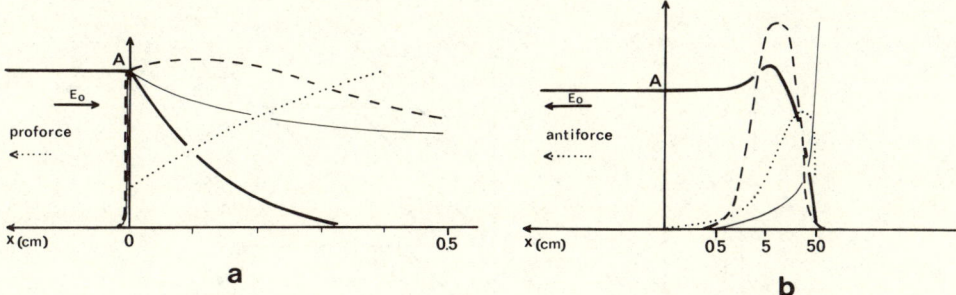

Fig. 23. Proforce (a) and antiforce (b) wave fronts. The propagation direction is from right to left. Plotted are the electrical field E(——), the electron temperature T_e (----), the electron density n_e (——), the electron mean velocity V_e(....). Units normalized at point A, are, for E(10^3 and 10^3 Vcm), T_e (6.1 and $9 \; 10^3$ eV), n_e (5.10^{11} and $9.2 \; 10^8$ cm^{-3}) V_e (10^9, $2.6 \; 10^9$ cm s^-) respectively for case (a) and (b).

that frame

$$j = e(n_e v_e - n_+ v_+) = 0 \qquad (47)$$

$$n_e = n_+ \frac{v_+}{v_e} \simeq n_+ , \qquad (48)$$

the space charge sign reverses too.

It follows that to the left of the maximum the space charge is negative and the field increases, while it decreases to the right of this maximum since the space charge becomes positive.

This is a very beautiful and interesting solution for antiforce waves; it shows that antiforce waves may theoretically exist without a supply of electrons in front of the wave by photoionization.

Figure 24 illustrates what is implied by this model. The wave front is sketched on (a), the space axis being either an axial position axis or a potential axis since $V(z)$ is a single-valued function of z. The problem is how an electron can be projected, in front of the wave, as indicated by the dashed line, since the field tends to carry the electrons along the positive direction. For example, let us consider an electron produced at position z_1 with no kinetic energy. This electron will transform its potential energy into kinetic energy, keeping its total energy constant, so that it follows

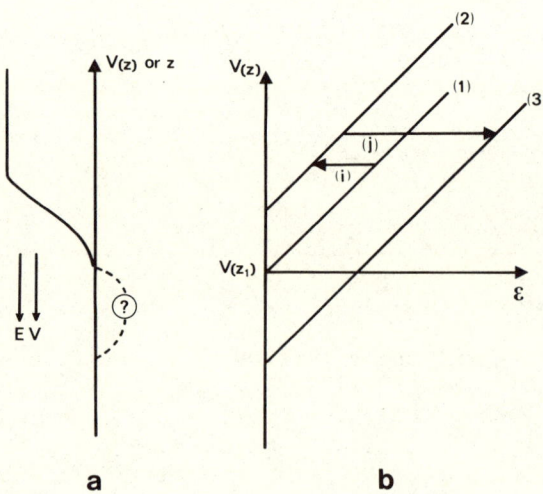

Fig. 24. Antiforce wave sketch with an associated plot of electron trajectory in the potential and kinetic energy space.

the line (1) on (b) in a $[V(z), \varepsilon]$ diagram. Elastic collisions, with negligible heavy species recoil, will not change these characteristics, i.e., may only make the electron go up and down, along (1), perhaps coming back to z_1, but never finding itself at $z < z_1$. An inelastic collision with a heavy species with loss of the electron energy will induce a transition like i, so that the electron, now on characteristics (2), will not even be able to reach z_1 anymore. Thus, the only way, that this electron can reach a position $z < z_1$ before the wave front, is through the existence of a collision like j, which communicates to the electron an additional kinetic energy that leads it to characteristics (3) where adequate elastic collisions may send it back to $z < z_1$. Sanmann and Fowler (1975) in fact suggested that a basic term, which was neglected in the Shelton and Fowler treatment, was the electron heat flux term which actually involves heat transfer by electron-electron collisions. However, for an electron density as low as $10^{10} - 10^{11}$ cm^{-3}, the e - e collision frequency seems to be too small ($\sim 10^5 - 10^6$ s^{-1}) to allow for a fast enough energy transfer by heat conduction.

It may be concluded, that even if the process may in principle be possible, the large value of the obtained temperature, the small value of the e - e collision frequency and the implicit assumption of a hydrodynamic regime, i.e., where the macroscopic collisions parameters are in equilibrium with the electrical field, require additional comparison with realistic situations.

Positive Streamer and Antiforce Electron Waves

The advancing tip of an ionized filamentary channel driven by an antiforce wave (a) or driven by a streamer process (b) is sketched on Fig. 25. An associated curve of the electrical field distribution at the head is given. A major difference appears: in front of the antiforce wave, the electrical field is large and remains constant, while in the streamer case the field rises from a small value to a large value. The one-dimensional assumption in case (a) implies:

$$\frac{dE}{dx} = 0 \tag{49}$$

in the space-charge free region in front of the wave, while in the nonunidimensional streamer case, the Poisson equation may be written as:

$$\frac{dE_x}{dx} + \frac{1}{r}\frac{\partial}{\partial r}(r\, E_r) = 0 \tag{50}$$

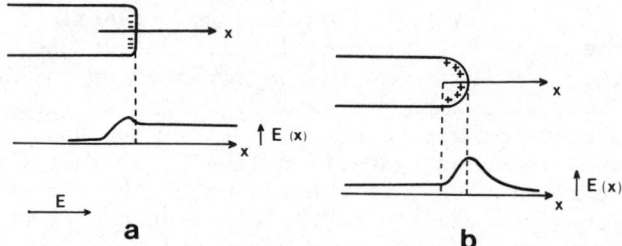

Fig. 25. Comparison between positive streamers and antiforce wave propagation.

in a cylindrical coordinate system. In that case dE/dx may be compensated by the radial term even with no space charge. Thus, in the antiforce wave process, the electrical energy $1/2(\varepsilon_o E^2)$ which drives the waves is already present in front of the wave and the wave space charge is negative because the electrons are pushed in front of the wave, while in the streamer process the electrical energy is carried with the cloud of the N_s positive charge contained within the positive head (Dawson and Winn, 1965).

However, the field in front of the streamer tip will not always rise from a small value to a high value. Figure 26 taken from Gallimberti (1972) shows that two sections can be distinguished during a streamer propagation. The first section extends from the positive electrode tip to a point x_m; in that section of the field in front

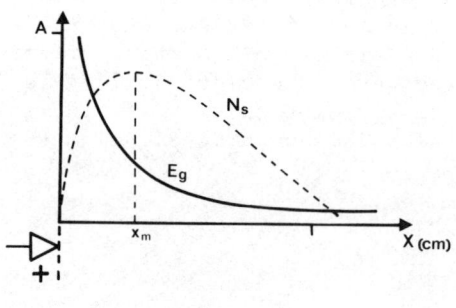

Fig. 26. An example of geometrical field distribution in the axis of a nonuniform field gap, and the corresponding change of the number N_s of positive changes situated, as sketched, in the streamer head. Typically A corresponds to 100 kV cm^{-1}, 1.5 10^9 charges (after Gallimberti, 1972).

of the streamer is the high geometrical field E_g imposed by the electrodes of small curvature, which gives at the same time the energy needed for the streamer avalanches and that stored as potential energy with the increase of number N_S of positive charges in the streamer head. After x_m, the applied field E_g, in front of the streamer head is no longer able to drive the streamer by itself, and the stored potential energy is used for the streamer advancement, leading to a decrease of N_S, i.e., the stored potential energy.

Could the wave propagation in the high field section $[0, x_m]$ be based on an antiforce wave mechanism rather than a streamer one? The observation that, in a nonuniform field, the wave never stops in the low field region, but propagates by itself largely inside this region seems rather to support a streamer-like mechanism since a storage of the potential energy must occur during the propagation in the high field region.

REFERENCES

Abbas, I. and Bayle, P., 1980, J. Phys. D, 13:1055.
Albright, N. W. and Tidman, D. A., 1972, Phys. Fluids, 15:86.
Badareu, E. and Popescu, I., 1965, "Gas Ionisés," Dunod, Paris.
Barreto, E., Jurenka, H., and Reynolds, S. I., 1977, J. Appl. Phys., 48:4510.
Bastien, F. and Marode, E., 1979, J. Phys. D, 12:249.
Brambilla, R. A., 1978, in: "Proceedings, 5th International Conference on Gas Discharges, Liverpool," IEE Conf. Publ. No. 165, p. 104.
Cavenor, M. C. and Meyer, J., 1969, Austr. J. Phys., 22:155.
Chalmers, I. D., Duffy, H., and Tedford, D., 1972, Proc. R. Soc., London A, 239:171.
Christophorou, L. D., 1977, in: "Proceedings, 13th International Conference on Phenomena in Ionized Gases, Berlin," Invited Lectures, VEB Export-Import, Leipzig, p. 51.
Davies, A. J., Davies, C. S., and Evans, C. J., 1971, Proc. IEE, 118:816.
Davies, A. J., Evans, C. J., and Woodison, P. M., 1975, Proc. IEE, 122:765.
Dawson, G. A. and Winn, W. P., 1965, Z. Phys., 183:159.
Doran, A., 1968, Z. Phys., 208:427.
Dupuy, J., Gilbert, A., Gallimberti, I., Bastien, F., Marode, E., and Hartmann, G., Les Renardières Group, 1981, Electra, 74:67.
Dutton, J., 1978, in: "Electrical Breakdown of Gases," J. M. Meek and J. D. Craggs, eds., Wiley, New York, Chap. 3.
Fowler, R. G. and Shelton, G. A., 1974, Phys. Fluids, 17:334.
Gallimberti, I., 1972, J. Phys. D., 5:2179.
Gallimberti, I., 1979, in: "Proceedings, 9th International Conference on Phenomena in Ionized Gases, Grenoble," Sup. No. 7; J. Phys., 40:C7-193.

Goldman, M. and Goldman, A., 1978, in: "Gaseous Electronics," M. N. Hirsch and H. J. Oskam, eds., Academic, New York.

Griem, R., 1964, "Plasma Spectroscopy," McGraw, New York.

Holt, E. H. and Haskell, R. E., 1965, "Plasma Dynamics," Macmillan, New York.

Jaeger, E. F., Oster, L., and Phelps, A. V., 1976, Phys. Fluids, 19:819.

Kekez, M. M. and Savic, P., 1974, J. Phys. D, 7:620.

Kline, E., 1974, J. Appl. Phys., 45:2046.

Klingkeil, E., Tidman, D. A., and Fernsler, R. F., 1972, Phys. Fluids, 15:1969.

Koppitz, J., 1971, Z. Naturforsch, A, 26:700.

Llewellyn-Jones, F., 1966, "Ionization and Breakdown in Gases," Methuen, London.

Loeb, L. B., 1965, "Electrical Coronas," Univ. of Calif., Berkeley.

Marode, E., 1975, J. Appl. Phys., 46:2005.

Marode, E. and Bastien, F., 1978, in: "Proceedings, 5th International Conference on Gas Discharges, Liverpool," IEE Conf. Publ. No. 165, p. 343.

Marode, E., Bastien, F., and Bakker, M., 1979, J. Appl. Phys., 50:140.

Mitchell, K. B., 1970, J. Chem. Phys., 53:1795.

Mitchner, M. and Krugger, H., 1973, "Partially Ionized Gases," Wiley-Interscience, New York.

Paxton, G. W., Fowler, R. G., 1962, Phys. Rev., 128:993.

Raether, H., 1965, "Electron Avalanches and Breakdown in Gases," Butterworth, London.

Les Renardières Group, 1974, Electra, 35.

Les Renardières Group, 1977, Electra, 53.

Les Renardières Group, 1981, Electra, 74.

Rogoff, G. L., 1972, Phys. Fluids, 15:1931.

Sanmann, E. and Fowler, R. G., 1975, Phys. Fluids, 18:1433.

Shelton, G. A. and Fowler, R. G., 1968, Phys. Fluids, 11:740.

Sigmond, R. S., 1978, in: "Electrical Breakdown of Gases," J. M. Meek and J. D. Craggs, eds., Wiley, New York.

Spitzer, J., 1956, "Physics of Fully Ionized Gases," Wiley-Interscience, New York.

Waters, R. T., 1978, in: "Electrical Breakdown of Gases," J. M. Meek and J. D. Craggs, eds., Wiley, New York.

Tholl, A., 1970, Z. Naturforsch., 259:420.

Von Engel, A., 1965, "Ionized Gases," Oxford.

Yoshida, K. and Tagashira, H., 1976, J. Phys. D, 9:485.

ARC DISCHARGE ELECTRODE PHENOMENA

G. Ecker

Ruhr-Universität Bochum
Bochum, West Germany

INTRODUCTION

The ASI was dedicated to the topic "Electrical Breakdown and Discharges in Gases," — a part of the general field called "Gaseous Electronics." There are physicists - more than a few - who have hardly heard about Gaseous Electronics. There are others - particularly, so called pure theoreticians - who consider this field the esoterica of physics filled with bizarre oddities. But there are also those experts who dared to look closer into this field and found that it has many fascinating and tempting aspects combining extreme features from various fields of physics.

It has become standard practice in Gaseous Electronics to order the manifold observations with a guideline of a set of standard discharges. One of these discharges is the "arc discharge."

The fact that for the arc — one of the best known and oldest standard discharges — one cannot find a generally accepted definition in the literature reflects the peculiar problems of Gaseous Electronics. Any discharge depends on a large number of parameters from various fields of physics: gas composition, gas pressure and motion, material and geometry of the walls, discharge current, discharge voltage, form of the electrodes, electrode material, impurities on the electrode surface, etc. The various model regions of the discharges more or less strongly influenced by sets of these parameters result in the great variety of phenomena.

THE ARC DISCHARGE

Without claiming that it has been or must be generally accepted, I will use the following definition of the arc discharge: An arc discharge is an ionized state in the gas, the main part of which can be described by a locally closed particle balance on the one hand and a locally open energy balance on the other hand. This definition is in agreement with the usual theoretical approach to the arc discharge.

The arc, burning freely, wall-stabilized, or otherwise — as long as it is not electrode-stabilized — can be subdivided in the "column" and the "electrode regions."

The column phenomena, I believe, are reasonably well understood and analyzed with a remarkable degree of mathematical sophistication.

The theory of the electrode regions, however, has gone only recently beyond the range of qualitative argumentation. My presentation here will be solely concerned with these complex and tempting arc electrode phenomena. In fact, since the decisive and most conspicuous processes occur at the cathode, I will concentrate on the cathode region. The methods described for the cathode can also be applied to the anode.

Although arc discharges frequently operate in an ambient gas, the basic processes governing the electrodic regions are most purely exhibited by the socalled vacuum arc. For this reason my discussions will also concentrate on this discharge. Let us first consider the phenomena. In arc discharges — typically with currents above the Ampere range — we observe at the cathode the current transition in small, highly luminescent areas, the so-called cathode spots. This is true for the arc operating in an ambient gas as well as for the vacuum arc operating in the metal vapor evaporated from the cathode.

The experiments show that the cathode spots burning in an ambient gas have qualities varying over a wide range but nevertheless, show some characteristic features. One observes three types of spots: one of very high current density ($O[10^7$ $A/cm^2]$), one of medium current density ($O[10^4$ $A/cm^2]$) and a spotless, unconstricted mode. There occur discontinuous changes from the spot of medium density to the spot of high current density at a certain critical pressure and to the spot free mode at a critical maximum current.

Experimental observation of the vacuum arc with quite sophisticated experimental methods produced no completely unique picture. Nevertheless, some people conclude that one can distinguish two typical spot types (I, II). Their qualities are compiled in the Table 1.

Table 1. Qualities of two modes of cathode spots.

	SPOT I	SPOT II
Appearance	Semi-spherical	diffuse
Autograph	craters	gross melting
Life time	μ sec	$> 200\ \mu$ sec
Current	10 A	20-30 A
Voltage	40 V	15 V
Velocity	10^2 m/sec	10 m/sec
Current density	average=(10^4 to 10^5)A/cm^2 crater = 10^8 A/cm^2	(10^4 to 10^5) A/cm^2
Electron Temper.	2 eV	1 eV
Electron Density	(10^{17} to 10^{18}) cm^{-3}	(10^{17} to 10^{18}) cm^{-3}
Erosion rate	$5 \cdot 10^{-7}$ gr/Coul. few droplets	$5 \cdot 10^{-5}$ gr/Coul. many droplets
	Order of magnitude only	

The above distinction of different spot types for the arc in an ambient gas and in the vacuum arc has also been critized and doubted. Moreover, the experimental information scatters over a wide range, as can be seen from Table 2 where, as an example, the data of various authors and for various experimental conditions, are compiled for the current density. The differences of the results do not necessarily indicate experimental discrepancies but may well be explained by the many parameter influences discussed above.

THE MODEL

Models make choices and assumptions about the essential constituent parts of the object. In the case of the cathodic arc component, we may — without claiming completeness — consider the following alternatives:

(a) planar surface — nonplanar surface

(b) smooth surgace — rough surface

Table 2. Compilation of experimental results of the current density (j) for various spot currents (I) with the year of observation.

metal	I/A	j/(A/cm^2)	year
Cu	3	10^5	1948
	200	$2 \cdot 10^3 - 10^4$	1963
	60	$3 \cdot 10^6 - 10^7$	1964
	2-200	$2 \cdot 10^5 - 8 \cdot 10^6$	1965
	1- 5	$4 \cdot 10^4 - 6 \cdot 10^4$	1967
	$5 \cdot 10^3$	$10^6 - 10^8$	1967
	$2 \cdot 10^4$	$8 \cdot 10^5 - 8 \cdot 10^7$	1967
	$3 \cdot 10^3$	$3 \cdot 10^4$	1968
	$2 \cdot 10^4$	$5 \cdot 10^6$	1969
	5-4000	$3 \cdot 10^4 - 10^5$	1972
Mg	3	30	1904
	10	$4 \cdot 10^3$	1922
	35	$2 \cdot 10^3$	1930
	3	$2 \cdot 10^5$	1948
	150	10^6	1949
	$8 \cdot 10^3$	$2 \cdot 10^6$	1957
	$2 \cdot 10^4$	$2 \cdot 10^6$	1969
	90-440	$3 \cdot 10^4 - 3 \cdot 10^5$	1972
W	3	$8 \cdot 10^4$	1948
	10^3	10^3	1956
	150	10^8	1958
	$3 \cdot 10^4$	$8 \cdot 10^3 - 2 \cdot 10^4$	1962
	50	10^5	1969

(c) homogeneous cathode material — inhomogeneous cathode material (grain, film, impurities)

(d) vacuum volume — volume with ambient gas

(e) spot at rest — moving spot

(f) stationary operation — nonstationary operation.

Obviously, these alternative choices in their combinations already yield more than fifty possible models. Fortunately, reasons can be given which reduce this number. Because of the small extension of the arc cathode spots, the planar surface is in most cases a good approximation. For the purpose of understanding the basic phenomena, we need not complicate the already complex issue by considering grain structures, film covering, and impurities. Since the vacuum arc produces the typical phenomena in pure form, we will in general study the spots of this type and only briefly consider arc spots in the presence of an ambient gas.

ARC DISCHARGE ELECTRODE PHENOMENA

Thus we study the following models:

(a) the stationary spot at rest on a planar smooth surface in vacuum,

(b) the spot in regular or irregular motion on a planar smooth surface in vacuum,

(c) the spot on a rough surface in vacuum,

(d) the spot at rest on a smooth surface in an ambient gas.

THE THEORETICAL CONCEPT

Let us recall that the arc discharge depends on many parameters, and particularly, the electrode components operate under conditions where laws of statistics, gas dynamics, electrodynamics and solid state physics are relevant simultaneously and sometimes in extreme ranges. This complex task cannot be solved without severe simplifications.

What we do not want to do is to pick out of the complete system one particular phenomenon like, e.g., the current density or the surface temperature and try to calculate just this quantity, while making assumptions about all the other parameters related to it. Under these circumstances one can prove practically anything by a suitable choice of the other parameters.

What we rather want to do is to give a complete and comprehensive description of the whole arc cathode region accounting for all processes of relevance in a reasonable approximation. Even in a simplified form, this is still difficult as we demonstrate for one example only, the energy balance at the cathode surface. There are many processes influencing energy gain and loss at the cathode surface. In the Fig. 1, we show the prevailing gain and loss processes as a function of the current density (j) and the cathode surface temperature (T_c). Obviously, all kinds of combinations may come into the picture; that means that in our general treatment all these processes must be taken into account. Even with numerical methods this is a hard problem to solve.

Tractability, however, can be reached, if one reduces the theoretical demands to a level that fits experimental knowledge. The appropriate tool for this is the method of the existence diagram (Ecker, 1971) which uses the available information completely, limits the demands sensibly, and gives a transparent presentation of the results. We must be content here with a somewhat naive but short and comprehensible picture of this method: the basic idea is to calculate characteristics from an incomplete set of equations apply-

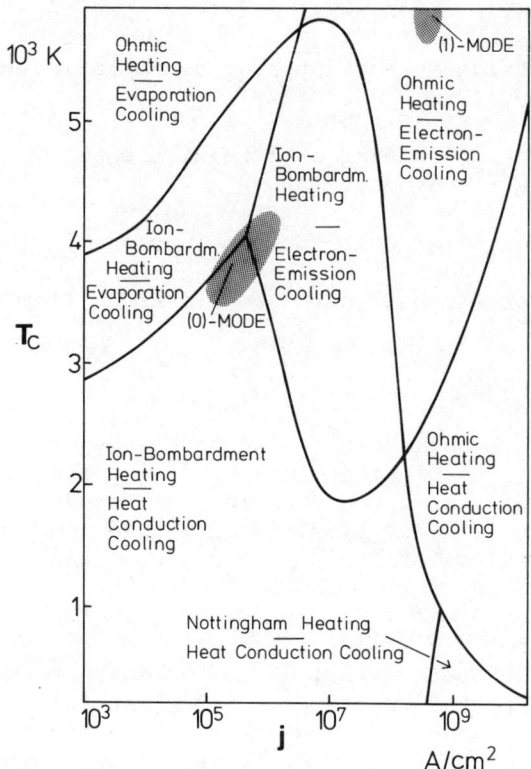

Fig. 1. Dominating heating and cooling processes as a function of the cathode surface temperature T_c and the current density j. The shaded areas designate the range of the spot modes.

ing arbitrary "unidirected" approximations. "Unidirected" means that the approximations are chosen so that they all shift the characteristic in the same direction. There may be as many approximations as we like and very serious ones, too. If we have several such shifted characteristics, then they limit an area to which a possible solution is restricted — the existence or uncertainty region (see Fig. 2). Of course, this uncertainty area will be larger the smaller our knowledge is and the more serious our approximations are. However, the statement that our solution is within this area is reliable. The size of the existence area allows us to judge our lack of knowledge.

In the special case of the cathode spot phenomenon, we find it suitable to choose the current density j and the temperature T_c at the cathode surface as variables of our energy diagram (E-diagram).

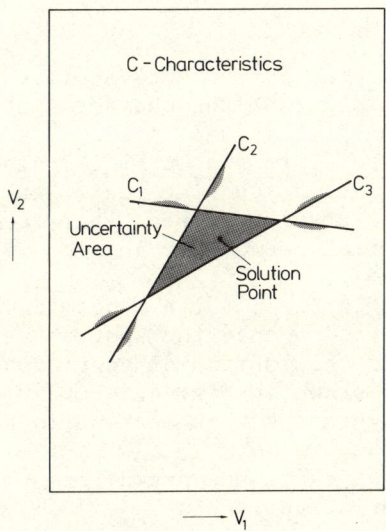

Fig. 2. Schematical example for three characteristics (C) in the diagram of the variables (V_1, V_2). The shaded area is the uncertainty range where the analysis does not exclude existence.

Then we calculate three characteristics defined by the following statements:

(C_o) The energy gain of the plasma ball of the cathode spot must always be greater than the energy loss of the same ball due to ion migration to the cathode surface.

(C_1) The number of ions returning from the plasma ball of the cathode spote to the cathode surface cannot be larger than the number of neutrals which have left the cathode surface.

(C_2) The energy loss at the cathode surface by heat conduction, Nottingham effect, and material loss must be smaller than the energy gain calculated from the kinetic and neutralization energy of the ions and the Joule heat under favorable assumptions.

The details of the numerical evaluation of these characteristics involve many laws of plasma physics, gas dynamics and solid state physics and cannot be presented here. We refer the reader to a comprehensive presentation (Ecker, 1980). In the following we show and discuss only the results.

THE SMOOTH-SURFACE SPOT

The existence diagram for a "spot at rest" on a smooth surface is given in Fig. 3. Note that the characteristics C_2 depend on the total discharge current (I) in the spot. In addition to the characteristics discussed above, we see in Fig. 3 also a characteristic C_3 which is relevant for the spot at rest. It excludes that range of the diagram where a stationary spot at rest can not exist because of the so-called "thermal runaway" (Hantzsche, 1972).

The corresponding diagram for a "spot in motion" was calculated under the concept of motion that the spot stays in a certain position for a certain time and then shifts discontinuously to a neighboring site. The diagram is shown in Fig. 4, where v designates the spot velocity. In this figure the spot current is assumed to be 200 A.

The conclusions which can be drawn from the results of Figs. 3 and 4 are very informative:

(1) On a smooth surface, a cathode spot at rest can exist only in one mode, the (o)-mode. For most materials this (o)-mode exists, but not for refractory materials like tungsten, molybdenum or tantal. For them a stationary spot at rest cannot operate. For experimentally observed values of I, the uncertainty regions are surprisingly small.

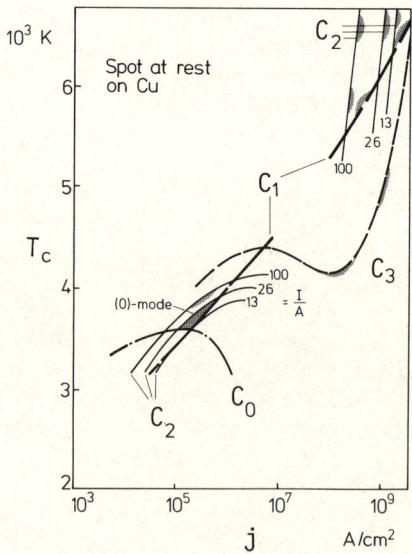

Fig. 3. E-diagram for the spot at rest on copper. The symbols and the characteristics (C) are explained in the text. Shading means regions admitted for solutions. The shaded triangle is the spot mode at rest with a current of 26 A.

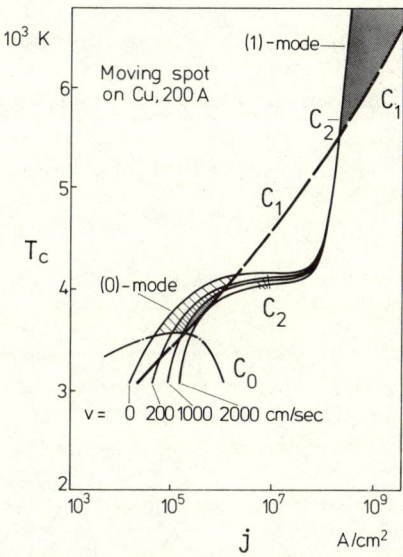

Fig. 4. E-diagram for the moving spot on copper with a current of 200 A. Note the additional (1)-mode and the variation of the allowed area for the (0)-mode with increasing spot velocity.

(2) Figure 3 shows that there exists a lower limit for the spot current I so that the requirement $I > I_o$ holds. I_o is the current for which the existence area disappears, but it is not necessarily the minimum current known from experiment, since we know that the spot can but must not necessarily be found in our existence area. Therefore the observed minimum current I_{min} may be larger than I_o ($I_{min} > I_o$).

(3) If the spot is in motion and its velocity v is larger than the critical velocity $v = 2\delta \, (I/j\pi)^{1/2}$, where δ is the thermal diffusivity, then the thermal runaway cannot develop, and the characteristic C_3 does not appear in Fig. 4. Under these circumstances the second mode, the (1)-mode, is accessible and consequently, spot solutions now exist also for refractory materials. Since our moving spot on the smooth surface corresponds probably to the experimental type-II-spot, we have chosen in Fig. 4 for the current, $I = 0(200)$ A, and for the velocity, $v = 0(10^2)$ cm/sec, and we find for the current density and the temperature, $j = 0(10^6)$ A/cm^2 and $T_c = 0(3\,900)$ K. These are reasonable values. With increasing velocity of the spot, the existence area decreases and the critical current I_o grows.

(4) The solution for the (1)-mode in the range $v < 2\delta \, (I/j\pi)^{1/2}$ shows that a spot with a regular motion, as described before, cannot exist. However, in this range a spot can operate on a irregular

basis in the sense that the thermal runaway develops at the point where the spot operates and produces an explosive evaporation as known from breakdown phenomena. This explosion itself destroys the operation mechanism at the spot and displaces it to a neighboring site.

THE ROUGH-SURFACE SPOT

For obvious reasons, a smooth surface is something appealing to the theoretician. A look with a microscope onto a realistic surface, however, will show that it has quite impressive structures, tips, mushrooms, groves and similar things. These structures can be destroyed by the operation of the arc spot, but they may also be created by the same process. Even on liquid surfaces, tip phenomena must be expected as already claimed by Tonks (1935).

Since the geometry of the surface is so bizarre, it is not surprising that the theoretical understanding of spots on rough surfaces is little developed in comparison to that of the smooth surface spots.

We distinguish two kinds of effects caused by the surface structure:

(a) <u>Average Surface Structure Effects</u>

Considering increasing surface structures starting from a smooth surface, one expects increasing local fluctuations around average values for all parameters. However, the main effect is in the space-charge layer and the corresponding electric field at the cathode surface which in turn invluences the electron emission.

If the space-charge layer is very thin, then it follows the surface structure like a skin, and the result with respect to our E-diagram is essentially a shift of the existence area to values of higher current density (see Ecker, 1980).

On the other hand, if the extension of the space-charge layer is large in comparison to the characteristic length of the surface structure, then there will be an enhancement factor β_E for the electric field determined via the effective electron emission. In this case, the existence area shifts to smaller current density values and smaller temperatures. The details can be found again in Ecker (1980). There one can also see the quantitative effect and recognize that the influences coming from the surface roughness may be essentially larger than the uncertainties in the theories of the smooth-surface spot.

(b) Individual Surface Effects

Here we consider the opposite case to (a). Average effects do not dominate but instead single fluctuations in very short time intervals and at localized tips. The first investigations of this problem came from scientists interested in the vacuum-breakdown problem. In this connection Mesyats and collaborators (1971) provided the concept of "explosive emission" which various authors including Mesyats himself tried to apply to the irregular arc spot motion. We have already discussed such a possibility in the case of a smooth spot operation.

The idea is that the spot operates at a preferential tip on the surface, heats it in nanoseconds to a very high temperature, and transforms the solid body into a highly ionized state before the material cloud can expand. When the plasma cloud starts expanding, it crosses another favorable tip in the neighborhood, and a new discharge ignites there, in this way propagating the spot over the surface.

As interesting as this principal concept is, it is as difficult to analyze. What has been tried is to calculate the temperature development in the tip under the influence of a given electric field. A number of interesting results were found, not without their engaging serious assumptions. One finds that current densities of 10^9 to 10^{10} A/cm^2 must be expected, as well as fields of the order of magnitude of 10^8 V/cm and characteristic times of the order 10^{-10} sec. However, the explosive transformation of the solid into the plasma state, the expansion of that plasma over the surface, and the ignition of a new tip discharge are not yet understood.

The application of the E-diagram method to this rough-surface phenomenon (Ecker, 1980) produced some interesting qualitative results. The requirements that

1) the tip size must be smaller than the size of the space charge region to produce individual effects, and

2) at the surface of the tip, temperatures high enough to ionize the material must be reached in a time short compared to the competitive destructive processes of surface tension and field rupture

showed that, with a potential drop of 15 V, individual explosive phenomena are unlikely to occur. However, with a voltage drop of order 200 V, they may be possible.

SPOTS IN AN AMBIENT GAS

The theory of cathode spots in an ambient gas becomes extremely involved if one allows simultaneously for the evaporation processes from the cathode. No such treatment is known.

Spots operating solely in the ionized ambient gas were treated however, a long time ago (Ecker, 1961). The method applied was that of characteristics but it was not applied to the extent that unidirected approximations were used such as the E-diagram. Without going into detail, we present in Fig. 5 the two diagrams which demonstrate the basic results.

Although the scheme is different from our T_c-j-diagrams, the background is similar. Here the points of existence are determined from the intersection of the ion current curve (I_+) with the electron defect current curve ($I - I_e$) plotted versus the parameter (j_c/j_{col}), where j_{col} is the current density in the arc column.

The diagrams prove the existence of three types of spots: the spotless mode (j_c/j_{col}) $\simeq 1$, the medium current density mode at (j_c/j_{col})=$0(10^4)$, and the high current density mode at (j_c/j_{col}) = $0(10^7)$. The spotless mode is governed by ion bombardment and conduction heating on the one hand, and heat conduction cooling on the other. The electrons are emitted by temperature emission. The medium current mode is governed by ion bombardment heating, heat

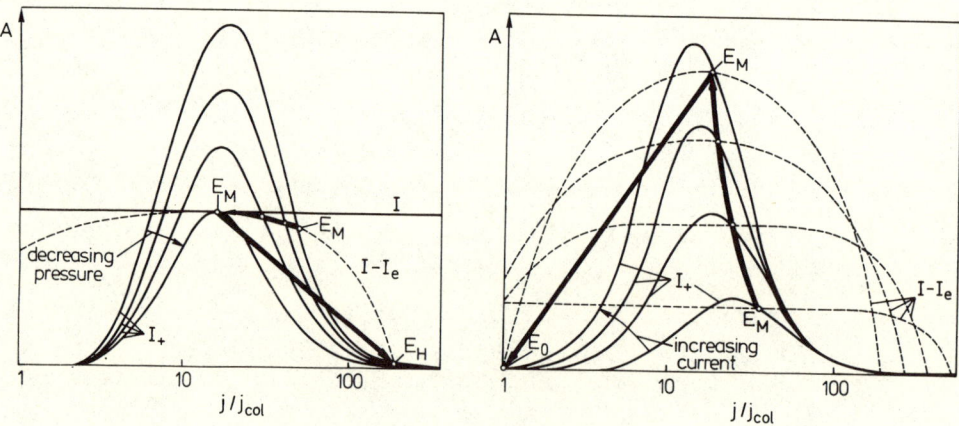

Fig. 5. Spot modes (E_0, E_M, E_H) in an ambient gas as calculated from the intersection of characteristics. I is the total, I_e, the electron-, and I_+, the ion-spot current. The heavy lines demonstrate the variation and the discontinuous mode change with decreasing pressure, respectively increasing current. (j/j_{col}) is the ratio of the current density in the spot and the column.

conduction and electron emission cooling: current transport is due to electrons by T-F-emission and plasma ion flow. Finally, the high current density mode is dominated by ion bombardment and Ohmic heating on the one hand and electron emission cooling on the other. Here field emission is the decisive current phenomenon.

Moreover, the diagrams show that there are transitions from the medium mode to the high-current density mode as well as to the spotless mode. The transition to the high-current density mode occurs with decreasing pressure at a critical pressure, the transition to the spotless mode with increasing current at a critical current.

CONCLUDING REMARKS

The cathode phenomena of arc discharges are very complex. Their theoretical treatment requires a number of different models, e.g. the vacuum smooth-surface model, the vacuum rough-surface model or the model with the ambient gas.

The application of the E-diagram method allows us to derive many qualitative, but also quantitative conclusions which are not in contradiction to say the least, with the experimental findings.

Nevertheless, the problem of the cathode spot is far from understood. In particular, those phenomena which are related to the radial distribution and the anode side conditions of the cathode spot are not understood. Therefore, further endeavours are required of those who are not afraid to tackle difficult tasks like the particle loss from the spots, the loss of liquid material, the velocity distribution of the particles, and, last but not least, the motion phenomena.

REFERENCES

Ecker, G., 1961, Ergeb. Exakten Naturwiss., 33:1.
Ecker, G., 1971, Z. Naturforsch., 26a:935.
Ecker, G., 1980, "Vacuum Arcs," Wiley, New York.
Hantzsche, E., 1972, Beitr. Plasmaphys., 12:245.
Mesyats, G.A., 1971, in: "Proceedings, 10th International Conference on Phenomena in Ionized Gases, Oxford," D.A. Davenport, ed., Parsons, Oxford, p. 333.
Tonks, L., 1935, Phys. Rev., 48:562.

DYNAMICS OF ARC DISCHARGES

J. Hackmann

Physics Institute II
University of Düsseldorf
Düsseldorf, West Germany

INTRODUCTION

The properties of steady arc discharges at atompsheric pressure are largely dominated by external and internal fluxes. Gradients of particle number density and temperature as well as the interaction of the plasma with external and internal magnetic fields can produce large flow velocities of the plasma particles. These fluxes may considerably influence the properties of the discharge, even in the absence of net mass transport.

In this paper, the theoretical treatment of arcs interacting with flows is considered. Although there exists a wide variety of phenomena related to flow effects in all kinds of arc discharges, we focus on two selected examples which provide a rather good insight into the physical background of the problems.

In particular the following topics are considered. In the first section, the arc interacting with a transverse magnetic field and a flow field is studied. The magnetic field interacting with the arc current produces a volume force which gives rise to a double vortex flow structure. If the arc is additionally exposed to an external transverse flow, two modes of operation can be analyzed. These modes are critically dependent on the operation conditions.

In the second section, we discuss the effect of internal diffusion fluxes driven by gradients of the partial pressure. The phenomenon is closely connected to the establishment of local thermodynamic equilibrium. For its investigation it is necessary to consider the elementary processes which dominate the interaction of the plasma particles.

For a theoretical treatment of the above mentioned flow phenomena, it is necessary to know the transport properties of the arc plasma. In a third section, we describe the evaluation of the electrical conductivity from arc measurements.

ARCS WITH TRANSVERSE MAGNETIC FIELD AND CROSS FLOW

Because of their importance in applications, high-pressure arcs in transverse flows and magnetic fields have been subject to a large number of experimental and theoretical investigations. For a general overview of this field, see the review articles by Uhlenbush (1976), Stäblein (1977), Jones and Fang (1980). In this discussion we follow the publications of Fischer and Uhlenbusch (1967) and Bartels and Uhlenbusch (1970), considered to be the most detailed analyses of the problem to date.

Basic Equations

For many practical cases, the interaction of the arc plasma at high pressure with external flows and magnetic fields can be described in the monofluid approach. The corresponding complete system of balance equations reads for stationary arcs:

Energy Balance.

$$\text{Div } K(T) \text{ Grad } T + \underline{J} \cdot \underline{E} - U(T) - \xi_m(T) \, C_p(T) \underline{v} \text{ Grad } T = 0 \qquad (1)$$

Viscous energy dissipation is neglected, and pressure variations are assumed to be small compared to the static pressure. Reabsorption of radiation by the arc plasma is also neglected.

Momentum Equation.

$$\xi_m (\underline{v} \text{ Grad}) \underline{v} = -\text{Grad } p + \mu_o \, \underline{J} \times \underline{B} + \text{Div } \underline{\tau}. \qquad (2)$$

$\underline{\tau}$ is the stress tensor with the elements

$$\underline{\tau}_{ij} = \eta(T) \left(\frac{\partial v_i}{\partial x_j} + \frac{\partial v_j}{\partial x_i} \right).$$

Mass Conservation.

$$\text{Div}(\xi_m \, \underline{v}) = 0. \qquad (3)$$

Ohms Law Without Convection-Induced Electrical Currents.

$$\underline{J} = \sigma(T) \underline{E}. \qquad (4)$$

Finally, we have the set of Maxwell's equations

rot $\underline{E} = 0$

rot $\underline{B} = \mu_o \underline{J}$ (5)

Div $\underline{E} = \rho/\varepsilon_o$

Div $\underline{B} = 0$.

The problem in solving this set of equations is the close coupling of the energy and momentum balance equations by the flow field and the thermal structure. The plasma flow gives rise to a displacement of the temperature distribution which in turn influences via the temperature dependence of the electrical conductivity, the current density distribution and the energy dissipation. In the following subsections, we give examples for methods of decoupling the energy and momentum equations for simplified cases.

The Wall-Stabilized Arc with Transverse Magnetic Field

The simplest configuration is the wall-stabilized arc in a transverse magnetic field as changes produced by the magnetic field take place inside the plasma column itself. The schematic of the geometry is shown in Fig. 1.

A qualitative description of the phenomena inside the arc plasma is easily done. The Lorentz force $J_z \times B_{ox}$ induces a mass flow in a plane perpendicular to the arc axis. The mass flow is accompanied by a convective transport of energy, and the original cylinder symmetric temperature distribution is shifted into the direction of the Lorentz force. Since the walls are impervious, the mass flows back through a relatively cold gas layer close to the wall before re-entering the plasma again at the opposite side. The resulting plasma flow field is a symmetric double vortex as observed experimentally by Rosenbauer (1971).

For a quantitative description the following simplifying assumptions are made:

(1) The plasma parameters do not vary along the arc axis.

(2) The electric field is constant over the arc cross section and parallel to the arc axis.

(3) The viscosity of the plasma is constant everywhere.

(4) The plasma is incompressible which suggests the introduction of a stream function ψ so that

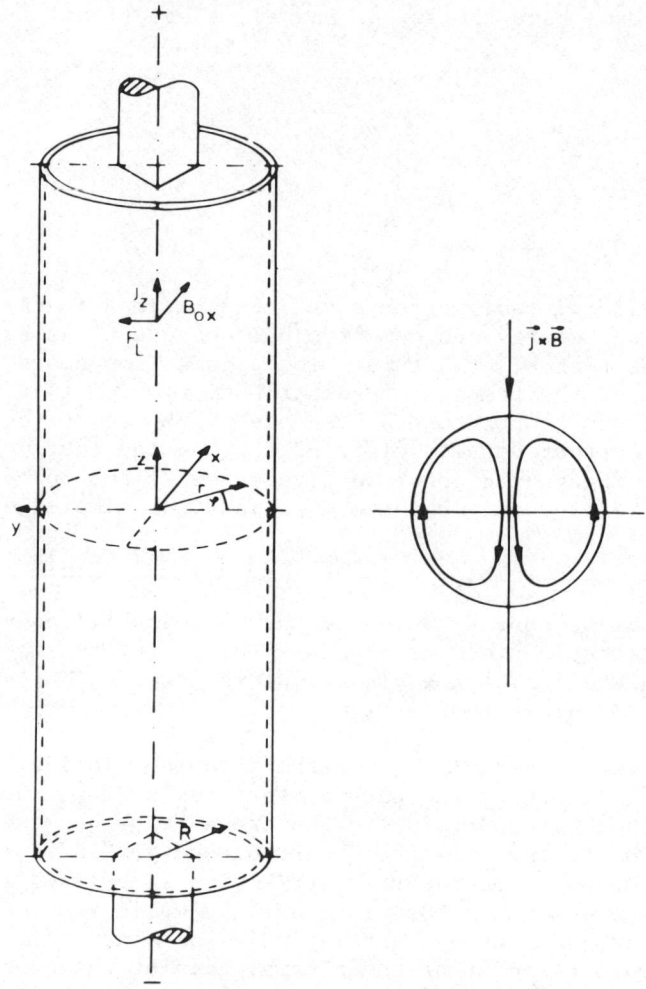

Fig. 1. Schematics of a cylindrical arc with crossed B-field and the double vortex induced by the j x B force.

$$v_r = \frac{1}{r} \frac{\partial \Psi}{\partial \psi} \quad v_\psi = - \frac{\partial \Psi}{\partial r} \tag{6}$$

(5) The curl of inertial forces is neglected.

The energy balance with these simplifications reads in cylindrical coordinates:

$$\frac{1}{r} \frac{\partial}{\partial r} \left(r \frac{\partial S}{\partial r} \right) + \frac{1}{r^2} \frac{\partial^2 S}{\partial \psi^2} = \frac{\xi_m C_p}{K(S)} \frac{1}{r} \left(\frac{\partial \Psi}{\partial \psi} \frac{\partial S}{\partial r} - \frac{\partial \Psi}{\partial r} \frac{\partial S}{\partial \psi} \right) - \sigma E^2 + U \tag{7}$$

S is the heat flux potential

$$S = \int_{T_{wall}}^{T} K(T)\, dt.$$

Eliminating the pressure from the momentum equation by applying the curl operation, we arrive at

$$\Delta\Delta\Psi = \frac{\text{Box } E_z}{\eta} \text{rot}(\sigma(S)\underset{\sim}{I}). \tag{8}$$

$\underset{\sim}{I}$ is the unit vector in the direction of the Lorentz force. Given the boundary conditions

$$S(\psi_1 R) = 0\ ;\ \frac{\partial\Psi(\psi,R)}{\partial\psi} = 0;\ \frac{\partial\Psi(\psi,R)}{\partial r} = 0 \tag{9}$$

and the symmetry of the problem with respect to the center plane between the electrodes, the solution of this system is carried out by an iteration procedure.

In the zero order, the energy balance is solved with respect to the heat flux potential S with zero magnetic field. Inserting the S - distribution into the right hand side of the momentum equation (8), a first order flow field is evaluated, and substituting this solution into the energy balance equation, a better approximation for S is derived. This process is iterated until the solution converges. The numerical problem of this procedure is that S and Ψ have to be expanded into Fourier series which converge very slowly. Therefore, for higher magnetic fields the computation becomes very time consuming. As typical results of these calculations, Fig. 2 shows flow profiles inside a nitrogen arc at atmospheric pressure displaying the double vortex. The plasma flow velocities are typically 10 - 20 m / s, as also observed in experiments. The corresponding lines of constant heat flux potential, i.e., constant temperature, are given in Fig. 3, showing the distortion of the temperature profiles.

This description of the interaction of stationary arcs with a transverse magnetic field is commonly adopted and the results confirmed by many experimental investigations (e.g., Rosenbauer, 1971; Sauter, 1969; Kollmar et al., 1969).

The Free Burning Arc with Transverse Magnetic Field and Cross Flow

This type of electrical discharge is of great practical interest, e.g., with circuit breakers, welding devices, and laser discharges. Experiments are performed in different ways: arcs moving along rails,

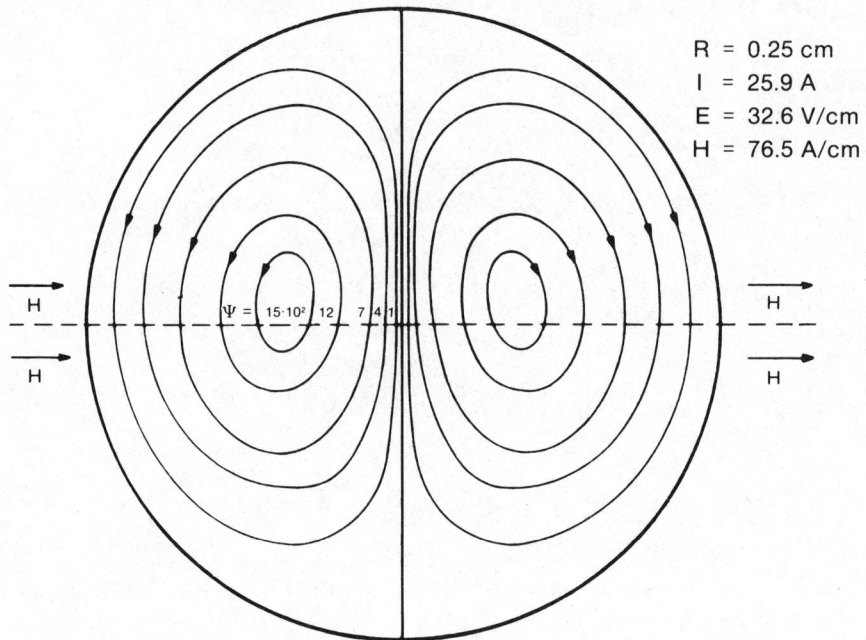

Fig. 2. Flow field inside the cylindrical arc with crossed B-field. According to Fischer (1967).

arcs fixed at one electrode and moving along the other, and the arc fixed at both electrodes (Sebald, 1980; Plessl, 1980). In this subsection we will discuss this latter case because of its well-defined geometry (see Fig. 4).

The treatment of this problem is, of course, more complex than the case of the wall-stabilized cylindrical arc discussed above. The major difficulties arise from the following facts:

(1) As experiments show, the arc is bent (see Fig. 4). Therefore, the electric field strength cannot be assumed to be independent from the radial coordinate.

(2) Energy and momentum balance again are closely coupled via the convection term. The distribution of the electrical field strength (i.e., Joule's heating per unit volume) in connection with the flow field inside and outside the discharge enforces an appropriate temperature distribution.

These difficulties can be overcome in the following ways:

(1) If the arc is only slightly bent, the distribution of the

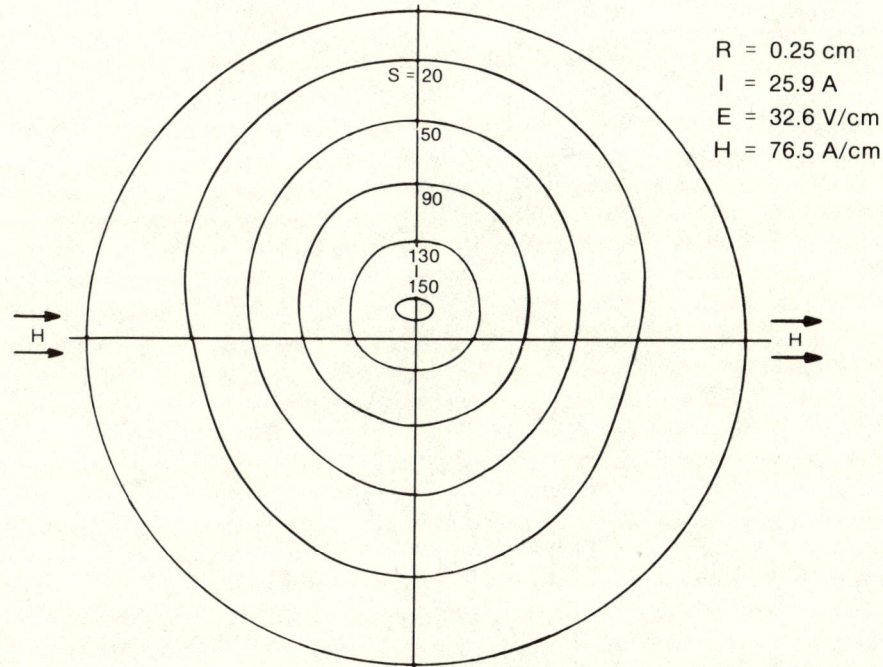

Fig. 3. Field lines of constant heat flux function inside the cylindrical arc shown in Fig. 2. According to Fischer (1967).

electrical potential can be calculated from the potential equation $\Delta\Phi = 0$, depending on the electrode configuration. For hyperpolic-shaped electrodes, the square of the $\underset{\sim}{E}$ - component in the axis value E_Z (0, 0) is given by the expression

$$E_Z^2(0,r) = \frac{E_Z^2(0,0)}{1+r^2/L^2} \tag{10}$$

where 2L is the spacing of the electrodes. Similar results follow for tip or plane circular electrodes.

(2) The temperature profile of an arc follows closely changes in the convection velocities, whereas small changes in the temperature distribution only slightly affect the flow field. Thus it is possible to calculate, first, the flow profile for an approximate temperature distribution and then, obtain a refined temperature distribution using the calculated flow field. The results can be improved by an iteration process.

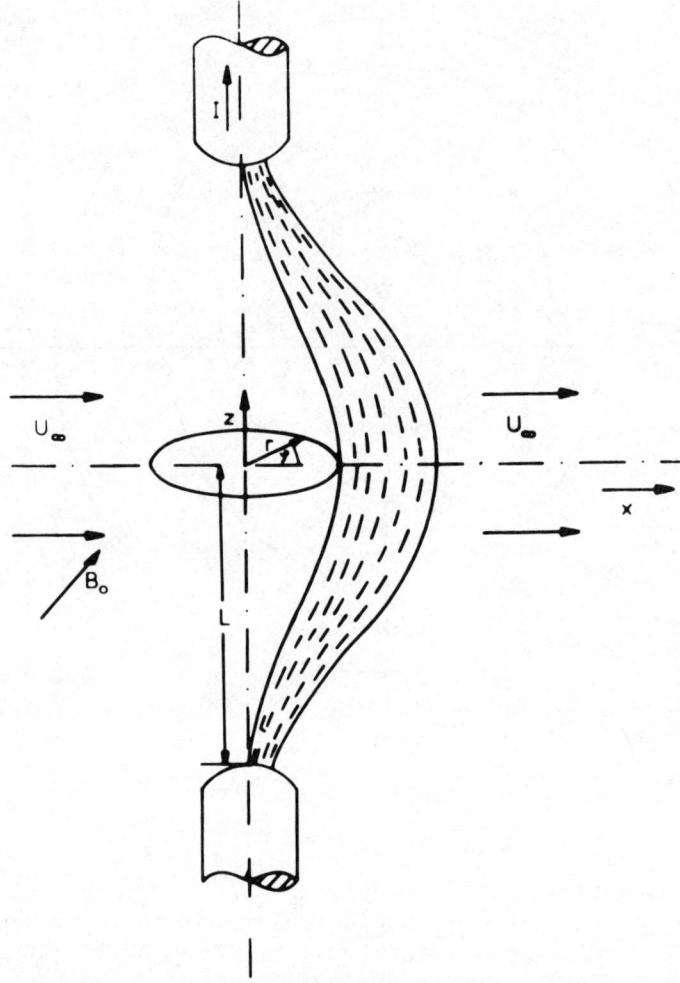

Fig. 4. Free-burning arc with crossed flow and magnetic field.

Let us first consider the case of an arc only weakly deflected, so that both temperature and current-density profile can be assumed to be radially symmetric. The B-field has the components

$$\underset{\sim}{B} = (B_o \sin\psi, \; B_o \cos\psi, \; 0) \tag{11}$$

with constant viscosity we write Eq. (12) for the stream function Ψ

$$\Delta \frac{1}{r}\left(\frac{\partial}{\partial r} r \frac{1}{\xi_m} \frac{\partial \Psi}{\partial r} + \frac{\partial}{\partial \psi} \frac{1}{r\xi_m} \frac{\partial \Psi}{\partial \psi}\right) = \frac{B_o}{\eta} \frac{\partial J_Z}{\partial r} \sin\psi. \tag{12}$$

A peculiar difficulty arises with the formulation of the boundary condition for this equation. With no magnetic field present we could assume an irrotational laminar flow upstream and downstream the arc. However, as shown above, vortices are produced by the magnetic field which propagate in the flow to infinity, and thus the assumption of a homogeneous flow is no longer valid. This problem can be solved by introducing a z-dependence of the flow and making the reasonable assumption that the flow is limited in the z-direction by plane walls at large distance from the center of the arc. To analize this problem, Fischer and Uhlenbush (1967) introduced a stream function of the type

$$\Psi = \xi_{m\infty} \cdot U_\infty \cdot r \cdot f(r) \, (1 - z^2/L^2) \tag{13}$$

which satisfies the boundary condition at the walls located at the electrodes (distance 2L). With Eq. 13 inserted into Eq. (12), the function f in the center plane between electrodes follows after integrating twice the following equation:

$$f'' + f'\left(\frac{3}{\rho} + \frac{T'}{T}\right) + \frac{f'}{\rho}\frac{T'}{T} = \frac{\sigma_o B_o E_o D}{\eta \, U_\infty} \left(\frac{L}{D}\right)^2 \\ \times \frac{1}{\sigma_o} \int_\infty^\rho \frac{1}{\hat{\rho} \, I_1^2(\sqrt{2}\,\rho)} \int_o^{\hat{\rho}} \frac{\hat{\rho} \, I_1(\sqrt{2}\,\hat{\rho})}{(1+\hat{\rho})} \frac{\partial \sigma}{\partial \hat{\rho}} d\hat{\rho} \, d\hat{\rho} \tag{14}$$

Here I_1 is the first order modified Bessel function, $\rho = r/L$, E_o the electrical field strength, σ_o the electrical conductivity on the arc axis, $D = 2 r_o$ diameter of the arc. Figure 5 shows the flow pattern and the vortex in the center of an argon arc with circular temperature distribution. The arc data are:

$T_o = 7500$ K; $U_\infty = 10$ m/s; $L = 2D = 2$ cm; $B_o = 4 \times 10^{-4}$ T; $p = 1$ atm

A very interesting result of these calculations is that the function $f(\rho)$ changes sign for a critical value of the ratio of magnetic force to friction force

$$\alpha = \frac{\sigma_o B_o E_o D^2}{\eta \, U_\infty} \left(\frac{L}{D}\right)^2 \tag{15}$$

indicating the generation of a vortex with a backflow in the center. This behavior is typical for the onset of instabilities in magnetized arc plasmas and also could be attributed to the phenomenon of retrograde motion.

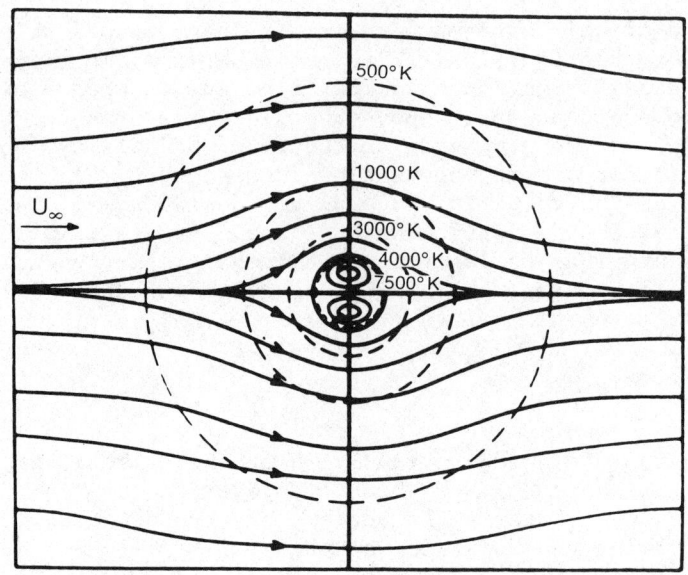

Fig. 5. Flow field inside an arc with backflow, calculated assuming a circular temperature profile.

In Fig. 6 the normalized diameter of the double vortex core of the arc shown in Fig. 5 is plotted versus the parameter α. Backflow occurs at α-values above $\alpha_{min} = 34$.

For noncircular and off-axis (shift δ) temperature and current-density distributions, α is replaced by

$$\beta = \frac{\mu_o I^2 \delta}{\eta U_\infty D^2} \left(\frac{L}{D}\right)^2 .$$

f(p) displays similar behavior with β as it did with α.

A realistic distribution of the heat flux potential obtained by iterating the solutions of momentum and energy balance is plotted in Fig. 7. In this example the backflow shifts the maximum of the temperature towards the incoming flow. Deflection in flow direction may be obtained for other arc parameters.

The calculations discussed above explain quantitatively the behavior of free-burning arcs in external magnetic fields and cross flow. At small values of the parameter, α (or β) introduced before the external flow passes through the hot arc. At higher values, a double vortex is generated. In this latter case, the arc behaves similarly to a heated, current-carrying, solid body; and aerodynamic models are applicable (see Roman and Myers, 1967).

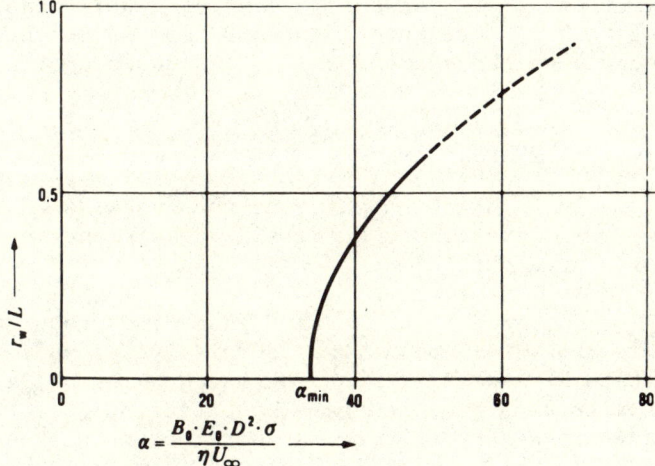

Fig. 6.
Normalized diameter of the double vortex core of the arc shown in Fig. 5 versus the parameter α.

INTERNAL FLUXES DUE TO GRADIENTS OF THE PARTIAL PRESSURE

As stated above, for many practical applications the interactions of electrical arcs with external flows and magnetic fields can be treated in the framework of the monofluid equations. This approach implies that the collision frequencies among the particles are large enough to establish locally thermal and reactive equilibrium of the plasma particles. A completely different situation arises if the mean free path for reactive processes, e.g., ionization by electron collisions, is large compared to distances over which the plasma parameters change considerably, say, the discharge radius. Then the local parameters of the discharge are dominated by particle fluxes driven by the gradients of the partial pressure. In this

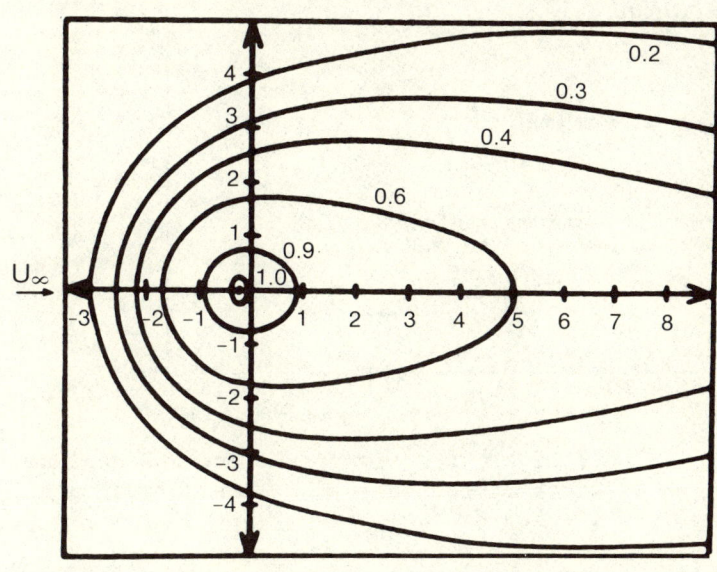

Fig. 7.
Realistic pattern of lines of constant heat flux potential in the plane z = 0 when backflow occurs.

section we will study the case of a wall-stabilized He – arc at atmospheric pressure. In this type of discharge the last conditions mentioned above are realized (Uhlenbusch et al., 1980; Richter, 1971).

Basic Considerations

To gain an insight into the dynamics of this type of discharge, we must investigate in detail the elementary processes between the discharge particles.

Compared to other rare gases the cross section for inelastic electron atom collisions is relatively small for neutral ground state He, the corresponding rates being small as well. Additionally the transport of neutrals by diffusion is rather efficient, since the cross section for charge transfer which determines the diffusion coefficient is small. Thus the mean free path for ionizing a ground state helium atom in a He – arc is of the order of 1 cm.

To elucidate the behavior of the neutrals in a He – arc, Fig. 8 shows a plot of realistic rates of the important reactions in this plasma. The rates are plotted as a function of the normalized

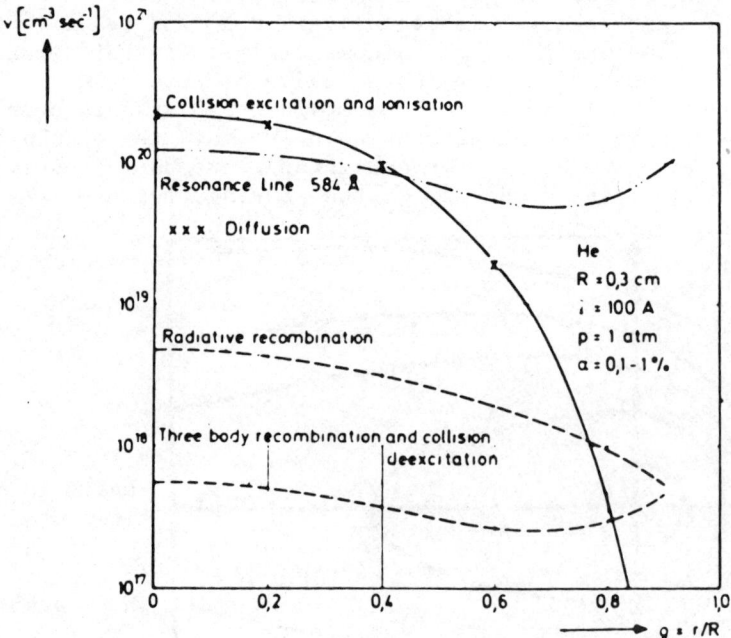

Fig. 8. Collisional and radiative rates for populating and depopulating the ground state of the He – atom versus reduced arc radius. R = 0.3, I = 100 A, p = 1 atm.

arc radius. The arc current is 100 A, and the radius 0.3 cm. The rates of collisional excitation and ionization are about 10^{20} cm^{-3} s^{-1} at the arc axis. The inverse de-excitation collision processes are three orders of magnitude smaller, as are the competing inverse processes of radiative recombination. The de-excitation rate into the groundstate by the transition at 584 Å is comparable to collisional excitation and ionization; however, its efficiency is completely cancelled by strong self-absorption.

Summarizing the situation, one must say that within the region of the arc axis, essentially all excited and ionized helium atoms are produced by collisions only, with the corresponding inverse process missing. On the contrary, close to the walls the rates of excitation processes are small, and ground state neutrals are generated by recombining radiative and collisional processes. To maintain a stationary discharge, an internal flow of neutrals are generated by recombining radiative and collisional processes. To maintain a stationary discharge, an internal flow of neutrals from the plasma boundary into the core has to be established, and vice versa, electrons and ionized atoms have to travel to the plasma edge by ambipolar diffusion. The driving force for both types of fluxes is the pressure gradient of neutrals, electrons and ions, respectively, as indicated in Fig. 9.

Mathematical Description of Internal Fluxes

Different from the monofluid approach considered previously, we now have to write the balance equations for each plasma constituent, assuming only a radial dependence of the parameters.

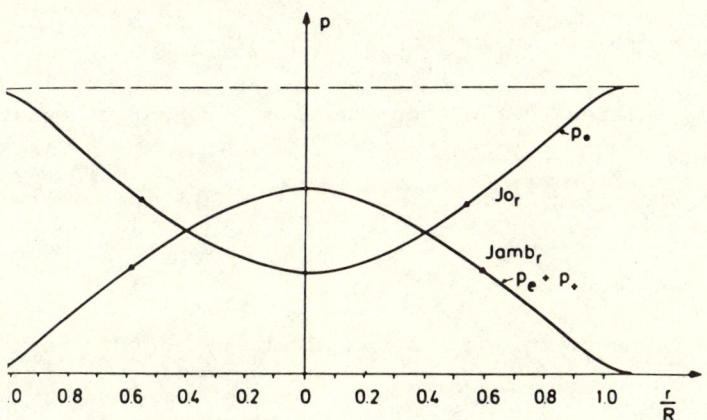

Fig. 9. Radial distribution of partial pressures and diffusion fluxes in a He discharge.

First, there is the equation of continuity for ground state neutrals including gain and loss terms:

$$\frac{1}{r}\frac{d}{dr} r j_{or} = (\frac{\partial n_o}{\partial t})_{coll\ rad} \qquad (16)$$

Here the gain and loss terms are:

$$(\frac{\partial n_o}{\partial t})_{coll\ rad} = \sum_{k \geq 1} n_k (n_e F_{ko} + A_{ko}) - n_o (n_e C_{ok} + \frac{n_k}{n_o}[1-\Lambda_{ok}]A_{ko})$$

$$+ n_+ n_e (R_o + n_e Q_o) - n_+ n_e (1-\Lambda_o) R_o - n_o n_e S_o \qquad (17)$$

with

C_{ok}, F_{ko} = rate coefficients for collisional excitation and de-excitation into and from level k, respectively;
A_{ko} = optical transition probability;
S_o = collisional ionization coefficient;
$R_o, n_e Q_o$ = rate coefficients for radiative and three-body collisional recombination into ground state;
Λ_o, Λ_{ok} = reduction coefficients due to absorption.

The diffusion of excited neutrals is neglected in the pressure range discussed here.

From kinetic theory, we have for the flux j_{or} the expression

$$j_{or} = -(3/8)[(n_o + n_+)\Omega_{a+} m_{He}]^{-1} \frac{dp_a}{dr} . \qquad (18)$$

Here Ω_{a+} is related to the cross section for charge transfer Q_{a+} by

$$\Omega_{a+} = Q_{a+} (\frac{kT_G}{m_{He} \pi})^{1/2} .$$

T_G is the gas temperature.

For the energy balance of the electron we arrive at

$$\frac{1}{r}\frac{d}{dr} r q_{re} = \sigma E_z^2 - \frac{4m_e}{m_{He}} k (T_{e'} - T_G) \nu_{ea} - U_{inel} . \qquad (19)$$

DYNAMICS OF ARC DISCHARGES

and the energy balance of the entire plasma:

$$\frac{1}{r}\frac{d}{dr} r q_r = \sigma E_z^2 - U_R \tag{20}$$

with

$$q_r = -K_G \frac{dT_G}{\partial r} - K_e \frac{dT_e}{dr} + K_D \frac{d}{dr}(n_a T_G)$$

(total heat flux),

$$q_{re} = -K_{eG} \frac{dT_G}{dr} - K_{ee} \frac{dT_e}{dr} + K_{eD} \frac{d}{dr}(n_a T_G)$$

(heat flux of electrons).

ν_{ea} = the electron-heavy particle collision frequency including collisions of electrons with neutrals and ions;
U_R = the optically thin, radiated power per unit volume;
U_{inel} = energy loss of the electron gas due to inelastic collision with atoms. Note that

$$U_{inel} \sim U_R - E_i (\frac{n_o}{t})_{coll\ rad} \quad (E_i = \text{ionization energy}).$$

Additionally we have the equation of state

$$p = (n_a + n_e)kT_g + n_e kT_e \tag{21}$$

and the equation of quasi-neutrality

$$n_+ \sim n_e . \tag{22}$$

If the transport coefficients are known from kinetic theory or experiments, the set of equations (16) through (22) can be solved with respect to temperatures, densities, and fluxes.

As a result, in Fig. 10 the flow velocities of neutral He-atoms in the ground state versus reduced arc radius are plotted for different arc currents (R = 0.3 cm). The velocities are rather high and of the same order of magnitude as for externally driven flows (see previous section). This diffusion flux has a very strong influence on the establishment of local thermodynamic equilibrium. This can be seen from Fig. 11, where the electron density is plotted versus electron temperature and compared to data evaluated from Saha's equation.

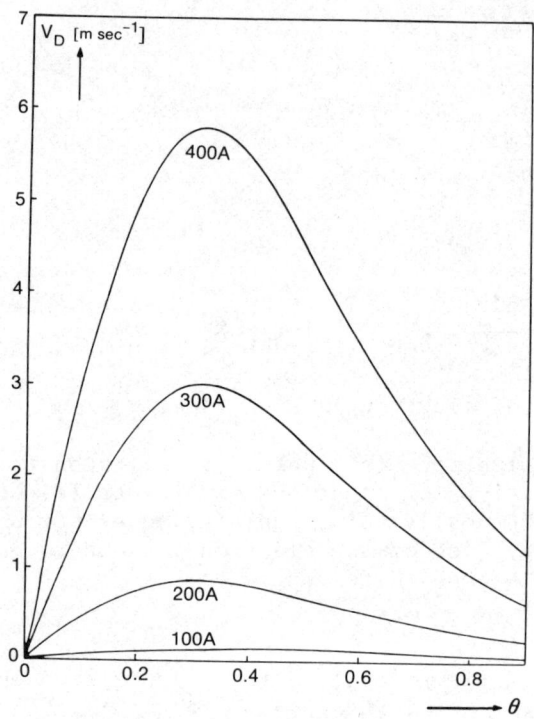

Fig. 10. Radial diffusion velocity of neutral He-atoms versus reduced arc radius.

Finally a criterion for the occurrence of diffusion fluxes can be derived from a simplified version of Eq. (16). Taking into account only excitation and ionization of ground state neutrals (rate coefficient Q_0) and replacing d/dr by $1/R$, we arrive at

$$\frac{1}{R^2} \frac{3}{8} \frac{k\, T_G n_a}{(n_a + n_+)\Omega_{a+}m_{He}} = n_a n_e Q_0. \tag{23}$$

$$\lambda_{Diff} = \left[\frac{3}{8} \frac{k\, T_G}{n_e(n_a+n_+)\Omega_{a+}m_{He}Q_0}\right]^{1/2}$$

defines a characteristic "diffusion length" for ground state neutrals

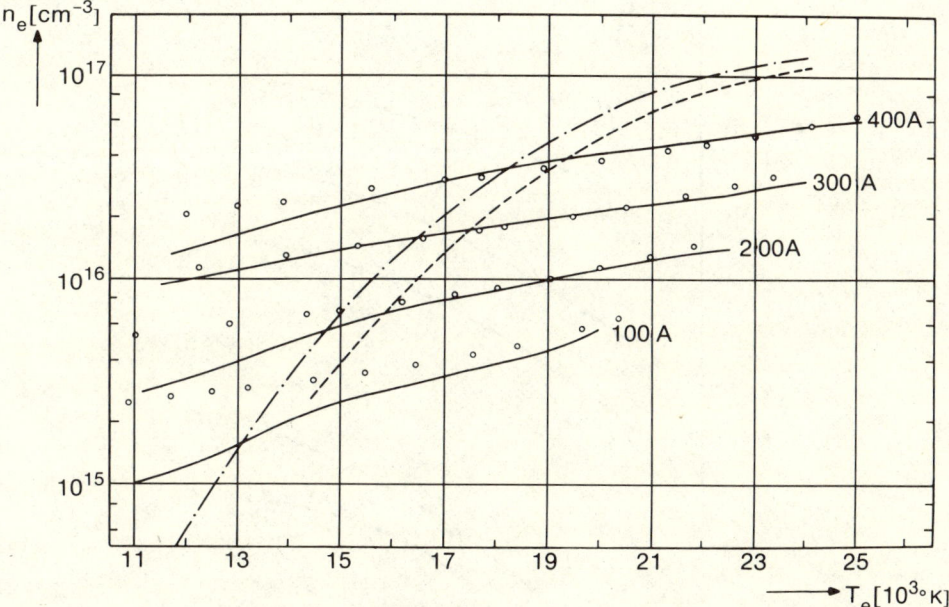

Fig. 11. Electron density versus electron temperature in He arcs of atmospheric pressure (R = 0.3 cm).
—·—·— complete LTE
— — — rate equations without diffusion
——— rate equations with diffusion
o o o o measurements.

which should be small enough compared to the discharge radius to suppress diffusion effects. This leads to the condition

$$\frac{R}{\lambda_{Diff}} = \left[\frac{3}{8} R^2 \Omega_{a} + Q_o n_e^2 (1 + \frac{n_a}{n_e}) \frac{m_{He}}{kT_G} \right]^{1/2} > 10 \qquad (24)$$

This parameter is plotted versus arc current for the rare gases in Fig. 12 (R = 0.3 mm, p = 1 atm).

DETERMINATION OF THE ELECTRICAL CONDUCTIVITY FROM ARC MEASUREMENTS

The determination of transport coefficients from arc measurements forms an important part of the diagnostic efforts in establishing the energy balance of an arc discharge. Besides the measurement of the radiative and particle heat conductivity (e.g., Hermann

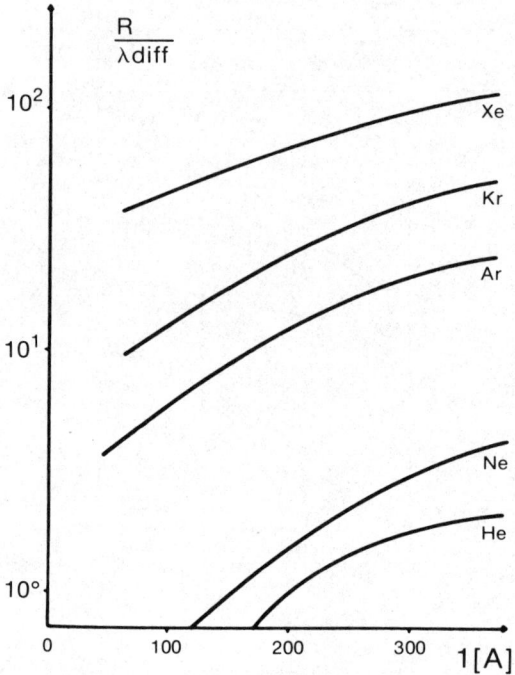

Fig. 12. Ratio of arc radius to diffusion length at various arc currents for rare gases. p = 1 atm., R = 0.3 cm.

and Schade, 1970), and of the viscosity (e.g., Asinovskii et al., 1978), the knowledge of the electrical conductivity is of importance for the description of the electrical properties of the arc. The problem in determining these quantities from arc measurements is that in general their evaluation requires the simultaneous solution of the energy balance, the radiation transport equation, and Ohm's law. However, this procedure causes problems concerning the absorption of radiation by the plasma. In the following we review a method for the evaluation of the electrical conductivity $\sigma(T)$ that avoids these difficulties (Hackmann, 1971 ; Hackmann and Uhlenbusch, 1971). This method is applicable to plasmas which are not too far away from LTE-conditions. The starting point is Ohm's law which in cylindrical coordinates reads

$$\frac{I}{2\pi R^2 E(I)} = \int_0^1 \sigma(T,p)\rho d\rho \; . \qquad (25)$$

Here I is the arc current; E (I), the electrical field strength; T and p, temperature and pressure; R, the arc radius; and ρ, the reduced radial coordinate r/R. Equation (25) can be transformed into

$$\frac{I}{\pi R^2 E(I)} = \int_{i_{axis}}^{i_{wall}} \sigma(i[T]) \frac{\partial \rho^2}{\partial i} di \qquad (26)$$

i (T) is an arbitrary but monotonously increasing function of the temperature, which can, for example, be the intensity of emitted radiation, the electron density, or the temperature according to what quantity is most accessible. The entries of Eq. (26) which have to be measured are E (I), I and $i(I,\rho)$. The evaluation procedure starts with the lowest measured arc current I_{min} and calculating from the integral (26) the value of σ at the axis, $\sigma(i_{axis}^{(1)})$. For a somewhat higher current I_2 this value of $i_{axis}^{(1)}$ and the corresponding value $\sigma(i_{axis}^{(1)})$ is already attained somewhere at the boundary of the discharge, and a new axis value $\sigma_{axis}^{(2)})$ can be calculated. Repeating this procedure finally for the highest current value I_{max}, we can construct a step-by-step profile of the conductivity $\sigma(I_{max}, \rho)$. Knowing the temperature profile for the maximum current $T(I_{max}, \rho)$, we can obtain the electrical conductivity as a function of temperature. Results for X_e are given in Fig. 13 (see Hackmann et al. 1972).

The values are compared to theoretical data reported by Devoto (1969). Values for other rare gases obtained by applying the method

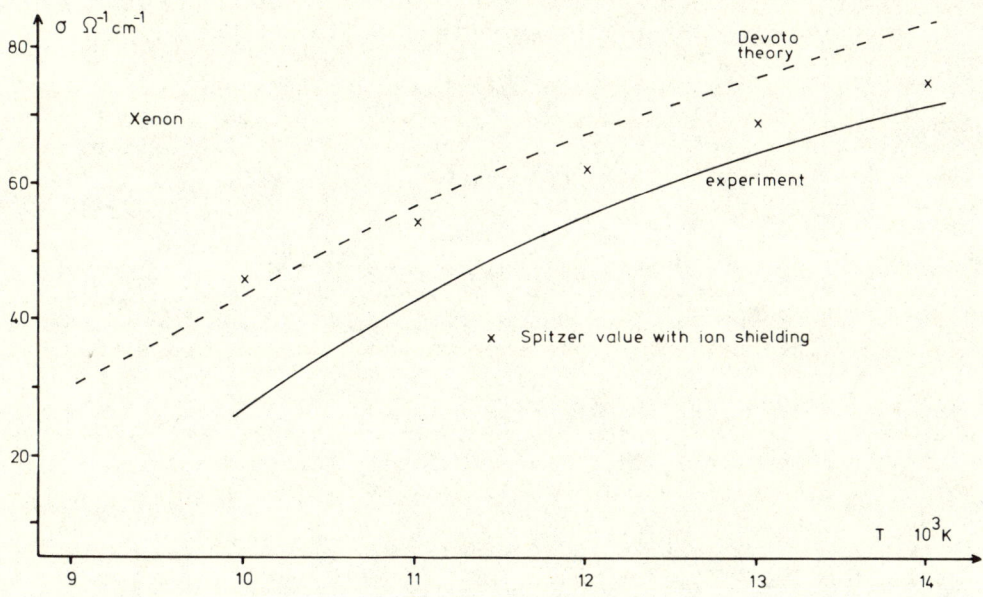

Fig. 13. Electrical conductivity of Xenon versus temperature.

described above have been reported by Hackmann and Uhlenbusch (1971).

SUMMARY

In this paper we have discussed flow effects in arc discharges by studying a few selected examples. Flows of plasma particles can be generated by external and internal magnetic fields and cross flows. We reviewed reaction of the discharge in these cases and their theoretical treatment by a monofluid approach. A different type of flow effect was investigated next. As shown, long mean free paths of neutrals and large gradients of the partial pressures drive internal diffusion fluxes. An adequate description of these effects has to involve the elementary collision and radiation processes occuring in the discharge plasma. The numerical evaluation of macroscopic plasma parameters then becomes somewhat more elaborate. For calculations of this type, the knowledge of the transport coefficients is essential. Finally, we presented a method for evaluating the electrical conductivity from arc measurements.

REFERENCES

Asinovskii, E. I., Pakhomov, E. P., and Yartsev, I. M., 1978, Tepl. Vys Temperatur, 16:28.
Bartels, K. and Uhlenbusch, J., 1970, Z. Angew. Phys., 29:122.
Devoto, R. S., 1969, AIAA J., 7:199.
Fischer, E. and Uhlenbusch, J., 1967, in: "Proceedings, 7th International Conference on Phenomena in Ionized Gases, Beograd," p. 725.
Hackmann, J., 1971, Thesis, Technical University Aachen, FRG.
Hackmann, J. and Uhlenbusch, J., 1971, in: "Proceedings, 10th International Conference on Phenomena in Ionized Gases, Oxford," Contributed Papers, p. 260.
Hackmann, J., Michael, H., and Uhlenbusch, J., 1972, Z. Physik, 250: 207.
Herrman, W. and Schade, E., 1970, Z. Phys., 233:333.
Jones, G. R. and Fang, M. T. C., 1980, Rep. Prog. Phys., 43:1415.
Kollmar, E., Rosenbauer, H.,and Seeger, G., 1969, in: "Proceedings, 9th International Conference on Phenomena in Ionized Gases, Bucharest."
Plessl, A., 1980, Appl. Phys., 21:377.
Richter, J., 1971, "Proceedings, 10th International Conference on Phenomena in Ionized Gases, Oxford," Invited Lectures.
Roman, W. C. and Meyers, T. W., 1967, AIAA J., 5:2011.
Rosenbauer, H., 1971, Z. Phys., 245:295.
Sauter, K., 1969, Z. Naturforsch., 24a:1694.
Sebald, N., 1980, Appl. Phys., 21:221.
Stäblein, H. G., 1977, in: "Proceedings, 13th International Conference on Phenomena in Ionized Gases, Berlin," Invited Lectures,

VEB Export-Import, Leipzig.
Uhlenbusch, J., 1976, Physica, 82C:61.
Uhlenbusch, J., Fischer, E., and Hackmann, J., 1970, Z. Phys., 238:404.
Uhlenbusch, J. and Fischer, E., 1971, Proc. IEEE, 59:578.

DIAGNOSTIC TECHNIQUES FOR DISCHARGES AND PLASMAS

R. T. Waters

University of Wales
Institute of Science and Technology
Cardiff, Wales, United Kingdom

INTRODUCTION

Many of the classical techniques developed by early workers for measurements of electrical discharge phenomena have found new application as a result of modern instrumentation. Thus, for example, both direct photography and schlieren techniques have benefited from the development of high-speed image converters and image intensifiers. Laser light sources have greatly extended applications of the classical Doppler and light-scattering methods, and again of shock-wave and schlieren photography and interferometry. Statistically-based measurements of breakdown voltages, time lags and pulse-to-pulse variations can be made by automatic test systems. Modern optics, photomultiplier tubes and analogue/digital electronics are a considerable aid to spectroscopic methods. Optical-fibre data links provide invaluable electrical isolation and noise avoidance. Digital computation of electric field distributions has application to high-voltage engineering, physical discharge studies, and modeling.

The present review aims both to recall the basis of a number of diagnostic methods and to examine some new developments. The application of techniques to discharges differing as widely as long sparks and glows, and high-current arcs and corona streamers, will be briefly illustrated.

A classification of the methods to be covered is summarized in Table 1.

Table 1. Optical and electrical measurements of electrical discharge phenomena.

Optical	Electrical

Schlieren techniques
  Schlieren photography
  Schlieren interferometry
  Direct-shadow methods
  Shock wave probes
	Current measurements at biased probes
	  Floating potential
	  Langmuir probe characteristics
	  Tassicker/Selim field probe

Light-scattering measurements
  Detection of microparticles
  Doppler anemometry
  Laser probes
	Induction probes for field
	  measurements
	  Nondiscrimatory capacitive probe
	  Field filter probes
	  Field mills and fluxmeters

Spectroscopic analysis
  Qualitative analysis
  Intensity measurements
  Line profiles
  Practical studies	

SCHLIEREN METHODS IN DISCHARGE STUDIES

Classification of Methods

Discharges which produce significant gaseous heating are especially suited to the well-known schlieren and related techniques. These methods are based upon the change of refractive index $(n - n_o)$ which arises when the density of the medium changes from ρ_o to ρ:

$$n - n_o = (n_o - 1)\left(\frac{\rho}{\rho_o} - 1\right). \tag{1}$$

There are three basic techniques (Holder and North, 1963):

(i) Schlieren (or strioscopic) methods (Toepler). Refraction of the illuminating beam in regions where there is a density gradient (and thus a gradient dn/dx of refractive index) is recorded.

(ii) Interferometry methods (Mach). A fringe pattern is formed between illumination passing through the discharge and a reference beam. The fringe pattern depends upon the value of n in

the discharge, since the velocity of light in that region is
$c(n) = c/n$.

(iii) Shadow schlieren methods (Dvorak). Here the optical arrangement is such that regions of constant density gradient are not visible, but only regions of large d^2n/dx^2.

Quantitative density measurements can be made by interferometry methods; schlieren and shadow methods are normally used to determine shapes and positions of discharge boundaries with significant density change.

Schlieren Photography

The basic schlieren method normally involves the formation of a parallel illuminating beam passing through the discharge; this beam is usually formed by placing the source at the focus of a mirror M_1 of focal length f_1 (Fig. 1(a)). Two image planes are formed by the receiving optical system:

(a) an image of the light source is formed in the focal plane f_2 of the second mirror M_2;

(b) an image of the discharge is formed by the optical combination of M_2 and a focussing lens L_3.

Any refraction of the illuminating beam arising in the discharge will cause

(c) a lateral movement of the image of the light source, but

(d) no movement of the image of the discharge.

However, intensity changes within the image of the discharge (corresponding to density gradients within the discharge) can be produced by placing a barrier in the image plane of the source so as to partially obscure the source image. This schlieren barrier is often in the form of a knife-edge (Toepler edge) which is set parallel to the expected density-change boundaries; then for one sign of density gradient the image brightness would be increased, and for the opposing sign it would be decreased. It is usually desirable that in the absence of the discharge, one half of the source image is allowed to pass the barrier. The range of density gradient which can be detected is defined by the maximum angular deflection θ_m from the normal which can be recorded; this in turn depends upon the ratio of the image half-height of the source (actual height, h, image half-height, $\frac{1}{2}(f_2/f_1)h$) to f_2:

$$\theta_m = \frac{h}{2f_1}. \tag{2}$$

It is usually inconvenient to reduce f_1 to small values. The range θ_m is thus increased by using a large source (either physically large or by use of a condensing lens or mirror in conjunction with the source).

The relative intensity or contrast with respect to the background of rays refracted at an angle θ is $C = \theta/\theta_m$. The image quality depends upon contrast sensitivity S, given by

$$S = \frac{dc}{d\theta} = \frac{1}{\theta_m}.\tag{3}$$

Sensitivity is thus inversely proportional to range.

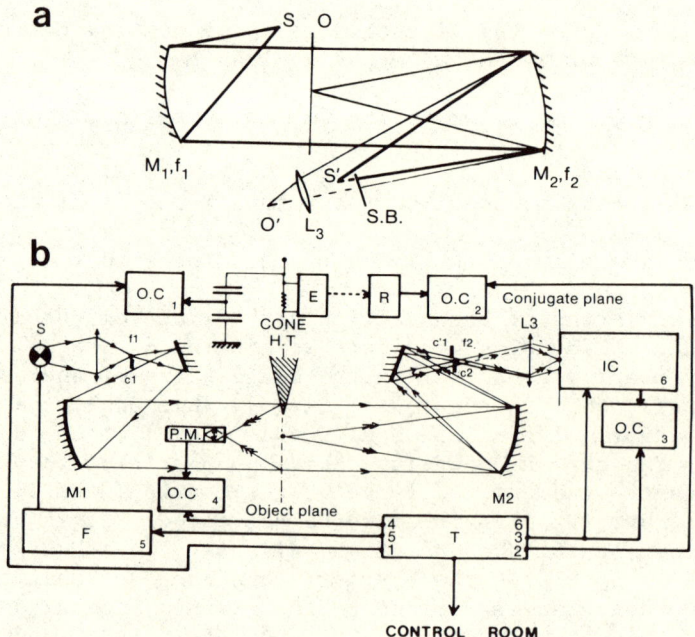

Fig. 1. Schlieren layout. (a) Basic arrangement: S,S' = source, source image; O,O' = object, object image; S.B. = Schlieren barrier. (b) Schlieren system for long spark leader channel study (Les Renardieres, 1981).

The size of schlieren mirrors M_1 and M_2 is clearly related to the area of the discharge to be studied. Lenses may be used, but their limited apertures and cost are restrictions. The tilt which is necessary in the optical layout introduces aberrations. Coma aberration is reduced by using an opposing tilt in M_1 and M_2 (i.e., the source and the camera on different sides of the illumination beam, not shown in Fig. 1(b)). Spherical aberration is reduced by adjustment of source and barrier positions so as to give uniform intensity attenuation as the barrier is moved across the source image.

Figure 2 shows schlieren time-resolved, image-converter photographs obtained with the system of Fig. 1(b). These are framing-mode sequences with an exposure time of 1μs, and a time interval between frames of 10μs. They represent a 45 mm diameter field of view just below the high-voltage applied to a 2 m cone/plane gap. The developing leader channel is observed to expand radially with time at a velocity of about 25 ms^{-1}. Weak spherical and cylindrical shocks of about mach 1 originate from the channel. The leader expansion is usually continuous in positive leaders (Fig. 3) and discontinuous (like the current and leader propagation) in negative leaders (Fig. 4). The precise measurement of thermal radius allowed by this method is valuable in modeling the hydrodynamic behavior of long sparks. Even more precise measurements can be obtained by a schlieren slit/streak arrangement. Figure 5 is the image of a 0.66 mm longitudinal section of the leader channel time resolved in a direction parallel to the leader channel. The leader inception (A) and associated shock (B) is followed by radial expansion (C); sparkover is at (D), followed by a rapid expansion in the arc phase (E). This causes a strong shock (F), which is attenuated to sonic velocity after some tens of millimeters.

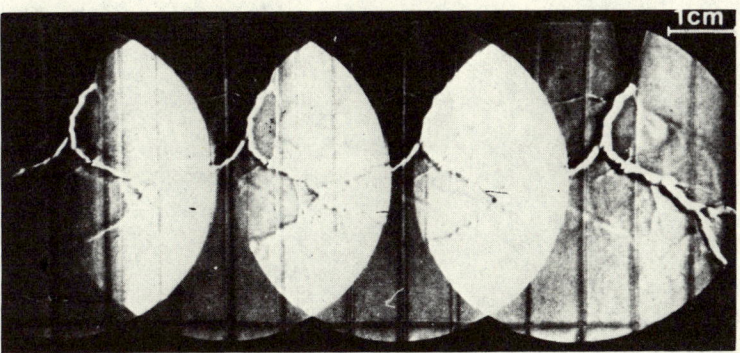

Fig. 2. Framing-mode schlieren photography. 2 m gap, positive polarity; framing interval 10 s; 45 mm field of view (Les Renardières, 1981).

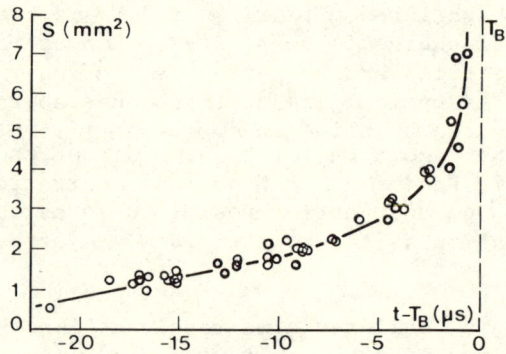

Fig. 3. Evolution of positive leader cross-sectional area. 60/3000 µs impulse shape (Les Renardières, 1981).

Schieren photography has also been useful in studies related to gas-blast circuit breaker applications. Farish and Lynch (1980) studied breakdown in air and SF_6/N_2 mixtures in supersonic streams at flows of up to mach 2.8. They found that breakdown is initiated in the low-density wake of electrodes, and the spark propagates within the higher density gas layers.

Where density gradients may appear in all directions as in a tortuous leader channel, there is some advantage in using a symmetrical shclieren barrier; thus instead of a knife-edge, a circular source and aperture (or disc) can be used, so that all regions of density gradient appear darker (or lighter) than the background.

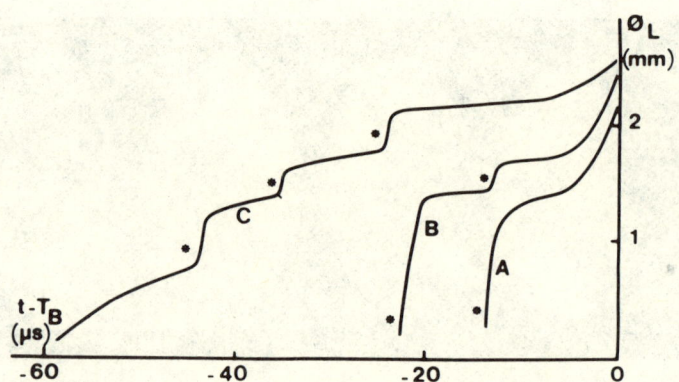

Fig. 4. Evolution of negative leader diameter. Impulse shape 20/1400 µs (Les Renardières, 1981).

DIAGNOSTIC TECHNIQUES FOR DISCHARGES

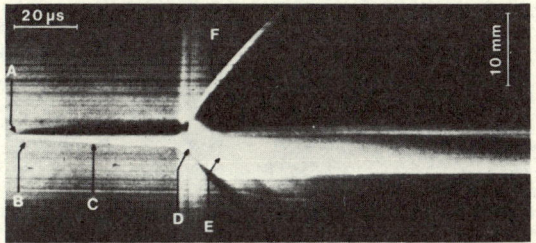

Fig. 5. Streak slit/schieren record showing growth of positive leader/spark channel diameter. Impulse shape 130/3000 µs (Les Renardières, 1981).

A further variant is to employ a radially colored circular filter; there are some visual advantages in discrimination between hues as opposed to monochrome shading. Ross (1977) used this technique to study positive impulse discharges in a 1.5 m cone/plane gap, with various impulse shapes. He used a 100 µs xenon flash, a single 300 mm diameter, 3 m focal length mirror, and a rotating-mirror camera recording up to 28 frames at a maximum speed of $1.25 \cdot 10^6$ frames sec^{-1}. Single-mirror systems involve double-pass rather than single-pass schieren layouts — that is, the illuminating beam is nonparallel and passes through the discharge both before and after reflection. Two schieren images are thus formed, of which either may be selected for study. Although image quality is slightly less good, the sensitivity S is greater since the source is located at the center of curvature of the mirror rather than its focus f_1. Figure 6 shows some frames from the sequence. The spatial resolution was about 0.1 mm.

An advantage of graded filters, whether neutral or colored, linear or circular, over opaque barriers is that a point light source can be used. This reduces aberrations and is useful for laser or high-speed flash sources. The use of laser sources avoids problems associated with discharge luminosity and also raises possibilities for holographic schlieren diagnostics.

<u>Schlieren Interferometry</u>

There are two possible approaches. Either a monochromatic source is used to produce a fringe pattern which is parallel in the absence of the discharge, so that in its presence the density gradient is constant along each fringe; or a white source is used with the optical layout arranged so that the zero fringe covers the field of observation. Color changes then correspond with gradient changes.

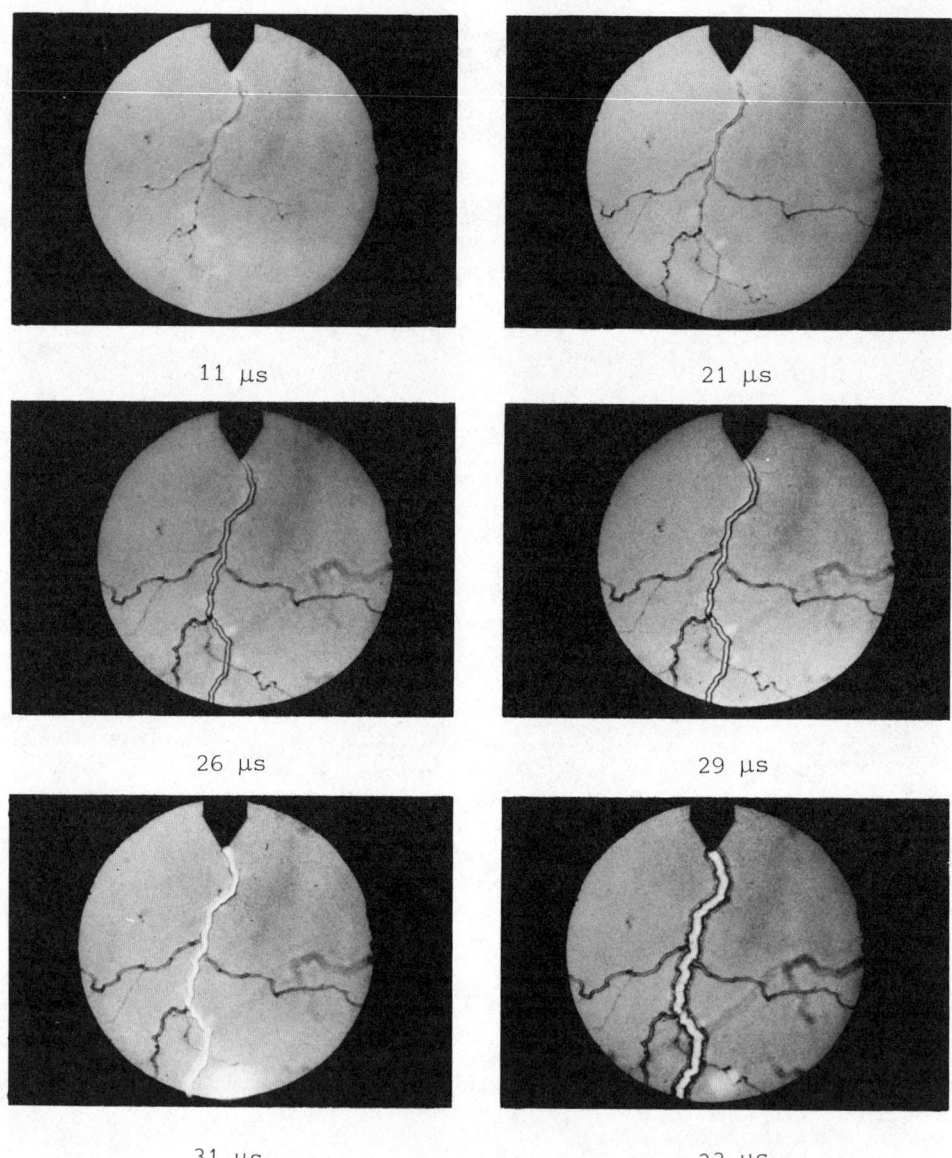

Fig. 6. Schlieren framing sequence of positive leader channel expansion using circular filter (Ross, 1977).

Quantitative studies of refractive index changes are thus possible. Figures 7 and 8 show a system for arc column analysis.

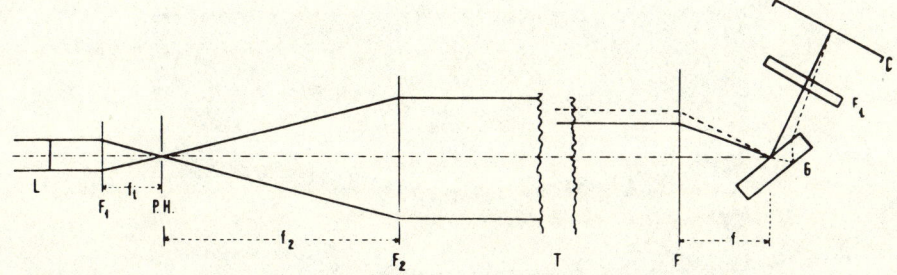

Fig. 7. Interferometric schieren layout (Tanner, 1965). L = Laser; F_1F_2F = Convex lens; P.H. = Pin hole; T = Test zone; G = Glass plate; F_i = Filter; C = Camera.

Kurimoto and Farish (1980) examined density variations in a negative d.c. corona discharge by means of a Michelson interferometer and a monochromatic (5451 Å) source. Conventional schlieren photography had shown a filamentary low-density column in both the glow and Trichel modes. The interferometric technique enabled quantitative estimates of gas density to be made (Figs. 9 and 10).

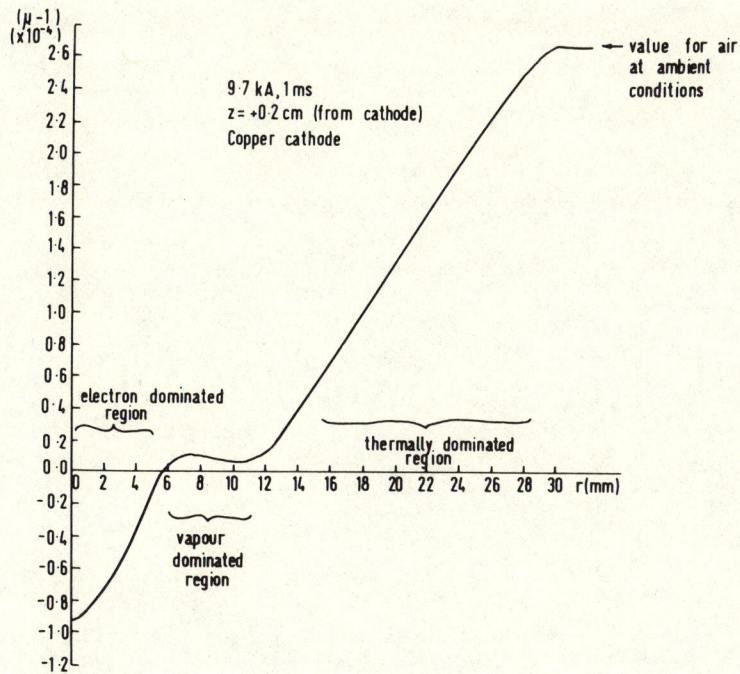

Fig. 8. Radially unfolded refractive index in free-burning arc (near cathode) (Rodriques, 1974).

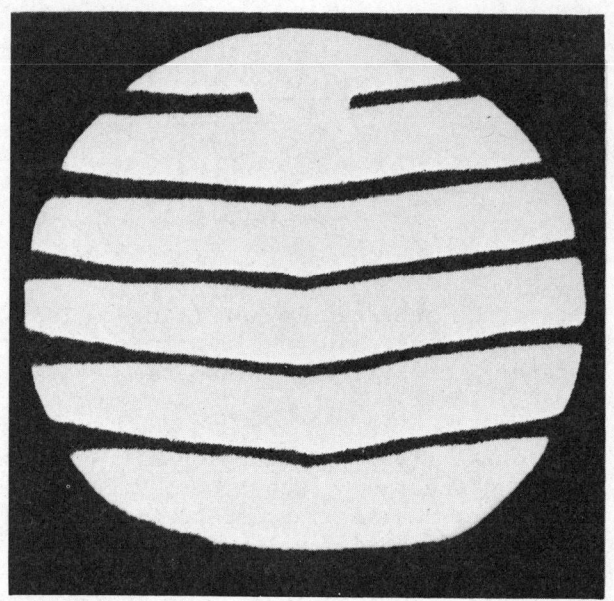

Fig. 9. Interferogram of 20 mm point/plane gap. Point radius 2 mm; negative corona (Kurimoto and Farish, 1980).

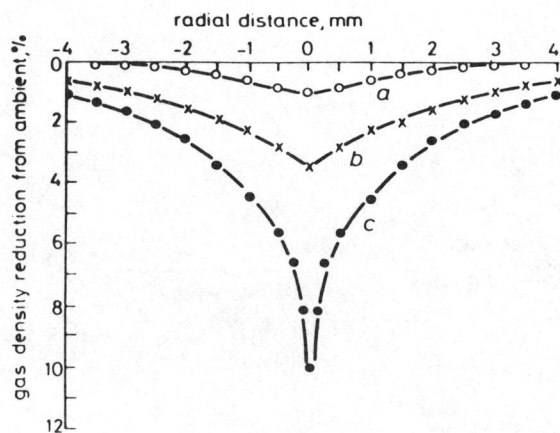

Fig. 10. Radial gas-density distribution at 6 mm from cathode. Negative corona (Kurimoto and Farish, 1980).
(a) Trichel-pulse (19 kV)
(b) Pulseless corona (28.5 kV)
(c) Near sparkover (36 kV)

Direct-Shadow Methods

Near strong shock fronts, a rapid change d^2n/dx^2 in density gradient occurs. Since closely adjacent rays are deflected by different amounts, an image of the shock is formed as a shadow at a given distance behind the shock. The magnification is unity, and the screen or plate usually must be fairly close to the working region. A second mirror can be used to give a focused direct-shadow record.

Shock Wave Probes

This is a method developed by Edels and Whittaker (1957) to determine temperatures within arc columns, following early work by Suits (1935). It is a particularly interesting diagnostic technique since the time-resolved records upon which measurements are based depend upon emission both from the schlieren light source and from the discharge itself. An externally generated shock wave is passed through the arc channel, and the shock velocity is measured as a function of time. Before entry into the channel, the shock is visible only by the schlieren technique; during its passage in the arc, it is visible also because of brightening of the arc luminosity.

Under assumptions of perfect gas behavior, the mach number of a shock front is given by

$$M = \frac{u}{s} = -\frac{u}{\sqrt{\gamma p \nu}} = \frac{u}{\sqrt{\gamma RT/m}} \qquad (4)$$

where u = shock front velocity and s = sound velocity under prevailing conditions. Determination of gas temperature within the discharge can be made if M (or s) is known:

$$T = \frac{mu^2}{\gamma RM^2} = \frac{ms^2}{\gamma R} . \qquad (5)$$

But the variation of M with u for any given gas is also a function of temperature. Edels and Whittaker (1957) determined the form of this function, which requires only a knowledge of a pair of u, T values at a single point during passage of the shock. This can be easily made outside the arc column by means of the Schlieren record.

The values of M so derived must also be corrected for the attenuation which occurs in the shock during its propagation. An exponential decrease was shown so that the change ΔM in mach number over a distance ℓ is

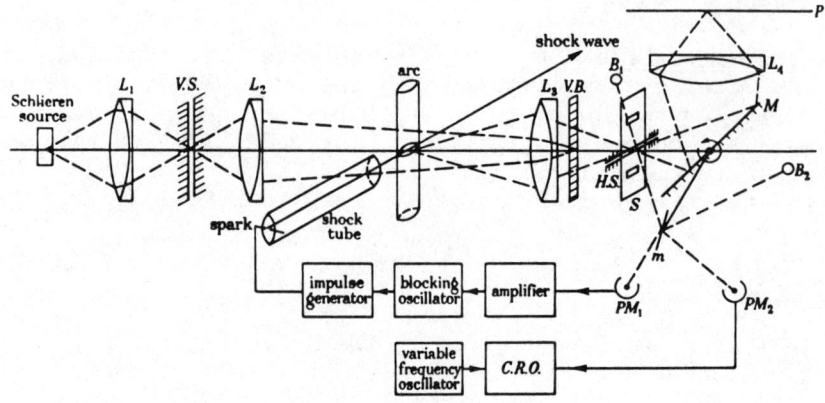

Fig. 11. Schlieren/shock-tube system (Edels and Whittaker, 1957).

$$\Delta M = (M-1)[\exp(\ell/d) - 1] \qquad (6)$$

where $d \simeq 20$ mm. The appropriate corrected value for M within the channel is obtained by interpolation between entry and exit.

Figure 11 shows the schlieren/shock-tube/rotating-mirror layout. Sufficiently strong shocks could be obtained by using a low-inductance impulse generator, and locating the shock-tube exit close to the arc under study. Arc temperatures in the range 6250 - 7700 K were found with an estimated accuracy of ± 200 K. The method is suitable for temperatures of up to 15000 K. Time resolution is limited by the shock transit time, which is typically a few microseconds. In highly ionized plasmas, the effect of electron density as well as neutral density becomes important, and temperature measurements may be less reliable.

LIGHT SCATTERING MEASUREMENTS

Laser light sources offer a variety of possibilities as probes for discharge studies, which are based upon (a) the high intensities available, which allow low-efficiency scattering processes to be detected, even in the presence of intense radiation from the discharge itself, (b) the coherence of laser light, allowing interferometric methods to be employed, and (c) the short light pulse durations which are possible, which allow a high degree of temporal discrimination and the achievement of high signal/noise ratios.

Detection and Sizing of Microparticles

The presence of such particles is known to be a cause of breakdown in vacuum gaps and in compressed gas-insulated systems. An analysis of microparticle sizing has been described by Grey Morgan (1979). A particle of cross section σ passing through a laser beam of power P which is uniformly distributed over a cross-sectional area A will scatter into a photomultiplier of acceptance area B placed at a distance L, a power P' given by

$$P' = \frac{P\sigma}{4\pi AL^2} . \tag{7}$$

The photomultiplier signal amplitude V_{pm} will be proportional to P', and thus to σ. Spherical particles of radius down to 2μm can be detected by this technique, and calibration tests with known microparticle sizes confirm the linearity of V_{pm} with r. Vacuum breakdown was observed to follow the detection of microparticles of size range 6 - 10μm within the discharge gap.

Doppler Anemometry

For low particle density, the particle velocity can be obtained from transit time measurements. Alternatively the Doppler shift due to motion of the scattering medium can be measured. Figure 12 represents the detection of light of frequency ω which has been scattered by a particle of velocity v illuminated by light of frequency ω_o. From the moving frame of reference of velocity v the frequency of the scattered light is

$$\omega' = \omega_o (1 - \frac{v}{c} \cos \gamma) . \tag{8}$$

At the fixed reference frame of the detector the frequency recorded is

$$\omega = \omega'/(1 - \frac{v}{c} \cos \beta) . \tag{9}$$

Hence the difference frequency between source and detection is

$$\omega - \omega_o = \frac{v}{c} \frac{\cos \gamma + \cos \beta}{1 - \frac{v}{c} \cos \beta} , \tag{10}$$

which may be measured by spectroscopic of interference techniques.

In arc discharges, the motion of microparticles will indicate the nature of the convective process. Correction for particle drag is possible in order to obtain gas flow velocities (Williams, 1973).

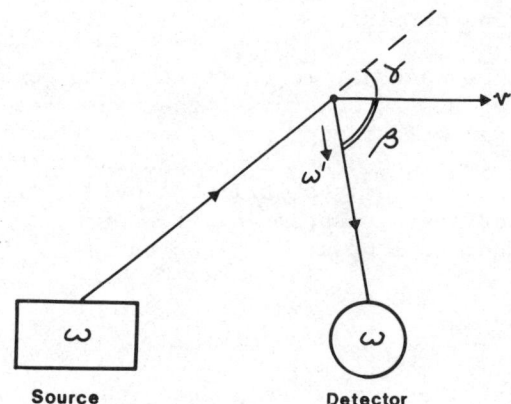

Fig. 12. Doppler anemometry — frequency change of scattered radiation.

Barrault and Jones (1974) describe an alternative technique. A narrowly converging (angled), argon laser split beam (wavelength λ) was used to form real fringes at right angles to an arc axis. Particles within the gas flow crossing these fringes at velocity v produced scattered light modulated at frequency $f = (2v \sin \alpha/2)\lambda$. Photomultiplier recording allowed measurement of the particle velocity (Fig. 13).

Laser Probing of Discharge Plasma: Introduction

The extent of the interaction between incident radiation and charged particles increases with increasing charge of the particles and decreases with increasing mass of the particles. The magnitude

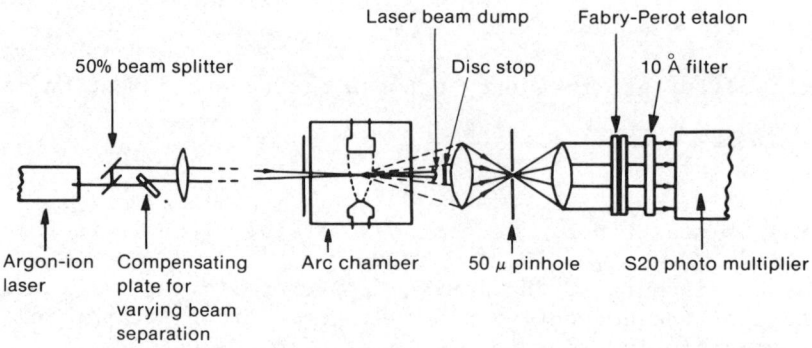

Fig. 13. System for laser anemometry (Barrault et al., 1974).

DIAGNOSTIC TECHNIQUES FOR DISCHARGES 217

of scattering effects (i.e., the total cross section) is therefore much greater for electrons than for ions. For low-density low-temperature plasmas, the interaction is defined by the classical Thomson cross section σ_T, and the angular distribution of the scattered light depends upon the angle γ between the direction of observation and the direction of polarization of the incident radiation. In dense plasmas the observed distribution of scattering will depend also upon the angle θ between the direction of the incident beam and the direction of observation, and the total cross section ($\sigma'_T = \beta \sigma_T$) will be less than the classical σ_T because of two effects: (a) the electron acceleration causing the scattered re-radiation will be reduced by the thermal velocity v_e of the electrons, and (b) the influence of the incident field upon the electrons will be limited by the Debye length λ_D of the plasma. The classical theory of Thomson scattering does not involve any change in wavelength between the scattered and incident light.

However, wavelength shift is observed in practice both because of Compton scattering and, more importantly, because of the effect of Doppler broadening.

<u>Thomson Scattering</u>. An electron vibrating due to an incident field will act like Hertzian dipole of moment p. Radiation field strength of the scattered radiation at angle γ to the dipole axis is

$$|E_s| = |H_s| = \frac{|\ddot{p}|}{c^2 r} \sin \gamma. \tag{11}$$

where r = distance from dipole (large compared with the wavelength).

The energy radiated/second is obtained from the Poynting vector magnitude

$$|S| = \frac{c}{4\pi} |E| |H| = \frac{e^2 |\ddot{x}|^2}{4\pi c^3 r^2} \sin^2 \gamma. \tag{12}$$

This gives the angular distribution of the scattered light.

The total scattered power is obtained by integration over a sphere of radius r:

$$P_s = \int_0^{2\pi} S \, 2\pi r^2 \sin \gamma \, d\gamma = \frac{e^2 |\ddot{x}|^2}{c^3} \int_0^{2\pi} \sin^3 \gamma \, d\gamma = \frac{2}{3} \frac{e^2 |\ddot{x}|^2}{c^3}. \tag{13}$$

Now the intensity of the irradiating beam is

$$I_o = \frac{c}{4\pi} E_o^2, \tag{14}$$

and the resulting electron acceleration is

$$|\ddot{x}| = \frac{eE_o}{m}. \tag{15}$$

Equations (13) - (15) give for the power of the scattered radiation

$$P_s = \frac{8\pi}{3} \frac{e^4}{m^2 c^4} I_o = \sigma_T I_o. \tag{16}$$

The Thomson scattering cross section σ_T of an electron is therefore

$$\sigma_T = 6.652 \; 10^{-25} \; cm^2. \tag{17}$$

This extremely small cross section indicates the usefulness of high-intensity laser irradiation for scattering experiments (Evans and Katzenstein, 1969). Two basic considerations are of importance: first, the denser the plasma, the higher the irradiating power required, since the degree of scattering is proportional to the electron density n_-, while in a highly ionized plasma the luminous emission is proportional to $n_- n_+ \simeq n^2$. To obtain a suitable signal/noise ratio, irradiating power densities of 0.1-10 MW/cm^2 may be needed. Furthermore, the low value of σ_T gives a ratio of P_s/P (incident) of only 10^{-12} even for $n_e = 10^{16}$ cm^{-3}. Stringent precautions against stray light effects are necessary. Second, in order to avoid unwanted absorption of the irradiating beam, its frequency must be much greater than the plasma frequency ω_p. For $n_e = 10^{14} - 10^{16}$ cm^{-3}, the beam needs to be in the visible or infrared. A ruby laser at 6943 Å and narrow line width 0.1Å is a suitable source.

The scattering cross section reduction factor $\beta = \beta_e + \beta_i = \sigma_T'/\sigma_T$ has been shown to be for a singly ionized plasma the sum of

$$\beta_e = \frac{1}{1 + \alpha^2} \tag{18}$$

and

$$\beta_i = \frac{\alpha^4}{(1 + \alpha^2)[1 + \alpha^2(1 + T_-/T_+)]}, \tag{19}$$

where

$$\alpha = \frac{\lambda_o}{4\pi \lambda_D \sin \theta/2}. \tag{20}$$

DIAGNOSTIC TECHNIQUES FOR DISCHARGES 219

and λ_0 is the irradiation wavelength and $\lambda_D = 7.40 \; 10^2 (T\_eV)/n\_)^2$ cm. The important parameter α is the ratio of the Bragg scattering depth $\lambda_0/4\pi \sin \theta/2$ to λ_D. We note that $\beta \to 1$ as $\alpha \to 0$, and $\beta \to 0.5$ as $\alpha \to \infty$. In high-pressure arcs, α is in the range 0.5 - 5 (Barrault and Jones, 1974).

Wavelength Change in Scattered Radiation. The Compton theory of quantum scattering shows the wavelength change resulting from momentum transfer to the recoiling electron. Consider a quantum of wavelength λ_0 which encounters an electron at rest, the quantum being scattered at a new (longer) wavelength λ_1 in a direction at angle θ, and the electron recoiling with velocity v at angle ψ.

Loss of photon energy = gain of electron K.E.:

$$\therefore \; hc\left(\frac{1}{\lambda} - \frac{1}{\lambda_1}\right) = m_o c^2 \left(\frac{1}{\sqrt{1 - \frac{v^2}{c^2}}} - 1\right) . \tag{21}$$

And for conservation of momentum:

$$\frac{h}{\lambda_o} = \frac{h}{\lambda_1} \cos \theta + \frac{m_o v}{\sqrt{1 - \frac{v^2}{c^2}}} \cos \psi \tag{22}$$

and $$0 = \frac{h}{\lambda_1} \sin \theta - \frac{m_o v}{\sqrt{1 - \frac{v^2}{c^2}}} \sin \psi . \tag{23}$$

Equations (21) to (23) give

$$1 - \cos \theta = \frac{m_o c}{h} (\lambda_1 - \lambda_o) . \tag{24}$$

The quantity $\lambda_c = h/m_o c = 0.0242 \text{Å}$ is known as the Compton wavelength. This wavelength change of scattered radiation is a function only of the angle of scattering θ:

$$\Delta \lambda = \lambda_1 - \lambda_o = 2\lambda_c \sin^2 \frac{\theta}{2} . \tag{25}$$

The maximum wavelength shift of 0.0484Å is a very small p.u. change in radiation of optical wavelengths. Momentum transfer in this case can often be ignored (the classical work used hard x-rays).

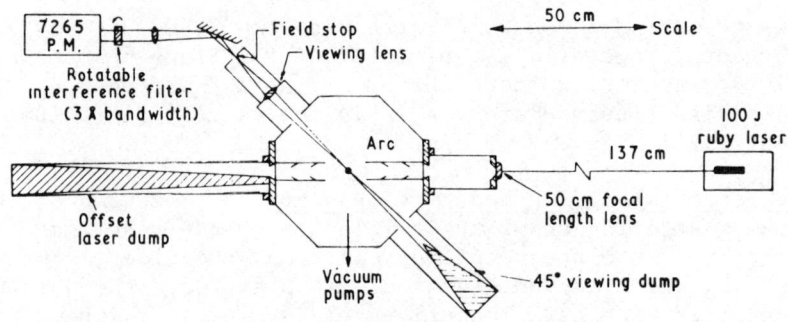

Fig. 14. Thomson scattering measurements in hollow-cathode argon arc (Gerry and Rose, 1966).

The Doppler broadening produced by electron and ion motion is much more significant. For large Debye length λ_D, the parameter α is small, say ≤ 0.5. Scattering is then by free electrons (Thomson domain), and the electron temperature is obtained from the line half-width $\Delta\lambda$:

$$T_- = 1.32 \; 10^{-3} \left(\frac{\Delta\lambda(\text{Å})}{\sin\theta/2} \right)^2 \text{ eV} . \tag{26}$$

An example of Thomson scattering measurements in a steady-state hollow-cathode argon arc is shown in Fig. 14 (Gerry and Rose, 1966). Reasonable agreement was found with Langmuir probe data.

For $0.5 \leq \alpha \leq 2$, splitting of spectral lines occurs because of electron scattering. If the wavelength shift is $\Delta\lambda$, then

$$\Delta\lambda(\text{Å}) = 4.218 \; 10^{-15} \lambda_o(\text{Å}) \sqrt{n_-} . \tag{27}$$

$\Delta\lambda$ thus depends only upon the electron density and is independent of scattering angle θ. For ruby laser light of $\lambda_o = 6943$ Å, an electron density $n_- = 5 \; 10^{15}$ cm^{-3} will cause a wavelength shift of 15 Å.

For $\alpha \geq 2$, most of the scattered light arises from the ions in the plasma. The line half-width due to Doppler broadening is then

$$\Delta\lambda = \frac{2\lambda_o}{c} \sin\frac{\theta}{2} \sqrt{\frac{K(T_- + T_+)}{m_+}} . \tag{28}$$

SPECTROSCOPIC ANALYSIS

The luminous emission from high-temperature, high-pressure arcs and plasmas is often of a complex spectral structure. Line spectra, particularly if time-resolved, can yield much information not only on the nature and concentration of the species present in the discharge medium but also on the gas and electron temperatures T and T_-, and the electron density n_-. Temperature can then be estimated from absolute or relative line intensity data, and electron density from line profile measurements. Molecular band spectra can provide values of vibrational and rotational temperatures (Bastien, 1981), the latter of the order of the gas temperature T. The relatively weak emission from radiative recombination

$$M^+ + e + \tfrac{1}{2}mv^2 \rightarrow M + h(\nu - \nu_i) \tag{29}$$

can sometimes be used to determine T_-. This involves relative intensity measurements in the continuum emission which extends from the series limit towards the UV.

Analytical Methods

Constituents of the Discharge Medium. Identification of the neutral and ionized species is usually straightforward, provided that the spectrum quality is adequate, even in the most difficult cases where the gas is a combination of molecular components such as air, and where the discharge characteristics are space and time dependent as in lightning, long spark, pulsed plasma and arc phenomena.

In air discharges, for example, the nature of the spectrum varies during the different stages of the discharge (Les Renardières Group, 1977):

(a) During the corona phases, when high electric field is present and little molecular dissociation has occurred, the spectrum consists of band and line spectra of N_2 (particularly the $2S^+$ second positive band) and N_2^+ (the $1S^{-1}$ first negative band) and some O_2^+ systems.

(b) In the ensuing leader stage, bands of N_2, NO, O_3, OH, CO and CO_2 appear.

(c) When the leader current increases in the final-jump phase, the spectrum consists mainly of NII and OII lines.

(d) These disappear during the follow-through arc, when a continuum with H_α, H_β and some OI and OII lines is seen. Basically similar spectra are found in the lightning discharge (Orville, 1968).

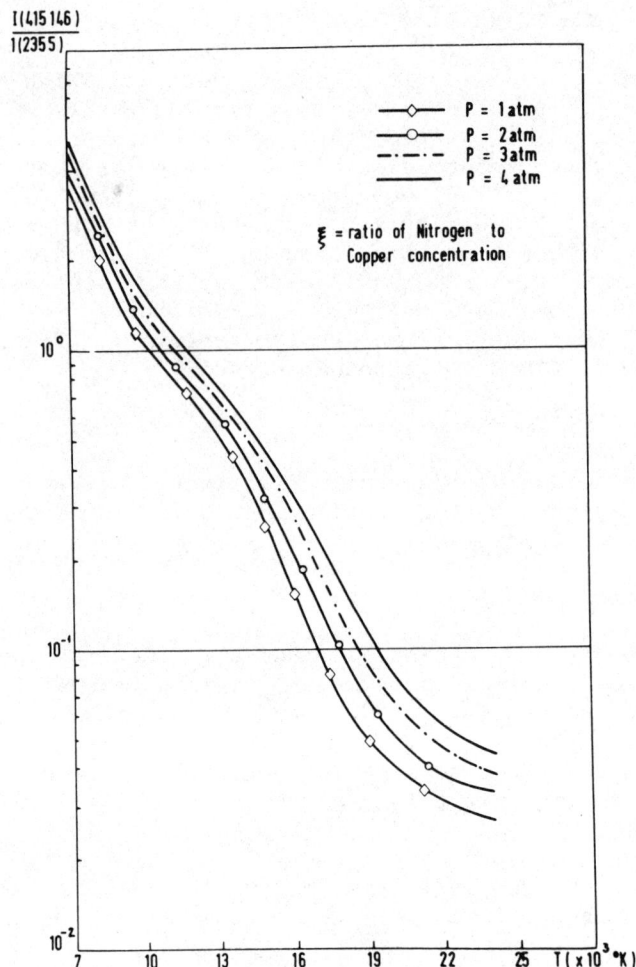

Fig. 15. Intensity ratio of NI to Cu II lines in arc discharge (Rodriques, 1974).

In arc discharges, the electrode material can have a profound influence on emission spectra. Here it is possible to employ the standard procedures used in spectrographic quantitative analysis, viz., the relative intensity of homologous line pairs (whose ratio is a weak function of excitation conditions) to determine impurity concentrations. An example (Barrault and Jones, 1974) is shown in Fig. 15 for copper-electrode impurities.

Intensity Measurements. The intensity I_x of an emitted line of frequency ν_x is expressed by the proportionality

$$I_x \alpha n_x A_x h\nu_x \tag{30}$$

where n_x is the number density of atoms in the upper energy level x, and A_x is the transition probability. If the statistical weights g_x, g_o of the level x and the ground level respectively are known, then the absolute line intensity is

$$I_x = \frac{G A_x g_x}{g_o} h\nu_x \, n \, \exp(-\varepsilon_x/kT) \tag{31}$$

where G is a geometrical factor, n is the density of the atom or ion, ε_x is the excitation energy, and T is the neutral, ion or electron temperature (depending upon the mode of excitation).

Measurement of relative intensity is simpler than absolute measurements. Since the ratio of number densities in two levels x, y is

$$\frac{n_x}{n_y} = \frac{g_x \exp(-\varepsilon_x/kT)}{g_y \exp(-\varepsilon_y/kT)}, \tag{32}$$

then the intensity ratio is

$$\frac{I_x}{I_y} = \frac{g_x A_x \nu_x}{g_y A_y \nu_y} \exp\left[(\varepsilon_y - \varepsilon_x)/kT\right]. \tag{33}$$

The assumption of Maxwell-Boltzmann distribution of energies is generally justified in thermally ionized plasmas such as arc discharges or lightning return strokes. Neutral temperatures of 24-28 10^3K have been estimated for lightning channels (Fig. 16), 20-24 10^3K in long (5m) sparks (Fig. 17), and 14-24 10^3K in high-current arcs.

Modification of these equations is required where excitation is by non-Maxwellian electron populations or occurs (as, for example, in optically thick plasmas at high electron densities) by cascade collisional excitation. In long-spark coronas, for example, the estimates of mean electron energy would be 2.5eV on the basis of a Maxwell distribution, and 0.5eV for a Druyvesteyn distribution. The local electric field responsible for electron excitation processes can be estimated from the spectroscopic evaluation of the Townsend energy factor η (Badaloni and Gallimberti, 1972):

$$\eta = 21 \, (E/p)^{0.49} \quad (E/P > 3). \tag{34}$$

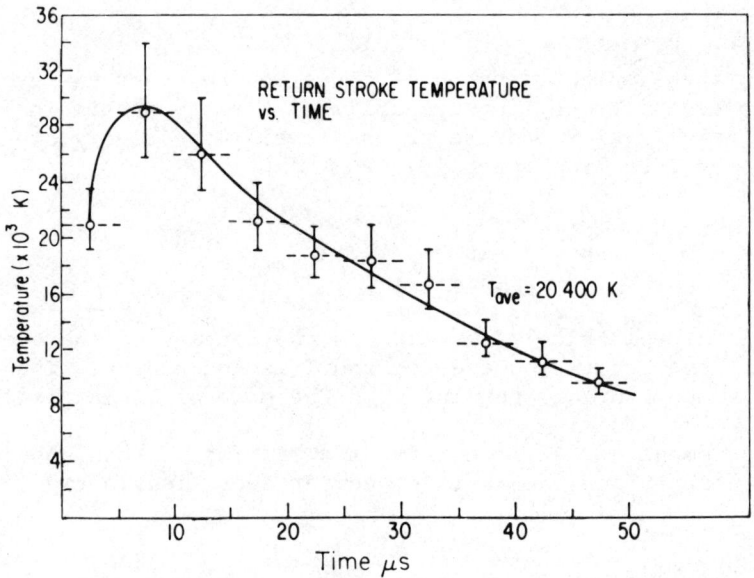

Fig. 16. Lightning return-stroke temperatures obtained from N II spectra (Orville, 1968).

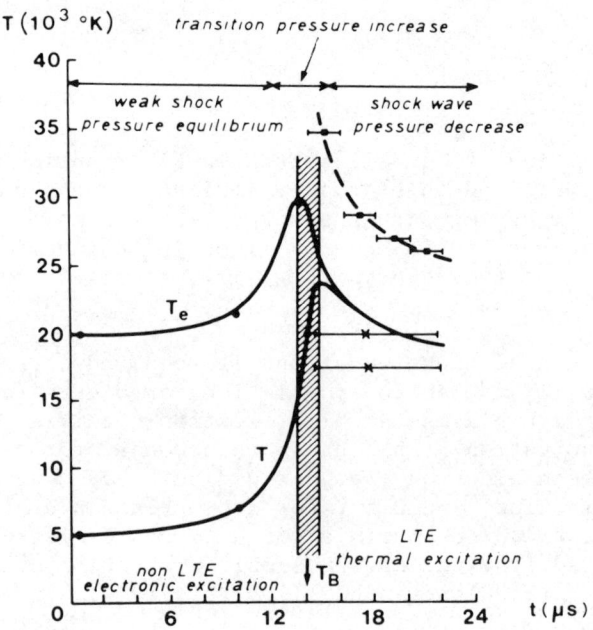

Fig. 17. Electron and gas temperatures in leader and return stroke in long spark (Les Renardières, 1977).

Line Profiles. The broadening of line spectra can arise from a number of causes.

(a) High neutral or ion temperatures lead to a Doppler line width of

$$\Delta\lambda = 10^{-6} \sqrt{\frac{T}{M}} \text{ Å} . \qquad (35)$$

(b) Pressure broadening is detected in high pressure sources. The line width is proportional to pressure, and is about 0.25 Å at 760 Torr.

(c) Self-reversal due to absorption processes can cause apparent broadening.

(d) The Stark effect can arise in high applied fields or from local space charge concentrations.

It is this latter effect that is valuable from a diagnostic viewpoint, since the electron or ion microfields are dependent upon particle densities. Most practical studies have been based upon line broadening in the Balmer series (particularly H_α and H_β) whose half-widths have been tabulated (Griem, 1964) in terms of a normalized function $S(\alpha)$ such that

$$\int_{-\alpha}^{+\alpha} S(\alpha) \, d\alpha = 1 \qquad (36)$$

where α is a reduced wavelength so that

$$\alpha = \Delta\lambda/F_o \qquad (37)$$

F_o is here the ionic (Holtzmark) microfield which depends upon the ionization density:

$$F_o = 3.735 \; 10^{-7} \; n_-^{2/3} \; \text{V/cm} \qquad (38)$$

(n_- in cm^{-3}) .

The profiles $S(\alpha)$ are also slightly dependent upon electron temperature T_- but an approximate estimate is sufficient for electron densities of 10^{15} cm^{-3} or more to be determined. For this purpose the actual line profile may be needed, but a simpler procedure is to measure the relative intensity within a given bandwidth $\pm\Delta\lambda$ of the line with respect to the total line intensity (which ratio can then be compared with those for the theoretical profiles).

Practical Studies

Long Sparks. Time-resolved spectra from a 5 m gap (Les Renardières Group, 1977) were obtained simultaneously with photomultiplier signals, streak and framing image-converter photographs, and electrical measurements (Fig. 18). The emission from a 3 mm section of the channel (0.15m from the anode) was analyzed by a

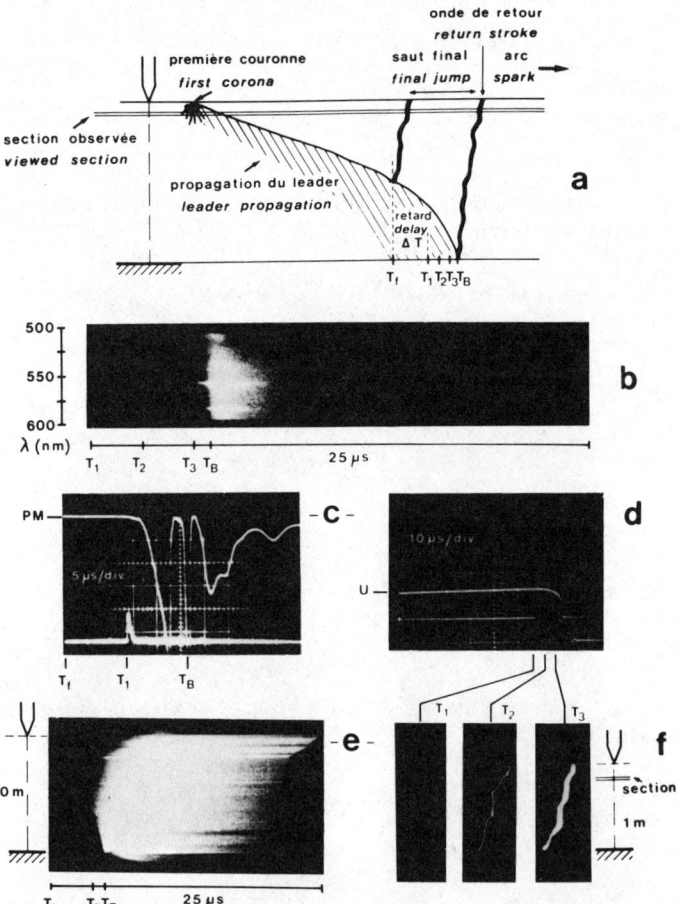

Fig. 18. Diagnostic records of long spark development (Les Renardières, 1977).
    (a)   Discharge structure
    (b)   Streak spectrum 500-600 nm
    (c)   Photomultiplier record
    (d)   Voltage
    (e)   Streak photograph
    (f)   Framing photographs

grating spectrograph in whose output plane was placed the photocathode of an image converter. The results from these studies are shown in Fig. 17.

Lightning. Spectra of single lightning discharges are successfully recorded by the slitless spectroscopic method (Salanave, 1965), which also provides sufficient line intensity for time-resolved studies. The time-resolved spectrum of Fig. 19 shows the stepped leader emission before the return stroke; the channel temperatures of Fig. 16 were obtained by this method. An alternative technique is to use a narrow bandpass filter or variable wavelength monochromator in conjunction with photomultiplier detectors. Both the lightning discharge (Krider, 1965) and the long spark (Gallimberti et al., 1977) have been studied by this method.

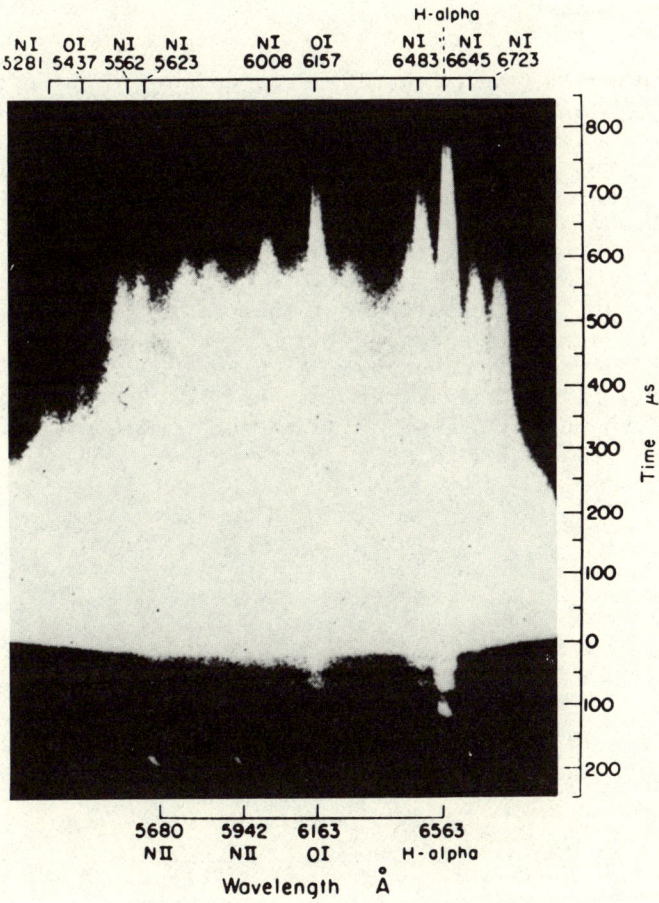

Fig. 19. Time-resolved slitless spectrum of lightning channel (Orville, 1968).

The intensity-ratio method was used to obtain the gas temperature estimates in the lightning return-stroke (Fig. 16, Orville, 1968). The Stark broadening of H_α has also been measured for these discharges to obtain electron density data (Fig. 20, Orville).

<u>Radial Variations in Discharge Channels</u>. The importance of radiation energy transport processes in arc channel structure has led to studies of the radially varying spectral components of high current discharges. It may be noted that for a symmetrical cylindrical channel of radius r_0, the real intensity $I(r)$ at radius r is related to the recorded intensity at the off-axis displacement $x = r$ by (Meek and Craggs, 1978)

$$I(x) = \int_x^{r_0} \frac{I(r) \, r \, dr}{\sqrt{r^2 - x^2}} \, , \tag{39}$$

An Abel inversion allows the recorded emission to be radially unfolded to obtain $I(r)$.

In work on wall-stabilized arc discharges, the confined nature of the discharge is an advantage. Wiese et al. (1972) have verified the Stark broadening method for hydrogen plasmas (using the H_β line), finding n_- in the range $1.5 \, 10^{16} - 10^{17}$ cm^{-3} and T_- of $9 - 14 \, 10^3$ k.

In the case of narrow spark channels, the random positional variations of successive discharges raises experimental difficulty. An interesting technique has been used by Tholl et al. (1970) (Fig. 21). A small slit allows light from the spark channel F to be time resolved by the spectrograph M and the photomultipliers PM for two

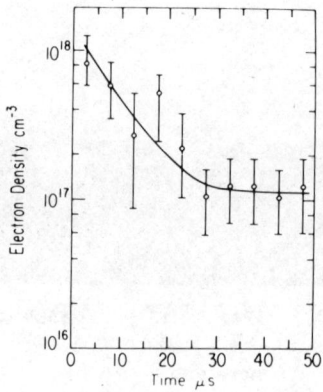

Fig. 20. Electron density in lightning return stroke obtained from $H\alpha$ line widths (Orville, 1968).

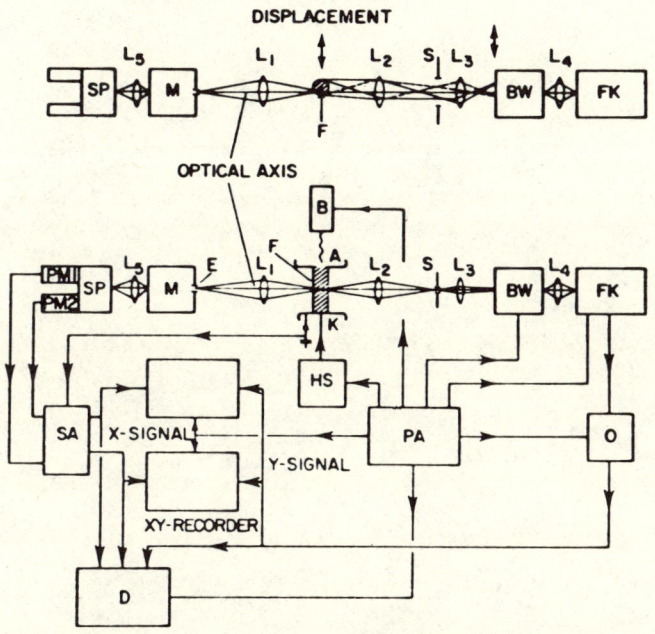

Fig. 21. System to obtain temporal and radial variation of relative line intensity in spark channels (Tholl et al., 1970).

wavelengths. Simultaneously a horizontal slit, combined with an image converter BW and an image orthicon FK, enables a streak record of the position and expansion process of the spark channel to be obtained. Combination of the photomultiplier signals with the scanning of the image orthicon by a discrimination technique then gives the time variation of a given wavelength at a known radial position. A large number of discharges are used to accumulate such data over the full diameter of the spark channel, the discharge system being successively laterally displaced with respect to the diagnostic apparatus. Figure 22 shows a typical time sequence of I(x) variations at 6000 Å, the I(r) variations then being obtained by solution of the integral equation. With such data Tholl succeeded in determining the thermalization time for each point in the channel (steep intensity rise, dissociation and the appearance of Balmer series, channel expansion due to gaseous heating), and to relate this time to electron density and gas temperatures obtained from Stark widths and relative line intensities.

Glow-to-Arc Transition. The development of the streamer and spark stages of 1 cm positive point-plane discharges in O_2 to which traces of H_2 were added have been studied by Bastien and Marode (1979). Extremely narrow time (25ns) and wavelength (0.17Å) segments were used in obtaining H_α and H_β profiles. The low luminous

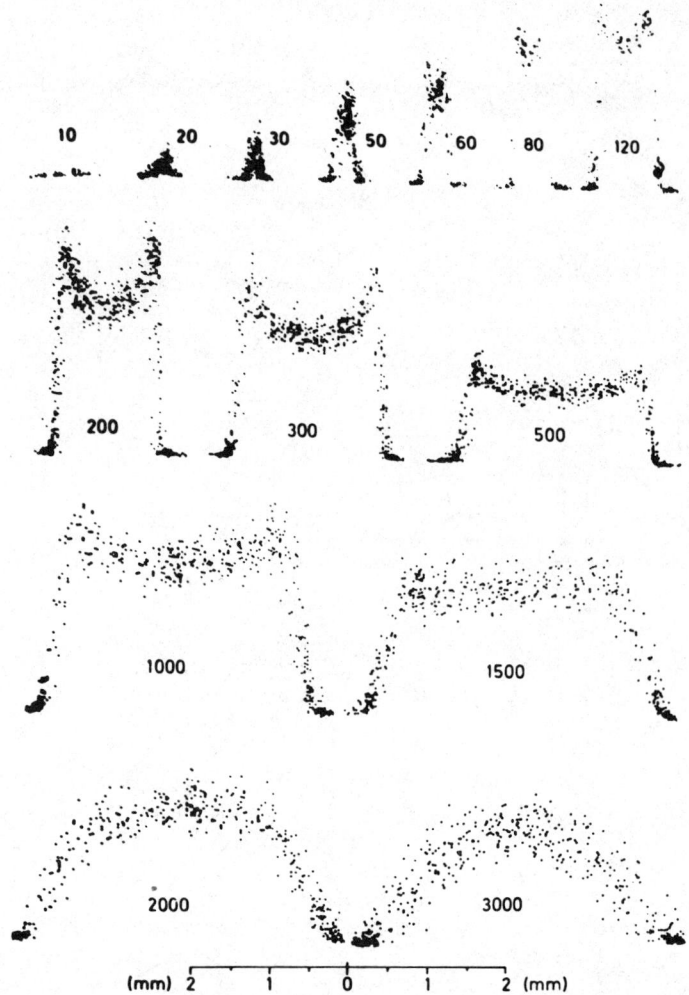

Fig. 22. Temporal and radial intensity variations in hydrogen spark channel $\lambda = 6000$ Å; Time in ns; p = 460 torr (Tholl et al., 1970).

flux in such narrow segments involved single-photo counting, multichannel analysis and long integrating times (15 min/segment) even at the high repetition frequency (10 kHz) of the discharge. Figure 23 shows line profiles at various times during discharge development. The authors emphasize that in weak plasmas such as channels, the channels, the macroscopic applied field F_c may be several times greater than the microfield F_o, and may be the greater cause of Stark broadening. Furthermore, such broadening would be noniostropic. It was shown that for F_c/F_o in the range 1 - 4 the

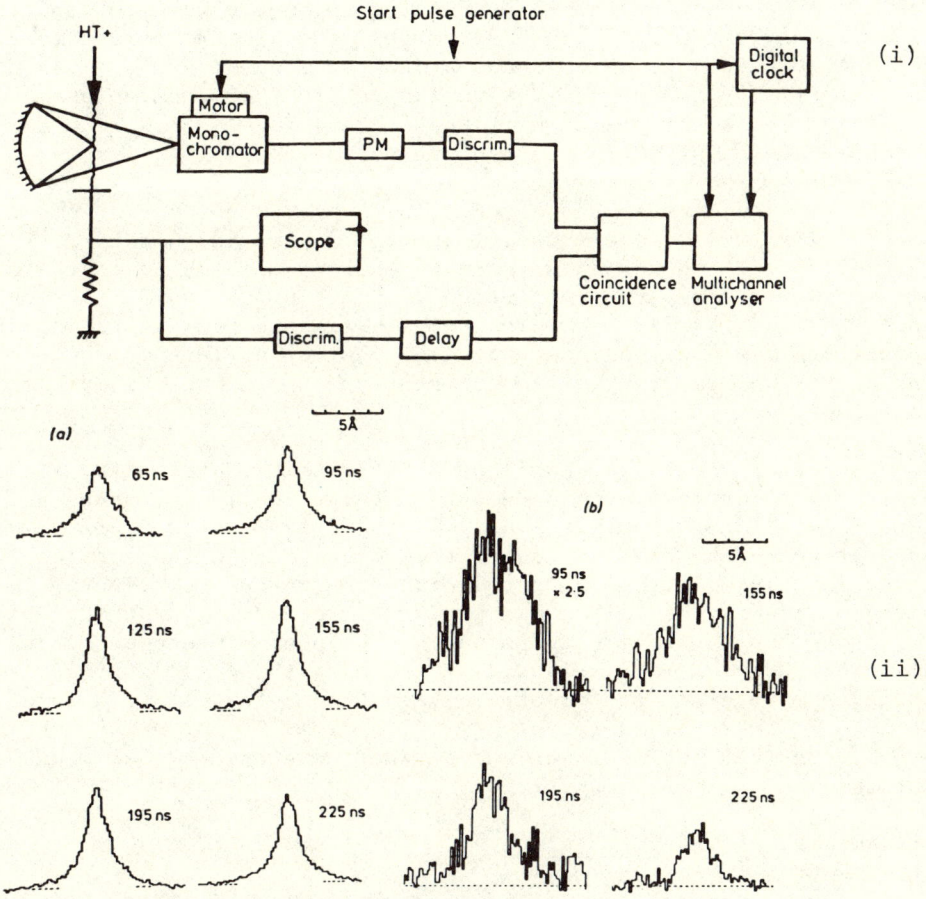

Fig. 23. Temporally resolved Stark broadening measurements in O_2 with added H_2 (Bastien and Marode, 1979).
(i) Test arrangement; (ii) (a) $H\alpha$
(b) $H\beta$ } p = 270 torr

profile depends on the product $n_- F_c$ rather than on n_- alone, and thus on the current density $j_- = e\mu_- n_- F_c$. Measurements for the primary streamer gave typical values of radius of 20 µm, F_c = 10kV/cm, n_- = 3 10^{15} cm^{-3} (thus F_o = 7.78 kV/cm).

CURRENT MEASUREMENT AT A BIASED PROBE

Introduction

An isolated probe inserted into a neutral plasma $n_- = n_+$ will take up a floating potential V_w (analogous to the wall potential of a discharge tube), which ensures that the electron diffusion current to the probe is balanced by the positive-ion drift current. The potential V_w is negative with respect to the actual plasma potential V_s at the position of the probe. $V_c = V_s - V_w$ can be regarded as the plasma/probe contact potential.

Whether a measurement of the floating or zero current probe potential V_w will provide useful information of the discharge parameters will depend upon the nature of the plasma region. In high-pressure discharges the reduced effect of diffusion is such that $V_w \to V_s$. Again, near electrodes or in the passive region of nonuniform field discharges where the charge carriers are predominantly unipolar, V_w will represent the space potential in a reasonably precise way.

In other circumstances in order to obtain useful data, it is necessary to arrange a measurement of the variation of probe current I_p with probe potential V_p. We shall examine two forms of such probes. In the first case, the collecting surface of the probe is aligned parallel to the direction of the net electric field. This is the well-known Langmuir probe configuration, where the probe current I_p is determined by the plasma/probe potential difference $V = V_s - V_p$. In the second case the probe surface is normal to the discharge electric field E_o, so that I_p is a function of $E_o - E_p$ where E_p is the probe field. This is the probe configuration described by Tassicker and further developed by Selim and Waters.

Langmuir Probe Characteristics

A probe whose potential V_p is strongly negative with respect to the plasma potential V_s will be screened from the plasma by a positive-ion sheath; between the sheath and the probe is a dark space of thickness d in which there is negligible emission by recombination or electron impact, since the probe repulsion field excludes even the fastest electrons from this region. In the steady state the positive-ion current density J_+ reaching the probe is equal to $(J_+)_s$, the random positive-ion current at the sheath. Thus the probe current density is

$$J_p = J_+ = (J_+)_s = \frac{n_+ e\, c_+}{4} \qquad (40)$$

where c_+ is the mean random positive ion velocity. Now the positive ion population is generally of a Maxwellian distribution with a temperature T_+ (of the order of 2 to 3 times the neutral gas temperature T) given by

$$T_+ = \frac{m_+ C_+^2}{3k} \tag{41}$$

(m_+, C_+ are the ionic mass and r.m.s. velocity and k is Boltzmann's constant). But $C_+ = (\sqrt{6\pi}/4)c_+$, which with Eqs. (40) and (41) gives for the probe current density

$$J_+ = n_+ e \sqrt{\frac{kT_+}{2 m_+}} \,. \tag{42}$$

This current density thus depends on the plasma conditions only, and is independent of the probe potential V_p. But since the condition current is controlled by the positive space charge in the sheath, it is also governed by the Langmuir-Childs law which for a plane probe is

$$J_+ = 5.4 \; 10^{-8} \frac{V^{3/2}}{m_+^{\frac{1}{2}} d^2} \; A/cm^2 . \tag{43}$$

The constancy of $J_p = J_+$ with varying V is maintained by a corresponding variation of dark-space thickness d. Under some conditions, simultaneous measurement of J_+ and d can yield values of V and hence of the plasma potential $V_s = V + V_p$.

If now the probe potential V_p is reduced to less negative a value, a proportion of the random electron current density at the sheath, given in analogy to Eq. (42) by

$$(J_-)_s = n_- e \sqrt{\frac{kT_-}{2 m_-}} \tag{44}$$

will succeed in reaching the probe surface. The proportion so succeeding will depend upon the electron energy distribution. We may note here that a Maxwellian distribution is certainly probable (a) at low E/n, in which case the electron mean free path and collisional energy loss are independent of energy and (b) at high electron densities (say 10^{12} cm^{-3}) where short-range electron-electron interaction is then possible. Many probe investigators suggest that the Maxwellian distribution is often valid even for n_- of about 10^9 cm^{-3}. In this case the electron current density J_- at the probe is governed by Eq. (44) and Boltzmann statistics so that

$$J_- = n_- e \sqrt{\frac{kT_-}{2\pi m_-}} \; \exp \; \left(-\frac{eV}{kT_-}\right) \,. \tag{45}$$

We note that as V_p becomes less negative (decreasing V) the electron current density J_- eventually considerably exceeds the positive ion current density J_+. This arises not only because $m_- \ll m_+$ but also because the electron temperature $T_- \gg T_+$ (the Townsend energy factor $\eta = T_-/T$ for molecular gases is about 10 for E/n=1 Td and about 100 for E/n=50 Td and is even greater for monatomic gases).

The measured probe current density is the net value of

$$J_p = J_+ + J_-. \tag{46}$$

J_+ will be little changed from the strongly negative V_p values of Eqs. (42) and (43) enabling J_- to be obtained from Eq. (46) and plotted against V_p in a logarithmic form of equation (45):

$$\ln J_- = \ln\left(n_- e\sqrt{\frac{kT_-}{2\pi m_-}}\right) - \frac{e(V_s - V_p)}{kT_-}. \tag{47}$$

Measurements indicating a linear $\ln J_-$ variation with V_p are evidence of a Maxwellian electron energy distribution and enable T_- to be determined. For an ideal probe function this linearity will be maintained through the $V_p = V_w$ value ($I_p = 0$) until a probe potential $V_p = V_s$ (i.e., V = 0) is attained.

We may note that at $V_p = V_w$ the equality of J_+ and J_- leads from equations (42) and (45) to an expression for the contact potential of the probe

$$V_c = V_s - V_w = \frac{kT_-}{2e} \ln \frac{m_+ T_-}{m_- T_+}. \tag{48}$$

When $V_p = V_s$ all of the random electron current density $(J_-)_s$ at the sheath is collected, and the probe current density due to electrons takes up the constant value of Eq. (44). As the probe potential V_p becomes positive with respect to V_s, the positive ion current density J_+ is reduced to zero and the probe current density $J_p \to (J_-)_s$.

The idealized current density/voltage characteristic for a plane probe is shown in Fig. 24. The electron temperature T_- obtained from the slope enables the r.m.s. and mean electron velocities C_- and c_- to be obtained from

$$C_- = \sqrt{\frac{6\pi}{4}} \quad c_- = \sqrt{\frac{3kT_-}{m_-}}. \tag{49}$$

DIAGNOSTIC TECHNIQUES FOR DISCHARGES

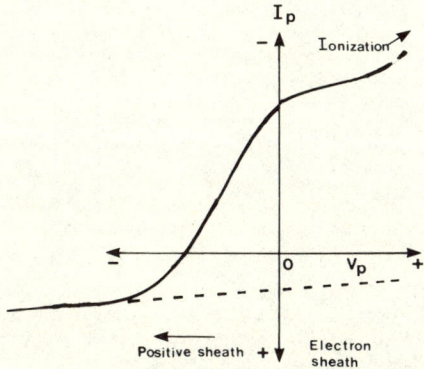

Fig. 24. Idealized Langmuir probe characteristic.

Identification of the probe potential at which the electron current density becomes constant enables the plasma potential V_s (= V_p) to be determined. The electron density n_- is found from Eq. (44). Finally, from the neutral plasma condition $n_+ = n_-$, the parameters T_+, C_+ and c_+ for the positive ions are obtained from Eqs. (41) and (42).

In ideal conditions the Langmuir biased probe offers a powerful technique for plasma studies. In practice, departures from the characteristic of Fig. 24 can arise from many causes. For example, the change in slope at $V_p = V_s$ can be obscured by such effects as secondary electron emission from the probe, electron reflection at its surface, or by ionization within the dark space d. Best function of the probe occurs when d is small, which also makes for a minimized disturbance of the plasma by the probe and good electron collection over the probe area.

If the probe action is satisfactory, a linear ln J_-/V curve indicates a Maxwellian distribution of electron energy, which justifies the above approach. In non-Maxwellian conditions, the Druyvesteyn relationship

$$f(C_-) = \frac{4m_-}{e^2} \frac{V d^2 J_-}{dV^2} \tag{50}$$

may be used to obtain the distribution of r.m.s. velocities. The second derivative is best measured in practice by the Emmeleus method of superimposing upon the probe voltage V_p a sinusoidal voltage of small amplitude V_{AC}. This results in a change ΔJ_p in the probe d.c. signal, so that

$$\frac{d^2 J_-}{dV^2} = \frac{d^2 J_p}{dV^2} = \frac{4}{V_{AC}} \Delta J_p \ . \tag{51}$$

Small spherical or cylindrical biased probes are frequently used in the Langmuir mode. For the positive ion current density J_+ in the strongly negative V_p region, Eq. (43) is simply modified to take account of the ratio r_s/r_p of the sheath radius to the probe radius. When V_p is such that electron current is being collected by the probe, then in general the value of r_s/r_p may be so large, or the probe/plasma potential difference V may be so small, that some electrons may orbit the probe without being collected. Correction factors for these effects have been evaluated by von Engel and Steenbeck.

Recent Examples of Langmuir Probe Applications

Hollow-Cathode Discharges. Hollow-cathode systems have had a recent application as electron sources for the neutralization of ion beams from r.f. ion thrusters for space propulsion. Groh et al. (1978) used a movable Langmuir probe along the axis of such a discharge in Hg vapor. Despite the small dimensions of the electron source, typical Langmuir characteristics and data were obtained (Fig. 25).

High Pressure Arc Discharges. The arc column consists basically of a high temperature, low gas-density core into which most of the input power occurs, and a surrounding cooler higher density zone (or thermal layer) to which energy is transferred from the core mainly by convection and heat conduction and by radiation. Although much of the core energy is transported to the electrodes by axial convection, the radial transport to the outer zone has important applications in nozzle design for gas-blast circuit breakers. The plasma within the arc core may be expected to be in thermal equilibrium, and temperature measurements may be made spectrocopically; the local variations of input power can only be effectively measured by means of potential probes to obtain electric field values, which often lie in the range 10 - 50V/cm. There are real difficulties in interpretation of the probe current/voltage characteristic at atmospheric pressure both because of the mean free path of the charge carriers and because of the local thermal interference by the probe. Nevertheless, probe characteristics are often of a form related to the Langmuir behavior, and may at the least be used to obtain plasma potential gradients. Three interesting forms of the probe should be mentioned:

(a) *Wall-stabilized arc probes.* Disc electrodes can be employed as wall supports used in stabilizing the arc to a symmetrical and axially uniform column. If the potential difference ΔV_w between

DIAGNOSTIC TECHNIQUES FOR DISCHARGES

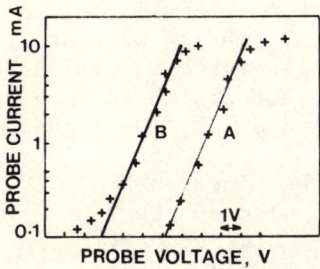

Fig. 25. Langmuir probe measurements in hollow-cathode Hg discharge (Groh et al., 1978).

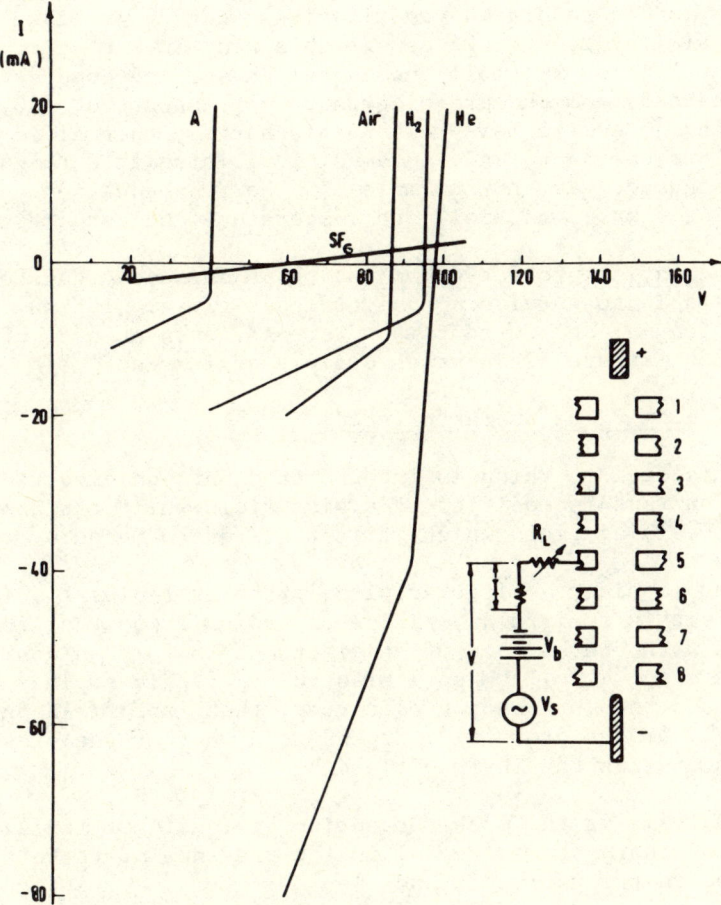

Fig. 26. Langmuir probe measurements in wall-stabilized arcs (George and Richards, 1968).

adjacent probes is measured, this may often give a reliable value of the potential gradient ΔV_s in the wall-stabilized arc since the contact potential V_c is unlikely to be very different for the two probes. The current/voltage characteristic for such probes (Kimblin and Edels, 1966; Barrault and Jones, 1974) shows ion current and electron current regions similar to the classical Langmuir probe (Fig. 26). For best precision in the measurement of ΔV_s a pair of biased probes operated in the steeper electron-current region of Fig. 26 is likely to minimize errors. For arc-recovery probe studies a useful technique consists of the passage of a low test current after interruption of the main arc (Hill, 1973).

(b) <u>Swept-wire probes</u>. Provided that a small probe collecting low current is moved through the arc column at a rate insufficient to disturb the convection process (although transit times of more than a few ms can result in vaporization), current/voltage characteristics similar to Fig. 26 can be obtained (Gick et al., 1973). Now for a sufficiently small sheath thickness, the product n_+c_+ can be obtained from the probe current by means of Eq. (40). In the arc plasma, unlike low-pressure discharges, thermal ionization processes are dominant, and ion density n_+ is related to gas temperature (and thus c_+) as, for example, in the Saha equation. The probe characteristic may thus yield arc temperature and conductivity values.

(c) <u>Ablation-protected probes</u>. PTFE shrouded tungsten rod probes are able to survive for significant times, with an exposed tip for potential and current measurements, even in arcs of many kA currents. Figure 27 shows an example of its use.

Tassicker/Selim Probe Characteristics

In this device, which is incorporated into an electrode surface, the probe current is modified by a bias field which can oppose or aid the prevailing field which it is required to measure.

We may consider a circular plane probe of radius r_p, located centrally within a circular orifice of radius r_a so that the probe lies flush with the surface of an electrode to form an annular gap of width $g = (r_a - r_p)$. When a bias voltage V_b is applied between the probe and the surrounding electrode, then the field pattern in the vicinity of the probe will depend upon whether the bias field E_b opposes or aids the surface field E_o.

An analysis of the probe operation by Tassicker (1974) gave for the ratio of the probe current I in the presence of V_b to the current I_o measured in the unbiased condition

$$\frac{I}{I_o} = \frac{\psi}{\psi_o} = 1 + \frac{C_o}{\pi r_m \varepsilon_o} \cdot \frac{V_b}{r_m E_o} \tag{52}$$

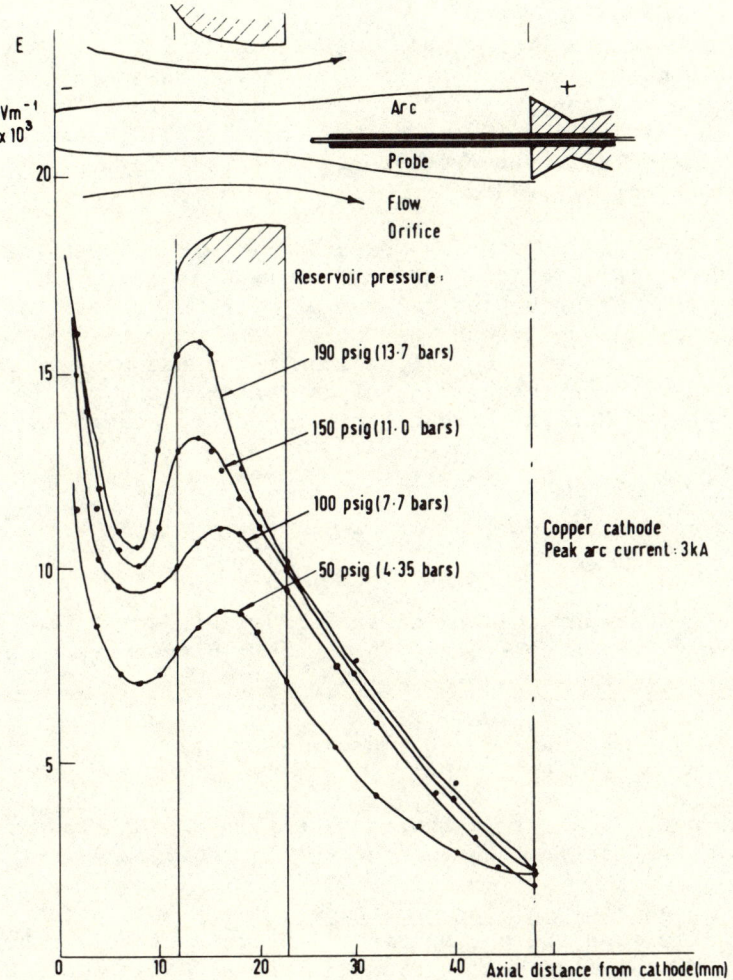

Fig. 27. Ablation probe measurements in high-pressure arcs.

where r_m is the effective radius of the probe, and C_o is the capacitance between the upper surfaces of probe and electrode. The unknown field can then be found if C_o is calculated.

Tassicker used an analytical expression derived by Spence (1970):

$$C_o = 4 r_p \varepsilon_o [1.07944 + 0.5 \ln (1 + \frac{r_p}{2g})]. \tag{53}$$

It was recognized that any reversal of the surface field at high bias voltage when $E_b > E_o$ would cause a deviation of the I/I_o characteristic from the linear dependence upon V_b and so invalidate the relationship. A more precise analysis of the probe function has since been carried out (Selim and Waters, 1980a) by computation of the bias field E_b at the probe surface, particularly in the important region near its edge.

(a) <u>Bias Field Opposing Measured Field</u>. Suppose that the net field $E_b + E_o$ is zero at a radius $r_1 < r_p$. At any plane boundary parallel with the probe and at several probe diameters distant, we can define a radius r_2 such that only the flux within r_2 associated with the measured field E_o will terminate on the probe. Any flux terminating upon the probe can only arise from the applied field E_o, and it is this flux which transports the charge carriers.

The total flux terminating on the probe is given by:

$$\psi_1 = 2\pi\varepsilon_o \int_0^{r_1} (E_o - E_b) \, r \, dr = \varepsilon_o E_o \pi r_2^2 \qquad (54)$$

since charge carrier collection is from within the radius r_2. The bias current I is thus related to the zero-bias current by:

$$\frac{I}{I_o} = \frac{r_2^2}{r_m^2} = \frac{r_1^2}{r_m^2} - \frac{2}{E_o r_m^2} \int_0^{r_1} E_b \, r \, dr, \qquad (55)$$

It can be noted that if we approximate $r_1 = r_m$ then this equation reduces to the earlier expression of I/I_o (Eq. (52)).

(b) <u>Bias Field Aiding Measured Field</u>. In this case the field reversal occurs on the surrounding electrode surface rather than on the probe surface, at some radius $r_3 > r_p$. We can again define a radius r_2 representing the boundary limit of the measured-field electric flux which terminates upon the probe.

This situation is best treated by considering the flux ψ_2 terminating upon the surrounding electrode, which is given by

$$\psi_2 = 2\pi\varepsilon_o \int_{r_3}^R (E_o - E_b) \, r \, dr = \pi\varepsilon_o E_o (R^2 - r_2^2) \qquad (56)$$

where R is the radius of the surrounding electrode. Thus the probe current ratio is given by:

$$\frac{I}{I_o} = \frac{r_2^2}{r_m^2} = \frac{r_3^2}{r_m^2} + \frac{2}{E_o r_m^2} \int_{r_3}^R E_b \, r \, dr \, . \qquad (57)$$

DIAGNOSTIC TECHNIQUES FOR DISCHARGES   241

For an approximation $r_3 = r_m$, and recognizing that the charge q_b on both the probe and the surrounding electrode are then of equal magnitude, this again reduces to the earlier form.

The computed probe characteristics can be presented in normalized form, i.e., I/I_o as a function of normalized bias voltage $V_b/E_o r_m$. In this case (Fig. 28) the nonlinear characteristic predicted by the present theory is seen to differ from the linear characteristic of the simplified theory. A more convenient form is shown in Fig. 29, in which the I/I_o characteristic is determined with E_o as the independent variable and V_b as the parameter. Correct functioning of the probe is then easily validated, since experimental points obtained for a given stable electrostatic situation but with varying V_b should lie upon a vertical line of constant E_o. This shows that the probe is a self-verifying device.

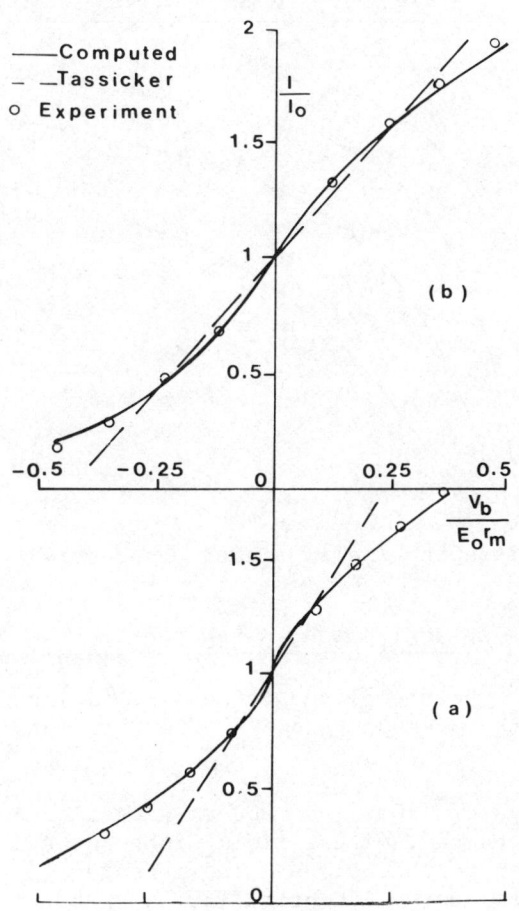

Fig. 28. Tassicker/Selim biased probe characteristic.

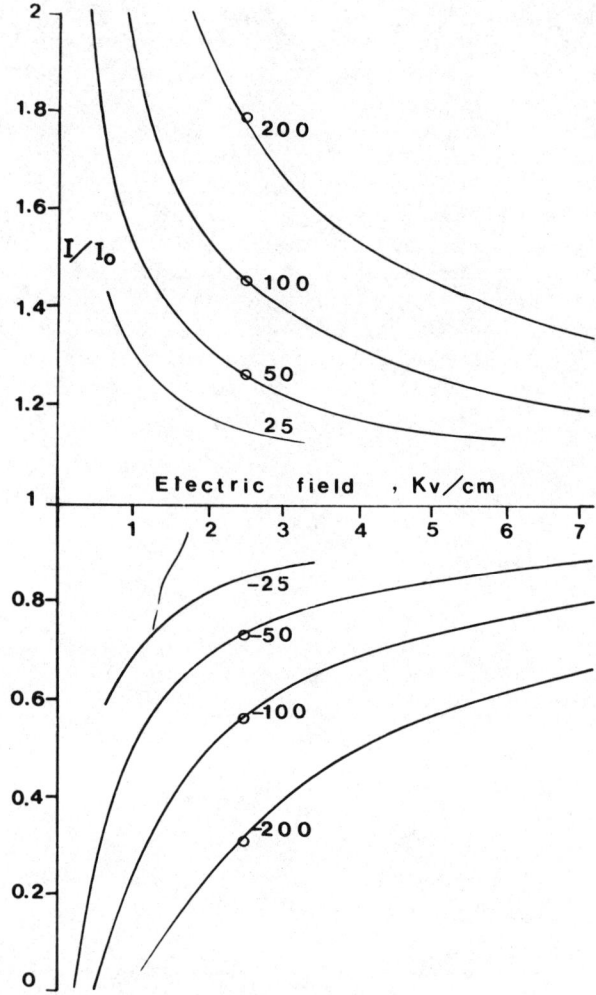

Fig. 29. Tassicker/Selim biased probe-working curves.

Application of Current/Voltage Probe to D.C. Corona Tests

This probe has been employed to measure fields in rod/plane corona in air in the pressure range 20-760 torr (Selim and Waters, 1980b).

Rod/plane gaps of 120 mm and 200 mm were incorporated into a large ionization chamber, with a field probe located centrally in the 900 mm diameter earthed plane. The measured I/I_o data from the probe confirmed its correct function (as shown by the experimental

DIAGNOSTIC TECHNIQUES FOR DISCHARGES

points in Figs. 28 and 29). Field strength E and average corona current I_c at the plane were measured for negative coronas at voltages of up to 100 kV. As shown in Fig. 30, the classical Thomson relationship $E \propto \sqrt{I_c}$ for coaxial systems is also valid for rod/plane coronas over the full pressure range for both gap lengths. A further useful generalization is that for the gap lengths d used, we have $E \propto \sqrt{I_c}/d$. Space-charge field enhancement causes E to become equal to the mean applied field V/d, and even to exceed that value for large I_c.

Two inferences may be drawn from the curves of Fig. 30:

(1) Classical theory predicts that space charge field at a given current depends upon carrier mobility μ so that $E \propto \sqrt{1/\mu}$; combining this with the experimental data gives $\mu \propto \sqrt{I_c/E^2 d}$. The per-unit change $\mu(p)/\mu(760)$ of carrier mobility with pressure p may be obtained for various E/p from Fig. 30. It is found that $\mu(p)$ is independent of E/p in the range 4-8 Td. A fivefold increase of $\mu(p)$ occurs as the pressure is reduced from 760 torr to 150 torr, which implies a normal inverse pressure dependence. At p < 150 torr, a faster increase of mobility which is proportional to $(1/p)^{1.75}$ is observed. Thus there is a tenfold mobility increase between 150 torr and 40 torr, and the mean carrier mobility is larger than for small ions. This indicates a mixed electron/negative-ion carrier population as a result of the free-electron lifetime becoming significant at low pressures.

(2) The probe data are represented by $E^2 \propto I_c/d$ and $E \simeq V/d$. Combination of these two relationships gives $I_c \propto V^2/d$, which is the generalized form for rod/plane gaps of the classical square-law expression.

INDUCTION PROBES FOR FIELD MEASUREMENT

Nondiscriminatory Capacitive Probes

A conducting probe exposed to an electric field will have a resulting charge separation induced over its surface of

$$q_i = \varepsilon_o \iint_S E_s \, dS \tag{58}$$

where E_s is the field strength into the surface element dS of the probe. The mean strength E over a surface area S is then obtained from $E = q_i/\varepsilon_o S$ or, if the time variation of the field is known, from the induced current to the probe.

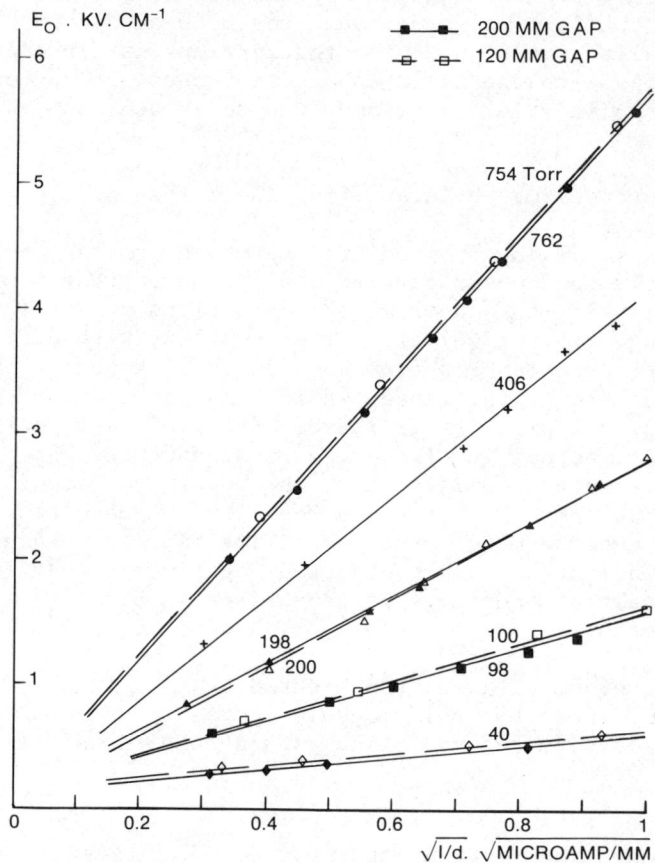

Fig. 30. Field at plane electrode for negative rod/plane coronas in air at various pressures.

However, the presence of conduction current arising from spatial ionization at the surface of the probe will result in an incorrect estimate of the induced charge q_i and thus of the field E. This is a major limitation in discharge studies, and for this reason simple capacitive probes may be described as nondiscriminatory. They therefore have to be located in a low-field region of the active electrode, and in consequence of this screening, can provide only a relatively insensitive measurement of field changes, particularly at large electrodes.

Furthermore, following the cessation of ionization activity in the gap, further field changes occur due to space-charge dissipation which can provide information on space-charge distributions. The ionic conduction current again prevents the use of capacitive probes for this phase of the discharge.

It was to augment the measurements available with the capacitive probe and to provide alternative techniques that discrimatory induction probes were developed as described below.

Discriminatory Electrostatic-Induction Probes

Field-Filter Probes: Principles and Design. The field penetration which occurs through orifices (circular or longitudinal) in the electrode surface is utilized, such field lines then terminating upon a sensor conductor within the electrode. The signal developed at the sensor is proportional to the electrode surface field; the maximum field at the sensor surface is also proportional to the external field but is attenuated by the geometry of the device.

When a reverse field is applied between the sensor and the external electrode which is higher than the maximum attenuated field at the sensor surface, conduction current is prevented from reaching the sensor. This reverse field does not affect the charge induction process, so that the effective action of the device is to filter the field changes (i.e., displacement current) from the conduction current components.

Field filters have been designed (Waters and Selim, 1979a; Stark et al., 1980) which have as small a disturbing effect as possible on the measured field, and optimize the opposing requirements of high sensitivity at a moderate applied bias voltage.

Circular orifices have the advantage that their axial symmetry simplifies digital computation of the electric field, and are easy to form mechanically with good precision.

The Single-Layer Circular-Orifice Field Filter. Typical profiles of the attenuated field E_a and bias field E_b for such a system are shown in Fig. 31. If the minimum value E_{bo} of the bias field is greater than the maximum value E_{ao} of the attenuated field, then no current conduction can take place to the sensor surface, and the signal registered by the sensor will be

$$V_s = \frac{2\pi\varepsilon_o}{C} \int E_{ar} \cdot r \, dr \, , \tag{59}$$

which is a purely induced-charge signal.

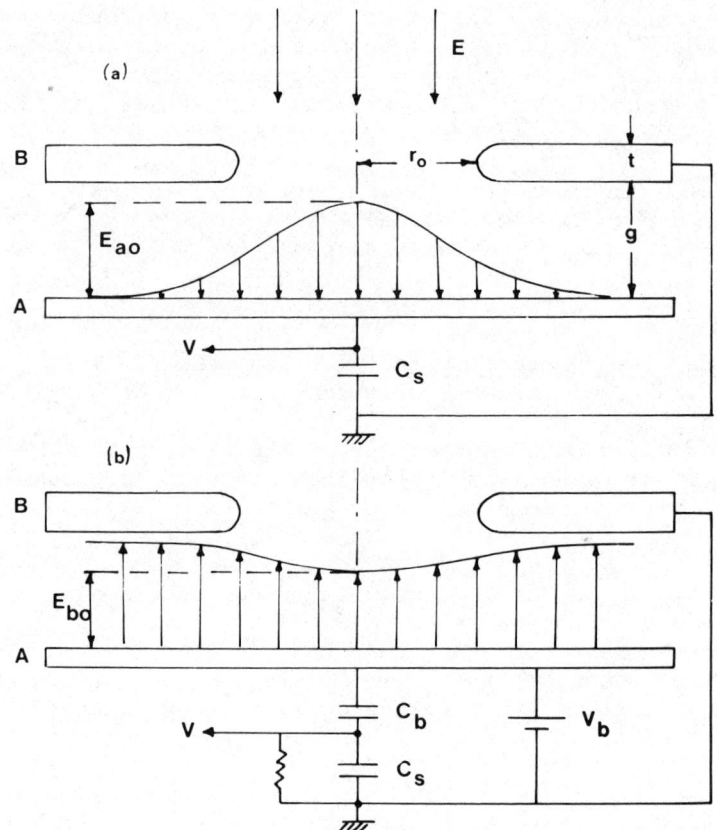

Fig. 31. Field-filter profiles of attenuated and bias fields.

The probe sensitivity is determined by the attenuated field integral and the total signal-circuit capacitance. Because the total sensor capacitance is restricted by the time constant requirements, the attenuated field integral is the main factor upon which sensitivity is dependent.

Parameters affecting sensitivity are the orifice radius r_o, the gap g between the sensor surface and the inner surface of the outer electrode, and the thickness t of the outer electrode. Figure 32 shows the curves of sensitivity and cut-off bias voltage versus r_o/g for single orifices of 3 mm and 2 mm diameter respectively in a wall of thickness 1 mm. The orifice edge was defined to be of semicircular section. These data were obtained from several computations of both attenuated field and bias field distributions using a charge-simulation field computation.

DIAGNOSTIC TECHNIQUES FOR DISCHARGES

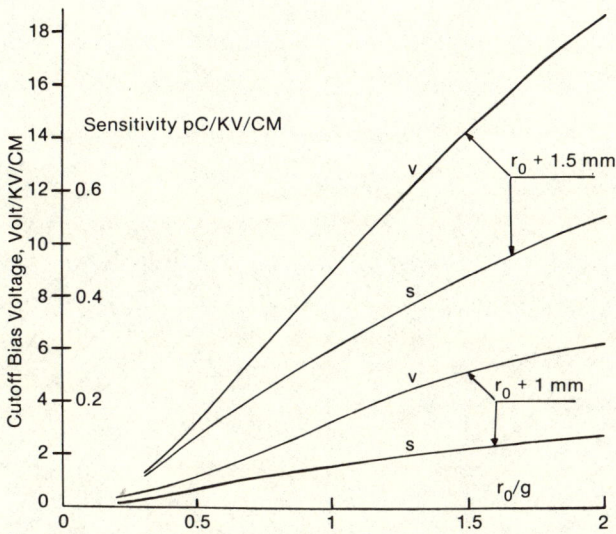

Fig. 32. Generalized characteristics for single-layer field filters.

The computed field distributions showed that for multiple orifice field filters, there will be a negligible effect of proximity between orifices if the orifice centers are spaced not less than two orifice diameters. Figure 33 shows a typical multiple-orifice design for a 75 mm diameter spherical electrode.

Double-Layer Circular-Orifice Field Filters. The single-layer field filter provides a satisfactory measurement for unidirectional fields. However, field reversal at the filter surface may occur in some applications such as a.c. coronas. It is then necessary to prevent ions of either polarity from reaching the sensor surface. The double-layer field filter system gives a bipolar-field measuring method, where between the electrode A containing the orifice, and the sensor B, there is an intermediate electrode C with a coaxial orifice. The biasing arrangement is such that the direction of the bias field between A and C is opposite to that between C and B.

Figure 34 shows an example for a design of a 14 orifice double-layer filter, having r_o = 1.5 mm, g = 1 mm and t = 0.4 mm.

Grid Field Filters. The basic construction of the grid field filter is shown in Fig. 35. A set C of parallel wires of separation $\underline{s}$ and radius $\underline{a}$ form a grid in the plane of the electrode surface B.

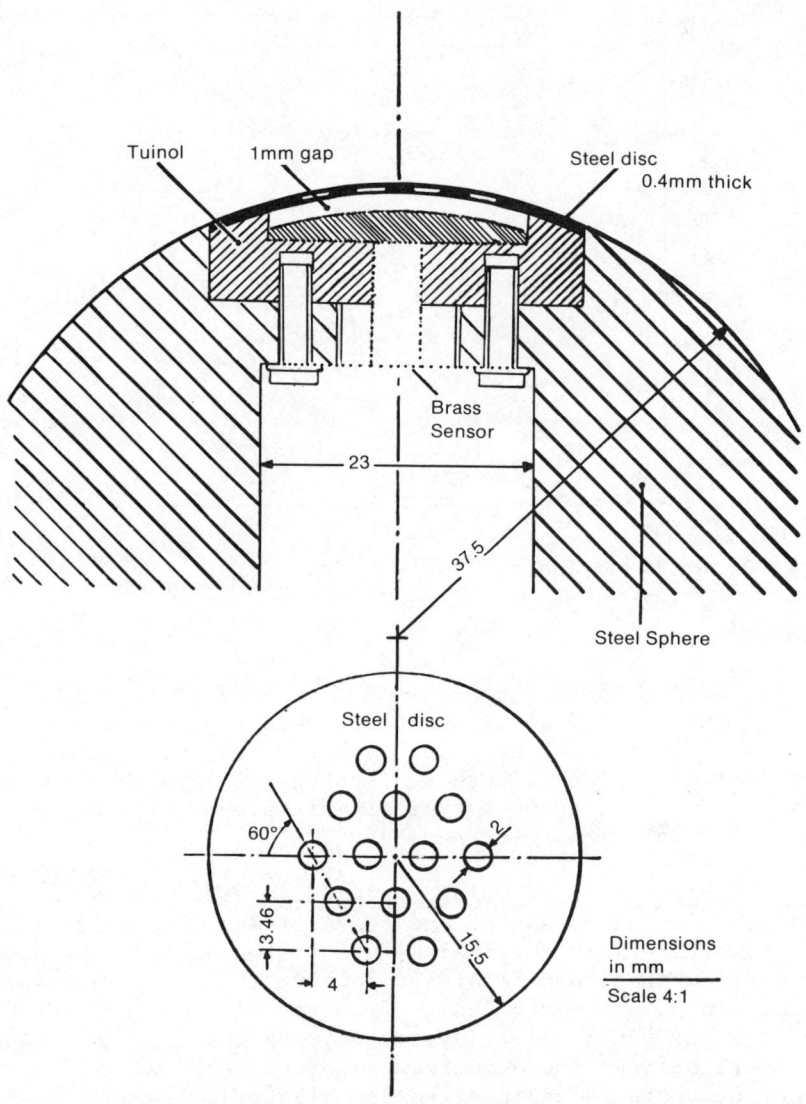

Fig. 33. Single-layer (multiple-orifice) field filter in 75 mm diameter sphere.

Alternate wires of the grid are connected to a symmetrical voltage supply $\pm V_b$. A circular sensor plate A with an outer guard ring C lies at a depth h below the grid C.

The sensitivity of this field filter is dependent upon the choice of the dimensions a, s, h and the sensor area S.

DIAGNOSTIC TECHNIQUES FOR DISCHARGES

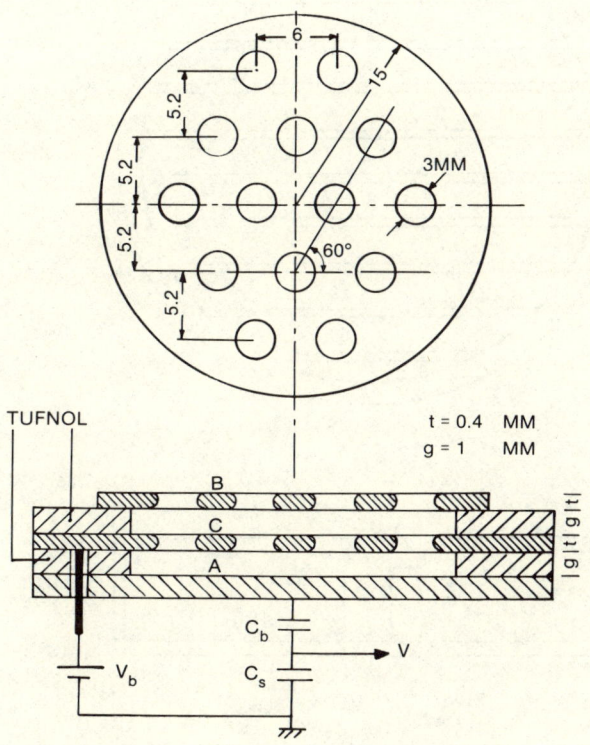

Fig. 34. Double-layer filter.

A typical construction had the following parameters: $a = 0.5$ mm; $s = 4$ mm; $h = 4$ mm; $S = 285$ mm^2; number of wires = 18; overall diameter = 76 mm.

Calibration procedures with a signal capacitor C_s of 4.00 nF and electrometer gain $g = 1.62$ gave a sensitivity of 14.8 MV m^{-1} V^{-1} of signal. The effective area of the sensor was therefore $S_e = C_s V/\varepsilon_o E g = 18.9$ mm^2, and the field penetration factor $F = S_e/S = 0.0663$. This gave the ratio of the grid filter sensitivity to that of a simple capacitive probe of similar area.

Applications of Field Filters

D.C. Discharges Studied with Field-Filter Probes. As a static

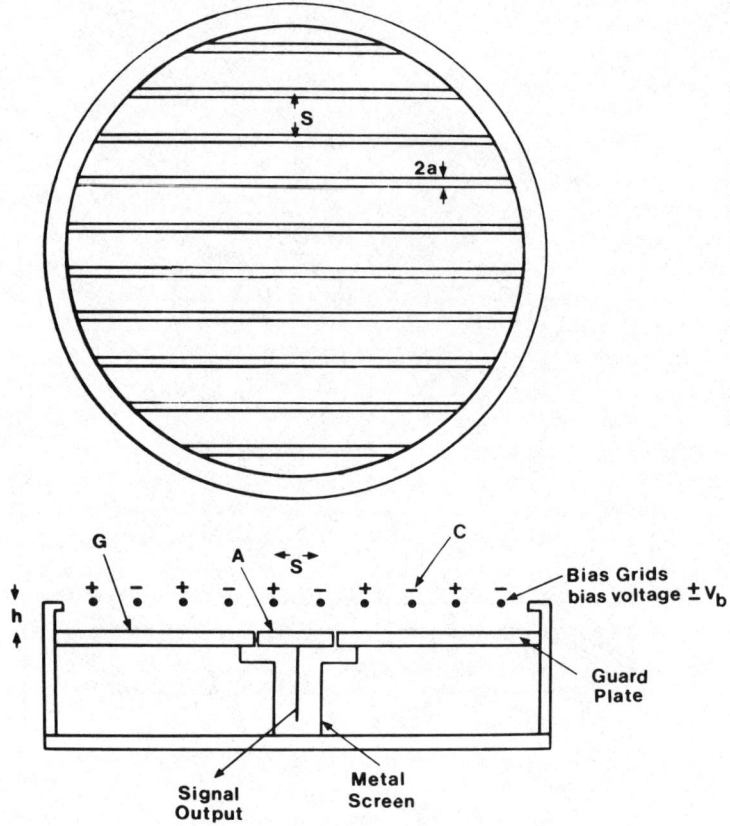

Fig. 35. Grid field filter.

induction device the field filter is best suited to the study of field-changes rather than steady-state values. A single-layer, single-orifice filter with $r_o = 1$ mm, $t = 0.5$ mm, and $g = 1.5$ mm was incorporated into the hemispherical tip of a 22.2 mm diameter rod cathode in a 83 mm gap (Waters and Selim, 1979b). A photomultiplier with telescope/slit optics recorded light output from the negative corona at the earthed rod.

Figure 36 shows field-changes and light output occurring in negative coronas. It is notable that both positive and negative field-changes can occur, with transition between these modes. This appears to indicate that while a preponderant negative space-charge will cause a periodic choking of the discharge as found in Trichel-pulse behavior, the positive ion space-charge can enhance the field close to the cathode.

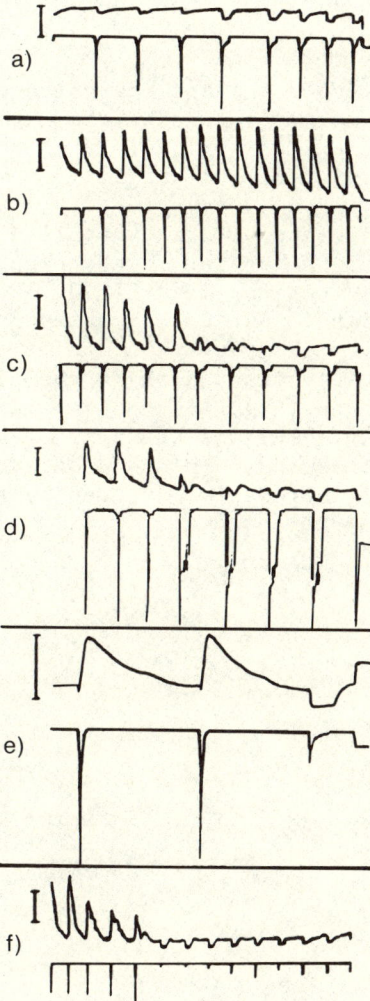

Fig. 36. Field-changes and photomultiplier signals in Trichel corona. Upper traces: field changes; Lower traces: photomultiplier; Full sweep time 800 µs (except (e), 200 µs). Bars indicate field changes of 100 kV cm^{-1}.

A.C. Discharges Studied with Field-Filter Probes. In order to examine the function of double-layer and grid field filters in corona discharges where charge carriers of both polarities were present, power-frequency (50 Hz) coronas in rod-plane gaps have been studied. A rod/plane gap of length up to 0.3 m was formed between a 6.35 mm-diameter hemispherically-tipped rod and a 0.9 m-diameter earthed plane. The field filters (Figs. 34 and 35) were incorporated into the earthed plane, and their location was variable in 50 mm radial steps. For comparison purposes, the field filters could be replaced by a rotating fluxmeter.

For a 0.2 m gap length and applied voltage 60 kV (rms), Fig. 37 shows the field signals recorded by the three devices, over one complete cycle of the applied voltage. This field variation was highly reproducible from cycle to cycle.

In order to examine the effect of applied voltage and gap length upon the total field at the center of the plane, tests were performed at voltages up to 72 kV rms for gaps of 50, 100, and 200 mm, using both field-filter designs. Agreement between the two filters was within 6% or better over all tests. At relatively low overvoltages above V_i, the electric field strength at the plane is below the mean field strength in the gap (Fig. 38) but becomes greater than the mean at higher voltages. This is the same behavior as has been found in direct-voltage coronas in rod-plane gaps and in coaxial-cylinder systems, using field measurements obtained by the biased static field probe.

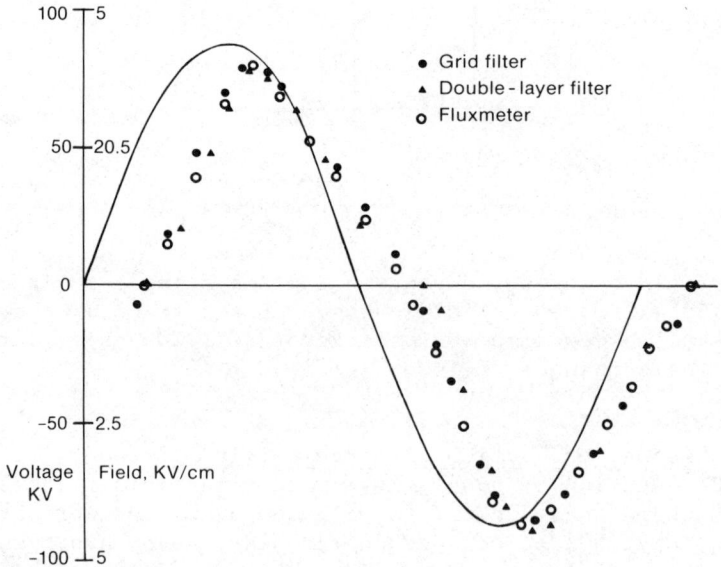

Fig. 37. Field at plane electrode of rod/plane gap in ac coronas. Gap length 200 mm; applied voltage 60 kV (rms).

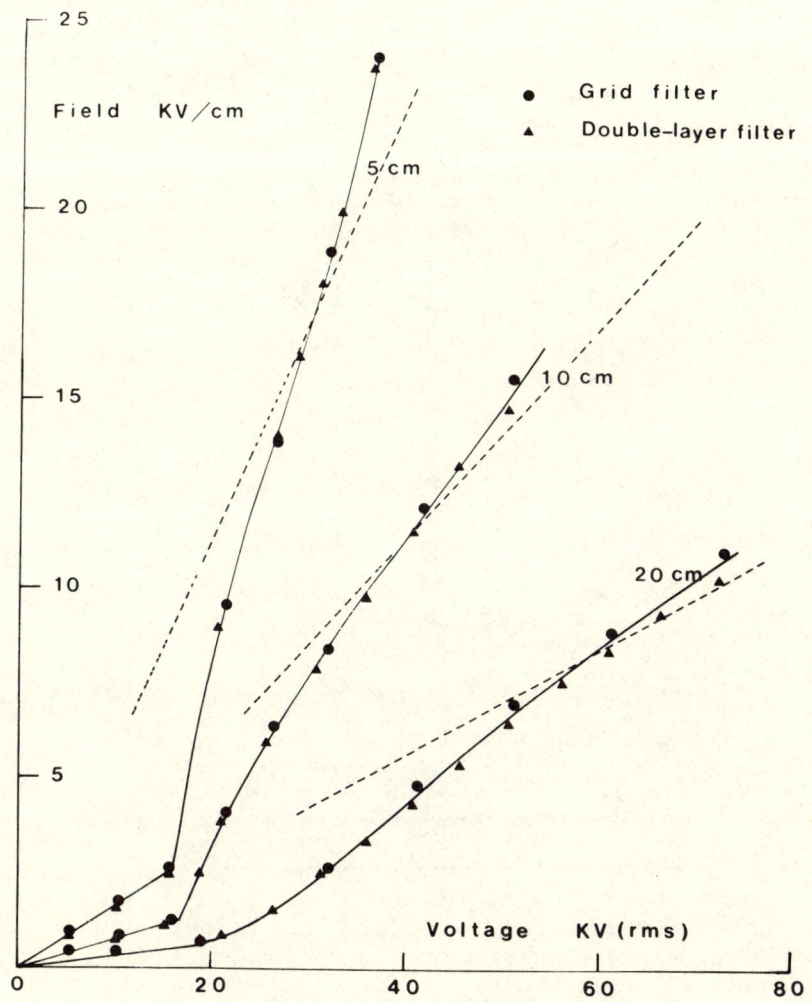

Fig. 38. AC corona field as function of applied voltage.

Impulse Discharges Studied with Field-Filter Probes. Field-strength measurements in impulse breakdown have been made under various conditions (Selim and Waters, 1981; Les Renardières Group, 1981; Pesavento, 1979; Baldo et al., 1979).

In short air gaps (0.15 - 0.45 m), positive and negative switching impulses were applied to a sphere-plane system (Selim and Waters, 1981). Electric fields at both the high-voltage sphere tip and the earthed plane were measured simultaneously, with charge flow to the sphere, with applied voltage, or light output from a defined spatial region in the gap, using the arrangement in Fig. 39.

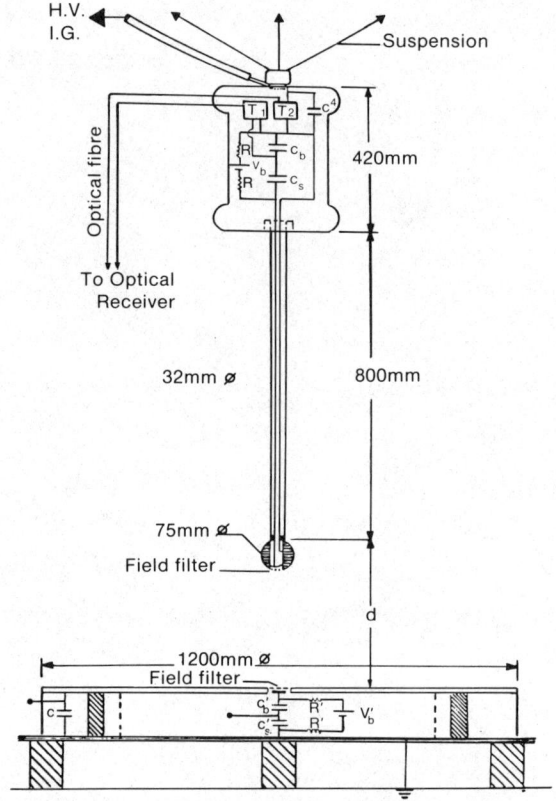

Fig. 39. Experimental arrangement for field measurement in impulse breakdown of sphere/plane gap. d up to 450 mm; T_1, T_2 optical transmitters.

Figure 40 shows a pair of simultaneous field oscillograms at the sphere (E) and plane (E_p).

In the study of negative impulse breakdown in gaps of up to 7 m (Les Renardières Group, 1981) the h.v. electrode of Fig. 41. was used. Simultaneous measurements of charge flow and spatial development were made.

DIAGNOSTIC TECHNIQUES FOR DISCHARGES

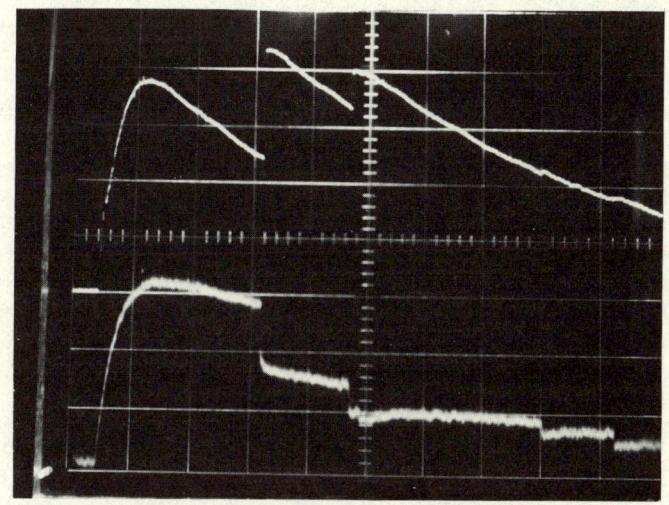

Fig. 40. Field changes at spheres and plane electrodes. Positive impulse crest 170 kV; 100 μs/div.; Upper trace — field at plane; Lower trace — field at 75 mm sphere.

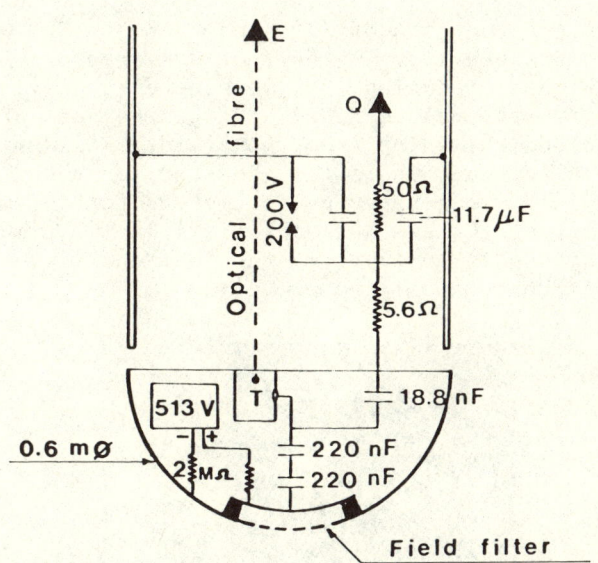

Fig. 41. Field and charge measurements at hemisphere tip of diameter 0.6 m; 7 m gap.

The field reductions at the h.v. electrode caused by the first corona sequence (Fig. 42 at 0.7 V_{50}) were much more variable than the corresponding charge flows, because of the variable surface location of the discharges.

At higher voltages ($V_{50} - \sigma$), the spatial progress of the leader is known to cause discontinuous restrike development of the negative discharge system. Major field changes were shown (Fig. 43) to occur at the electrode tip, mainly synchronized with the restrike processes. Field reversal was sometimes indicated.

Pesavento (1979) has described tests performed with positive nonstandard lightning impulses in a 1 m rod/plane gap. Again field-filter records obtained simultaneously with image-converter photographs showed the effect of space-charge field both before and after arrival of the corona streamers at the plane. Figure 44 shows the field variation observed with an oscillatory impulse. It was also demonstrated that when the leader/leader-corona system bridges the gap, the field in the leader-corona system is nonuniformly distributed; this follows from the observation that the field strength E_p at the plane is greater than V/d-L) where V and L are the instantaneous values of applied voltage and leader length in a gap of length d.

Field-Mills and Fluxmeters

<u>Discriminatory Function of Rotating Probes</u>. The well-known field-mill consists usually of a pair of sectored discs on a common axis, one of which is rotating; the sensing disc provides an alternating signal proportional to field strength because of its successive exposure to and shielding from the field. Field-mills have been in continual use for many years for the study of thunderstorm phenomena (Anderson, 1977) and in generating voltmeters for the measurement to high direct voltages (Craggs and Meek, 1954). Their main advantage in such applications is to overcome the inherent time-constant limitations of simple capacitive probes.

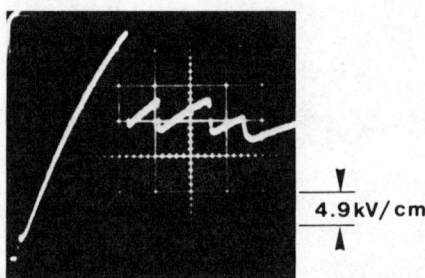

Fig. 42. Field-changes at hemisphere tip at 0.7 V_{50} due to negative-corona sequence (Les Renardières, 1981).

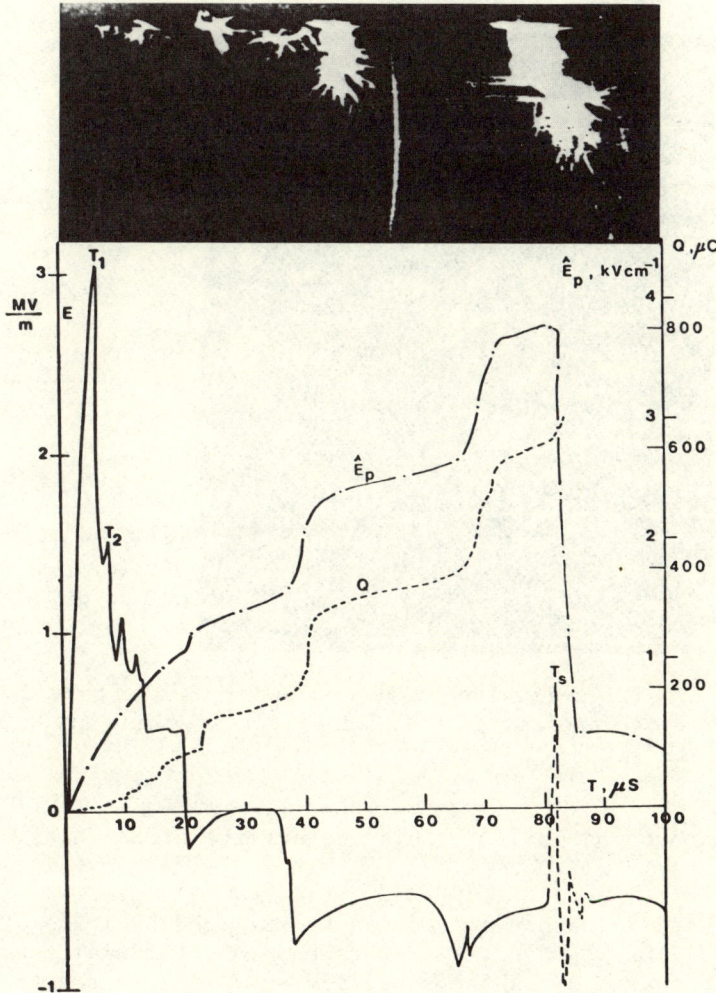

Fig. 43. Simultaneous E, E_p, and Q measurements with image converter record of negative impulse breakdown. 7 m gap; T_1: first corona; T_2: second corona; T_s: sparkover; (Les Renardières, 1981).

The more recent application of rotating probes to the measurement of electric fields associated with discharge phenomena in impulse breakdown (Waters et al., 1970, 1979; Les Renardières Group, 1974, 1977, 1981; Bye et al., 1978), d.c. coronas (Waters et al., 1972a, 1975; Bogdanova et al., 1976, 1978; Dallaire and Maruvada, 1979) and a.c. coronas (Waters et al., 1972; Bogdanova et al., 1978) has arisen because of an appreciation that such probes also offer a means of discrimination between conduction-current and displace-

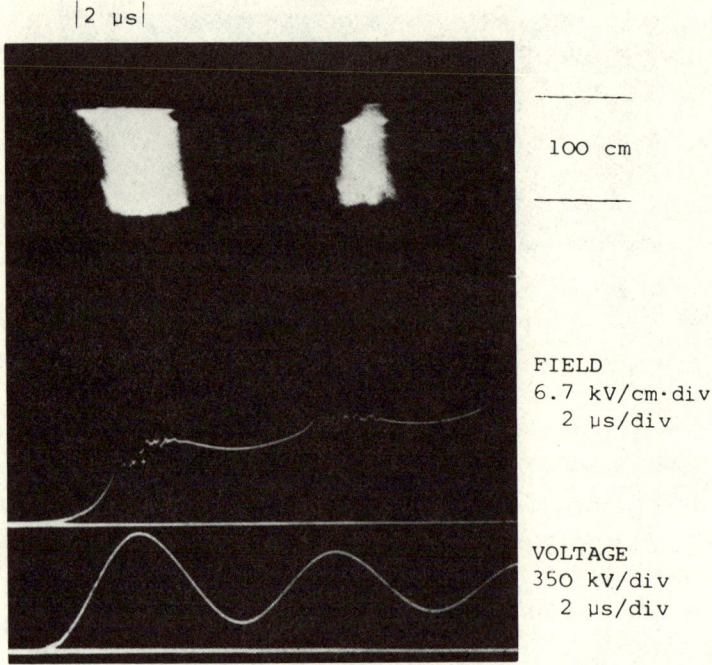

Fig. 44. Positive-impulse breakdown under oscillatory impulse. 1 m rod/plane gap; field-filter at plane (Pesavento, 1979).

ment-current components. In this context the probes have been known as electrostatic fluxmeters.

This discriminatory action can be examined by means of Fig. 45 (Waters, 1972). If the disc sectors consist of radial segments, then the induced charge q_i on the sensing disc varies in the triangular form of Fig. 45 (i), the dashed line representing a changing field. The conduction current i_c to the disc which arises from ionization will also be modulated because of disc rotation (Fig. 45 (ii)). A signal capacitor C will provide a signal voltage

$$v = (q_i + q_c)/C = (q_i + \int i_c \, dt)/C . \qquad (60)$$

The integration performed by the capacitor C produces the phase shift which discriminates between q_i and q_c (Fig. 45 (iii)). The envelope of the modulated signal remains of an amplitude proportional to field strength even in the presence of conduction current (Fig. 45 (iv)).

The modulated signal of the rotating fluxmeter lacks the high-speed response to rapidly varying fields which is a valuable feature

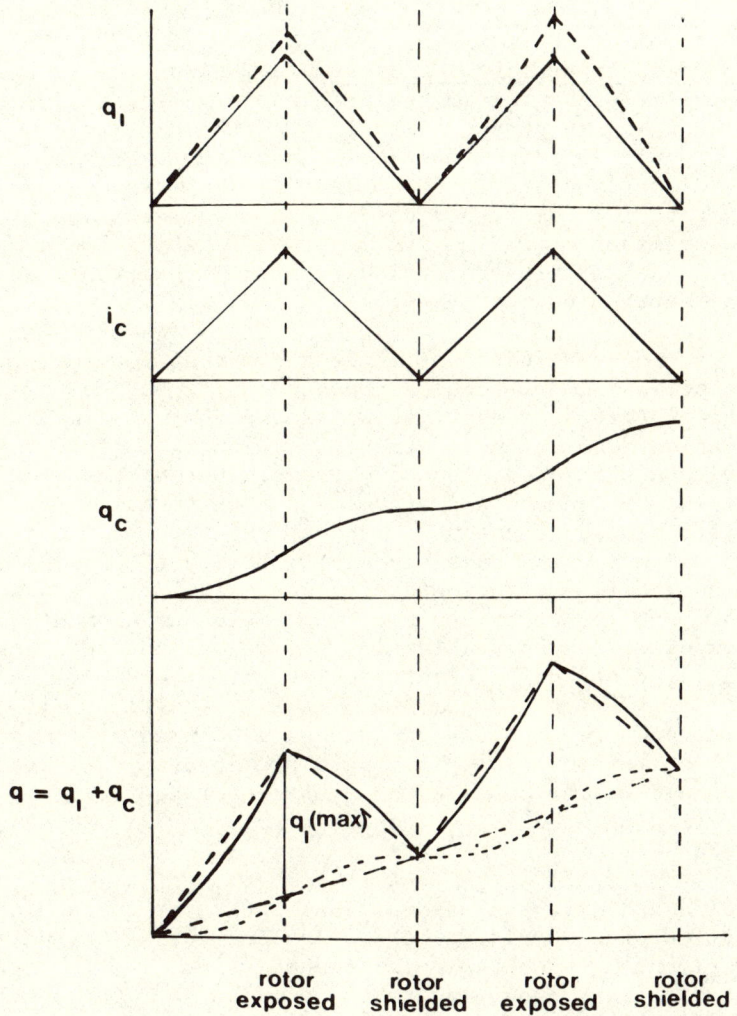

Fig. 45. Fluxmeter probe — principle of discrimination.

of capacitive and field filter probes, although a miniaturized version has been developed to operate at 5 10^4 rev/min, which for 24 sectors corresponds to a time resolution of 25 μs (Les Renardières Group, 1981). The nonshielded disc of a fluxmeter (normally the stator) can, in any case, be used as a nondiscrimatory capacitive probe. The main application of the fluxmeter is in the study of steady-state or slowly varying discharges such as d.c. or a.c. coronas, or in the examination of space-charge resulting from impulse coronas. For this purpose fluxmeters of both plane and cylindrical geometry have been used.

Applications of Fluxmeters

Field Measurements in D.C. Systems.
The surface field strength at a cylindrical conductor in the presence of coronas is an important boundary condition in physical modeling of the discharge and in analysis of power loss from overhead lines. Cylindrical fluxmeters of various diameters between 10 mm and 25.5 mm were devised (Fig. 46) for incorporation into earthed conductors which were coaxial with an outer h.v. cylinder of radius 291 mm; these enabled field measurements to be made with the central conductor with coronas (Waters et al., 1972a; Waters and Stark, 1975).

For d.c. positive coronas, it is known that streamer formation at the inception field E_i can be replaced at higher voltages by a glow discharge (ultracorona). It was found that streamer formation caused transient reductions of the average surface field to about $0.85\ E_i$, with a subsequent recovery to E_i. During the glow regime, which can also be induced and stabilized by external irradiation, the surface field was constant at about $0.95\ E_i$. It was shown that this is consistent with a positive-ion mobility of $1.8\ 10^4\ m^2\ V^{-1}\ s^{-1}$ and a secondary ionization coefficient of 10^{-4}. It was also clear that there was no evidence of the field enhancement which would be expected if the stable glow regime was due to a negative-ion sheath adjacent to the electrode (Hermstein, 1960).

In d.c. negative coronas the discharge consists of discrete sites at which Trichel-type discharges are located. Consequently, transient field variations are found, the average field strength being a maximum between sites (Waters et al., 1972a; Bogdanova et al., 1978).

Dallaire and Maruvada used a plane fluxmeter to obtain transverse profiles of electric field and ion current at ground level under experimental $\pm$ 900kV bipolar transmission lines. Figure 47 gives results of an indoor laboratory simulation. Outdoor tests showed considerable modification because of wind effects upon ion flow.

Field Measurements in A.C. Systems.
The successive positive and negative coronas at line conductors in a.c. systems are significantly influenced by the discharge in the previous half-cycle.

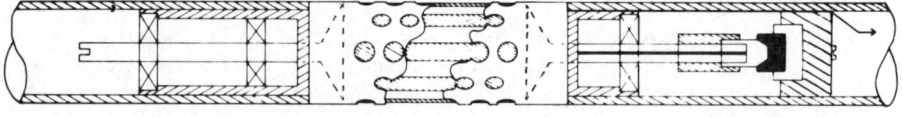

Fig. 46. Cylindrical fluxmeter.

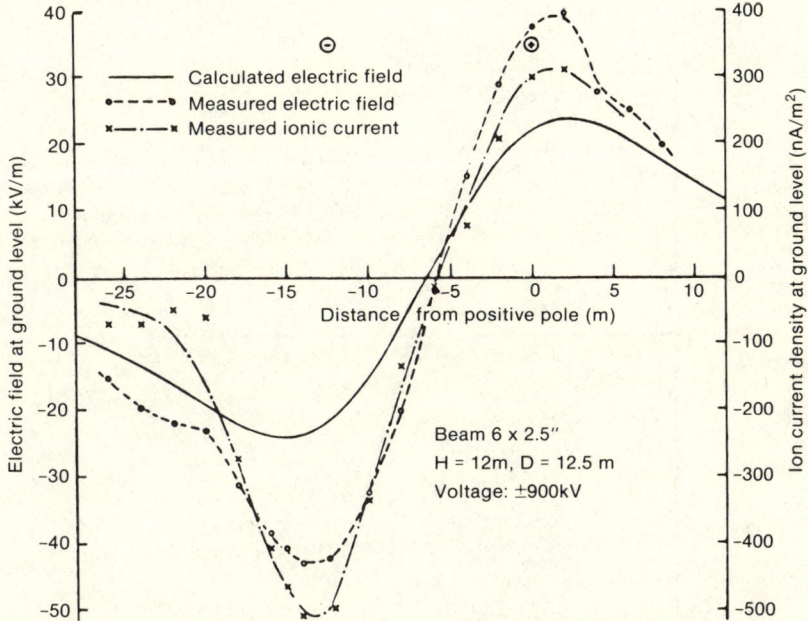

Fig. 47. Electric field and ion current profiles at ground level below bipolar ± 900 kV dc test arrangement (Dallaire and Maruvada, 1979).

Fluxmeter studies (Waters et al., 1972b; Bogdanova et al., 1976) have shown that this remanent space charge causes a field enhancement on the subsequent half-cycle, so that corona inception occurs earlier than if the geometrical field acted alone.

Figure 48 shows a typical surface field variation during the voltage cycle.

<u>Field Measurements in Impulse Coronas</u>. In impulse tests the fluxmeter probe has been employed with two main objectives:

(a) To examine the ionic currents and remanent space charge fields during the dissipation period which follows discharge extinction in long air gaps. From these data, information can be obtained of spatial charge distributions and electric fields.

(b) To detect the arrival of corona streamers at the electrode containing the fluxmeter. This provides an estimate of the macroscopic field (as enhanced by space charge) which is necessary for streamer propagation.

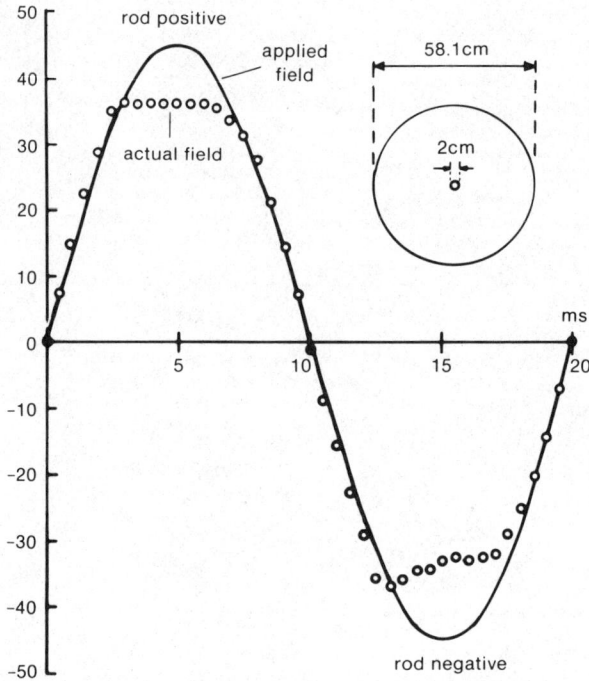

Fig. 48. Field variation during ac cycle at surface of cylindrical conductor.

Figure 49 shows the form of a typical fluxmeter oscillogram at the earthed plane electrode of a 10 m gap (positive impulse, 1.74 MV (Les Renardieres, 1977). Following a withstand event, in which the leader can propagate several metres before quenching, the space-charge field at the plane is a maximum when the drift field causes ions to reach the plane. From measurements of this time of arrival, the ions can be shown to have originally resided near the extremities of the corona growth. In 10 m gaps the space charge field subsequently decays to half-value in about 1 s. In short gaps (0.4m) the decay constant is about 10 ms (Waters et al., 1970). Recent measurements in 7 m gaps under negative impulses (Les Renardières, 1981) have shown times of about 0.1 s. In all cases it has been verified that during this initial period the main cause of decay of space charge density, from an initial value of $\rho(o)$ to a value of $\rho(t)$ after time t, is due to the self-repulsion space-charge field so that

$$\rho(t) = \frac{\rho(o)}{1 + (\mu/\varepsilon_o)\rho(o)t} \quad . \tag{61}$$

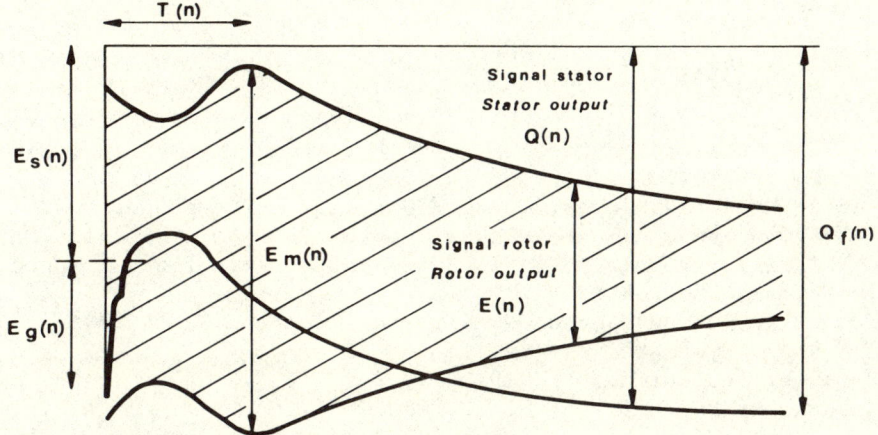

Fig. 49. Fluxmeter signals following withstand impulse in long gaps. $E_s(n)$ = space-charge field after leader arrest; $E_g(n)$ = peak applied field; $E_m(n)$ = space-charge field at time of arrival of space-charge; $T(n)$ = time of arrival; $Q_f(n)$ = final charge/unit area.

Values of $\rho(o)$ can then be reconstructed from the fluxmeter data, and spatial densities of up to 100 μC m^{-3} have been estimated. These measurements have been used to determine leader channel gradients, energy dissipation, and storage in long sparks.

Use of a miniaturized high-speed fluxmeter in the h.v. cathode of the 7 m gap has recently indicated that reversed fields can sometimes occur (as found also by field-filter probes).

Measurements of the field strengths necessary to support streamer propagation have been made both for positive coronas (Bye et al., 1978) and negative coronas (Waters et al., 1979); the respective values found were 0.5 MV m^{-1} and 1.8 MV m^{-1}. The former value is consistent with theoretical estimates based on corona streamer modeling. The value for negative coronas corresponds to the field necessary to support series cascade avalanches near the streamer tip. A progation field of 1.8 MV/m has also been deduced by indirect methods for negative long-gap coronas (Les Renardières, 1981).

REFERENCES

Anderson, R. B., 1977, in: "Lightning," Vol. 2, R. H. Golde, ed., Academic, New York, Chap. 13.
Baldo, G., Marchesi, G. and Lalot, J., 1979, in: "Proceedings, 3rd International Symposium on High Voltage Engineering, Milan," Paper 52.19.
Barrault, M. R. and Jones, G. R., 1974, in: "Proceedings, 7th Yugoslav Symposium and Summer School on the Physics of Ionized Gases, Rovinj, Croatia," V. Vujnović, ed., Inst. of Phys., Univ. Zagreb, Zagreb, p. 701.
Barrault, M. R., and Jones, G. R., and Blackburn, T. R., 1974, J. Phys. E, 7:663.
Bastien, F. and Marode, E., 1979, J. Phys. D, 12:249.
Bogdanova, N. B., Pevchev, V. G., and Polevoy, S. W., 1976, in: "Proceedings, 4th International Conference on Gas Discharges, Swansea," IEE Conf. Publ. No. 143, p. 231.
Bogdanova, N. B., Pevchev, V. G., 1978, Izv. Akad. Naut SSSR, Energ.: Transport, 16:83.
Bye, P., Davies, A. J., Dutton, J., and Waters, R. T., 1978, in: "Proceedings, 5th International Conference on Gas Discharges, Liverpool," IEE Conf. Publ. No. 165, p. 312.
Craggs, J. D. and Meek, J. M., 1954, "High Voltage Laboratory Technique," Butterworth, London.
Dallaire, D. and Maruvada, P. S., 1979, R. Can. Génie Elec., 4:22.
Edels, H. and Whittaker, D., 1957, Proc. Roy. Soc. London, A, 240:54.
Evans, D. E. and Katzenstein, J., 1969, Rep. Prog. Phys. 32:207.
Farish, O. and Lynch, B. R., 1980, "Proceedings, 6th International Conference on Gas Discharges, Edinburgh," IEE Conf. Publ. No. 189, p. 213.
Gallimberti, I., Hartmann, G., and Marode, E., 1977, Electra, 53:123.
George, D. W. and Richards, P. H., 1968, Br. J. Appl. Phys., 1:1171.
Gerry, E. T. and Rose, D. J., 1966, J. Appl. Phys. 37:2715.
Gick, A. E. F., Quigley, M. B. C., and Richards, P. H., J. Phys. D, 6:1941.
Griem, H. R., 1964, "Plasma Spectroscopy," McGraw, New York.
Groh, K. H., Walter, S. E., and Loeb, H. W., 1978, in: "Proceedings, 5th International Conference on Gas Discharges, Liverpool," IEE Conf. Publ. No. 165, p. 53.
Hermstein, W., 1960, Arch. Electrotech., 45:279.
Holder, D. W. and North, R. J., 1963, NPL Notes App. Sci. No. 131, H.M.S.O., London.
Hill, R. J., 1973, Ph. D. Thesis, University of Liverpool.
Jones, G. R. and Naidu, S. R., 1974, J. Phys. D., 7:2254.
Kimblin, C. W. and Edels, H., 1966, Br. J. Appl. Phys., 17:1607.
Krider, E. P., 1965, J. Geophys. Res. 70:2459.
Kurimoto, A. and Farish, O., 1980, Proc. IEE, 127:89.
Les Renardières Group, 1974, Electra, 35:47.
Les Renardières Group, 1977, Electra, 53:31.

Les Renardières Group, 1981, Electra, 74:67.
Meek, J. M. and Craggs, J. D., eds., 1978, "Electrical Breakdown of Gases," 2nd Ed., Wiley, New York and Chichester.
Morgan, C. G., 1979, Chem. Soc. Rev., 8:367.
Orville, R. E., 1968, J. Atmos. Sci., 25:827.
Pesavento, G. C., 1979, in: "Proceedings, 3rd International Symposium on High Voltage Engineering, Milan," Paper 51:13.
Rodriques, F. A., 1974, Ph. D. Thesis, University of Liverpool.
Ross, J. N., 1977, CEGB Research Report RD/L/N55/77.
Salanave, L. E., 1965, "Problems of Atmospheric and Space Electricity," Elsevier, Amsterdam, p. 371.
Selim, E. O., and Waters, R. T., 1980a, IEEE Trans., IA-16:458.
Selim, E. O. and Waters, R. T., 1980b, in: "Proceedings, 6th International Conference on Gas Discharges, Edinburgh," IEE Conf. Publ. No. 189, p. 146.
Selim, E. O. and Waters, R. T., 1981, in: "Proceedings, 15th International Conference on Phenomena in Ionized Gases, Minsk."
Spence, D. A., 1970, Proc. Cambridge Philos. Soc., 68:529.
Stark, W. B., Selim, E. O., and Water, R. T., 1980, IEEE Trans. 1A-16:46A.
Tanner, L. H., 1965, J. Sci. Instrum., 42:834.
Tassicker, O. J., 1974, Proc. IEE, 121:213.
Tholl, H., Sander, I., and Martinen, H., 1970, Z. Naturforsch., 25a:412.
Waters, R. T., 1972, J. Phys. E, 5:475.
Waters, R. T., Rickard, T. E. S., and Stark, W. B., 1970, Proc. Roy. Soc. London, A, 315:1.
Waters, R. T., Rickard, T. E. S., and Stark, W. B., 1972, in: "Proceedings, 2nd International Conference on Gas Discharges, London," IEE Conf. Publ. No. 90, p. 188.
Waters, R. T. and Stark, W. B., 1975, J. Phys. D, 8:416.
Waters, R. T. and Selim, E. O., 1979a, in: "Proceedings, 3rd International Symposium on High Voltage Engineering, Milan," Paper 44.05.
Waters, R. T. and Selim, E. O., 1979b, J. Phys.(Paris), 40:C7-245.
Waters, R. T., Allibone, T. E., Dring, D., and Allen, N. L., 1979, Proc. Roy. Soc. London, A, 367:321.
Wiese, W. C., Kelleher, D. E., and Paquette, D. R., 1972, Phys. Rev. A, 6:1132.
Williams, M. M. R., 1973, J. Phys. D, 6:744.

SPECTROSCOPIC DIAGNOSTICS IN

GAS DISCHARGES

>F. Bastien

>Laboratoire de Physique des Décharges (CNRS)
>Ecole Supérieure d'Electricité
>Plateau du Moulon - 91190 Gif Sur Yvette, France

EMISSION SPECTROSCOPY AND TEMPERATURE

This chapter is specially devoted to emission spectroscopy of gas discharges. At first glance emission spectroscopy seems to be a very good diagnostic method. There is a witness without the medium disturbed. If we could get the light emitted from a given atom at a given time, the situation would be very good indeed, but, in fact, we measure only an average emission from which we cannot get a complete knowledge of the plasma. In other words, from the microscopic state we could theoretically compute the light emission, but not the other way around. We can also deduce macroscopic variables from the microscopic state, but the problem is to deduce macroscopic variables from the measured light emission.

In order to make the connection between light emission and the macroscopic variables, we have, in fact, to consider some equilibrium. The most complete equilibrium corresponds to an ionized gas completely enclosed in an ideal box; then the macroscopic state of the plasma can be described without any detailed knowledge of the nature of the interaction between plasma particles with each other and between the plasma particles and the photon field. In that case, which is called complete thermodynamic equilibrium, we can use a set of well-known laws (see Table 1). Unfortunately, this complete thermodynamic equilibrium (C.T.E.) is never achieved in electrical discharges. But it is possible to consider cases in which only some of the previous laws can be applied. An important case is the so-called complete local thermodynamic equilibrium (C.L.T.E.) in which the Maxwell, Saha, and Boltzmann laws apply. In such a case, it is relatively easy to deduce macroscopic parameters from spectroscopic measurements. The question is: Can the

Table 1. Some laws valid for plasma in complete thermodynamic equilibrium (C.T.E.) and complete local thermodynamic equilibrium (C.L.T.E.) (after Drawin, 1970).

	C.T.E.	C.L.T.E.
Temperature	$\nabla T = 0$	$\nabla T = 0$
Density	$\nabla n = 0$	$\nabla n = 0$
Photons	Plank law	No Plank law
Energy velocity	Maxwell $T = T_{kinetic}$	Maxwell $T - T_{kinetic}$
Population of discrete energy levels	Boltzmann $T = T_{excit.} = T_{kinetic}$	Boltzmann $T = T_{excit.} = T_{kinetic}$

C.L.T.E. be achieved in electrical discharges? If we consider the diagram of Fig. 1 after H. W. Drawin (1970), we see that, apart from the arc, where C.L.T.E. is sometimes achieved, gaseous discharges do not reach C.L.T.E. Since in this chapter we do not consider arcs, we have to answer the following question: What kind of emission spectroscopy diagnostic can be used without C.L.T.E.?

First, before presenting diagnostic techniques, let us consider some examples concerned with the Boltzmann law application. For this purpose, we consider a given set of particles or excited particles or a particular excited state of excited particles. Boltzmann's law applies if the energy gained or lost from, or to, the external medium is much smaller than the total energy exchange between the particles or the excited states under consideration (see Fig. 2).

Let us give some examples. In most cases, Boltzmann's law does not apply to the electronic excited states due to low exchange rates between these electronic states compared with electronic excitation rates and radiation losses.

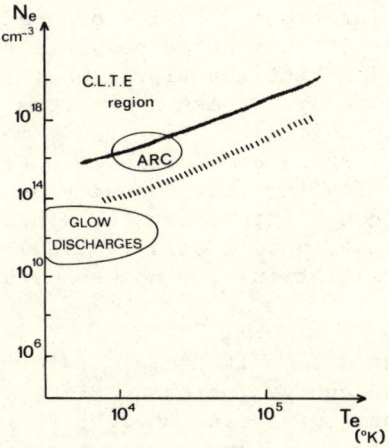

Fig. 1. Region of L.T.E. in a N_e (electron density), T_e (electron temperature) diagram (after Drawin, 1970).
　　　　▬▬▬ lower limit of complete L.T.E.
　　　　ııııı {lower limit for partial L.T.E. for the third quantum level in the hydrogen atom approximation.

On the other hand, if we consider rotational excited states, the exchange between these excited states is very fast; consequently Boltzmann's law applies generally and it is possible to define a rotational temperature T_r. Furthermore, due to rapid rotational relaxation (for O_2 at 1 atmosphere the relaxation time is less than 1 ns) the rotational excited states are in equilibrium with the kinetic energy of the molecules. It is often possible to define a gas temperature T_g approximately equal to T_r.

Now let us consider the vibrational excited states. If the electronic excitation is not too fast, due to the large transition

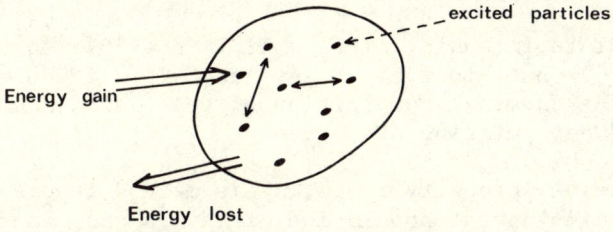

Fig. 2. Set of particles (or excited particles or particular excited state of excited particles) with energy exchanges.

probability between vibrational levels compared with the vibrational translation probability, it is often possible to define T_v from Boltzmann's law. Notice that the vibrational translation transition is the most important decay process apart from the wall deactivation for homopolar diatomic molecules. For these molecules, radiational decays of vibrational levels are forbidden. For example, in N_2 at $T = 300$ K the probability for vibrational-vibrational transitions is 5 orders of magnitude greater than the vibrational-translational transition probability (Campbell et al., 1980). Furthermore, in N_2 at 1 atmosphere the relaxation time between vibrational states is about 1 µs.

Besides Boltzmann's law, if the electric field is not too large, that is to say, if the energy increase during a mean free path is much smaller than electron average energy, it is possible to define an electron temperature (T_e) from an electron distribution which is often assumed as Maxwellian.

In conclusion, it is possible to say that in many cases we can define an electron temperature (T_e), vibrational temperature (T_v), rotational temperature (T_r), and neutral gas temperature (T_g) with the relation

$$T_e \gg T_v \gg T_r \simeq T_g. \tag{1}$$

We shall present now some methods to determine T_e and, in more detail, methods to obtain T_v and T_r.

After that, we shall tackle another topic, namely the determination of electron density from Stark broadening, especially the limitation of this method in the presence of a large external or charge space field.

T_e DETERMINATION

Now we shall examine very briefly three different methods of obtaining T_e when Boltzmann's law does not apply.

(1) Absolute intensity lines. It is possible to use absolute intensity lines. But, in fact, this method is difficult because we have to know the chemical composition of the plasma and all the populating and depopulating processes.

(2) Relative intensity lines. This method is easier, but we need equations giving the population of levels and, consequently, use a lot of macroscopic parameters.

(3) Free-bound emission recombination. It is theoretically possible to use emission from recombination processes ($\chi^+ + e \to \chi + h\nu$). The emission coefficient computed with L.T.E. can be used without L.T.E. if it is possible to assume a Maxwellian electron distribution.

MOLECULAR SPECTROSCOPY

Introduction and Notation

Before examining in some detail the determination of T_r and T_v, we discuss some results concerned with molecular spectroscopy. More detail on molecular spectroscopy can be found in Pecker-Wimel (1968), Marr (1968), and Herzberg (1950).

In Fig. 3, we show the energy diagram of a molecule. As a first approximation, the total energy E can be considered as the sum of three terms: E_e (electronic energy), E_v (vibrational energy), and E_r (rotational energy), with generally $E_j \ll E_v \ll E_e$. X, B, and C refer to electronic states and the dotted arrows represent examples of transitions between levels.

Figure 4a shows some transitions between vibrational levels. It should be pointed out that there are no selection rules for vibrational transitions. It is different for rotational transitions where the general selection rules are (1) J being the quantum number associated with a state, (2) the allowed transitions corresponding to $\Delta J = 0$ or ± 1, (3) the $J = 0 \to J = 0$ transition being forbidden. The lines corresponding to $\Delta J = 0, + 1, - 1$ are designated respectively by Q, P, and R (see Fig. 4b).

In Fig. 5, a schematic spectrum of molecular emission is given. With a low resolution (Fig. 5a) only the bands corresponding to different transitions between vibrational levels can be recorded. Increasing the resolution allows us to obtain sequences of vibrational transition (Fig. 5b). A further increase of the resolution gives a separation of different rotational lines. Figure 5c represents a case in which only R and P branches exist. ν_o is the so-called band origin.

Before we examine coupling between various angular momenta, let us recall some notation used in order to classify molecular states. First, different upper case Greek letters are used to classify the states as a function of Λ respectively Σ, Π, Δ and ϕ for $\Lambda = 0, 1, 2$ and 3, $\Lambda = M_L$, M_L being the component of orbital momentum along the molecular axis. M_L can have $2L + 1$ values from $-L$ to $+L$.

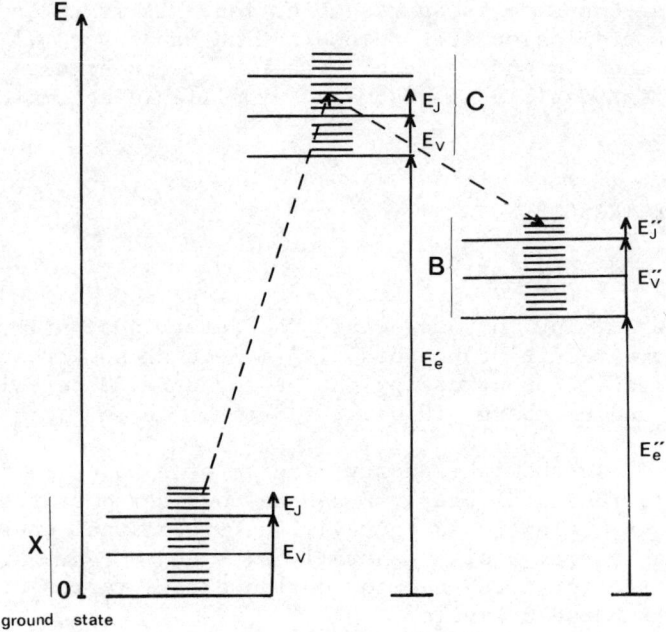

Fig. 3. Molecular energy level diagram.

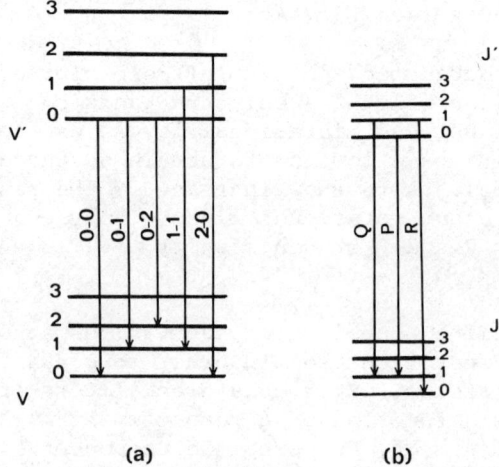

Fig. 4. a) Diagram of vibrational transitions.
b) Diagram of rotational transitions.

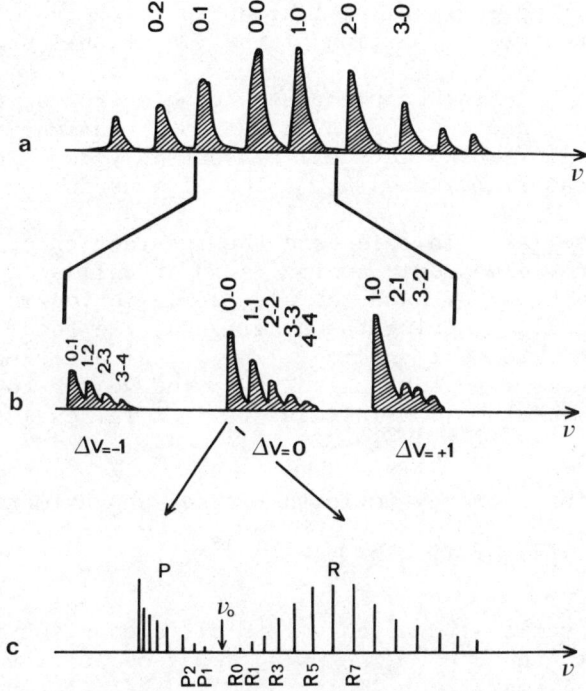

Fig. 5. Schematic spectrum of molecular emission is given with increasing resolution from a to c.

Furthermore, if it is defined, the total angular momentum component along the internuclear axis is given by

$$\Omega = |\Lambda + \Sigma| \tag{2}$$

where $\Sigma = M_S \cdot M_S$ is the component of the resultant spin vector S along the internuclear axis. Σ can have $2S+1$ values from $-S$ to $+S$.

The quantum number Ω appears as a subscript to the term symbol. Ω can have integral or half-integral values. The multiplicity of the states is written as a superscript. For example, the ground state of nitric oxide is written $X^2\Pi_{1/2,3/2}$ where X denotes the ground state. Other upper case letters are used for excited states. An example of a transition in O_2 is $B^2\Sigma \rightarrow X^3\Sigma$.

Coupling Between Angular Momenta

Now we shall briefly discuss the coupling between various angular momenta. The interaction effects of vibration and rotation can be, at first approximation, introduced in the rotational constant B which becomes a function of the vibrational state.

We examine only the coupling between electron spin, orbital angular momentum, and the angular momentum of nuclear rotation. Applying the vector model we shall present only two limiting cases, namely Hund's cases, (a) and (b), which give useful examples.

Hund's Case (a). In this case the interaction of the nuclear rotation with the electronic motion (spin as well as orbital) is very weak. On the other hand the electronic motion is coupled very strongly to the line joining the nuclei. Ω, the total electronic angular momentum, is well defined. This total electronic angular momentum interacts with the nuclear rotation vector to produce the total angular momentum J. (This interaction is schematized in Fig. 6).

The rotational energy is in that case, approximately

$$F_v(J) = B_v [J(J+1) - \Omega^2], \qquad (3)$$

B_v being the rotational constant. $B_v \Omega^3$ is a constant for a given electronic state and $J \geq \Omega$ (the energy must be positive). An example of $^3\Pi$ rotational energy levels for Hund's case (a) coupling is given in Fig. 7. Missing rotational levels are shown as dotted lines, and integer numbers correspond to the quantum number associated with the total angular momentum J.

Hund's Case (b). When $\Lambda = 0$ and $S \neq 0$, the spin vector S is not coupled to the internuclear axis, and consequently Ω is not defined. Sometimes, especially with light molecules, even if $\Lambda \neq 0$, S may be only very weakly coupled to the internuclear axis.

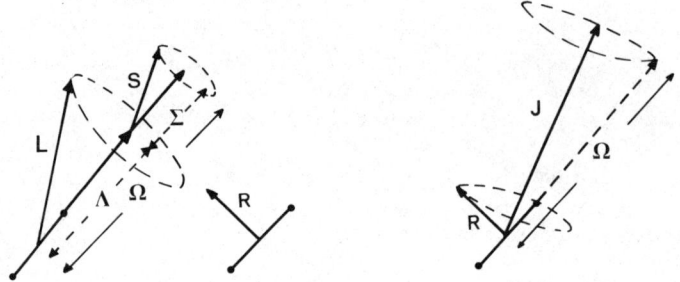

Fig. 6. Vector diagram for Hund's case (a).

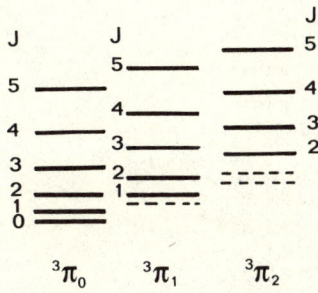

Fig. 7. Diagram of energy levels for Hund's case (a) coupling in $^3\Pi$ levels.

In that case, it is possible to consider the interaction of Λ and R (angular momentum of nuclear rotation) to form K which is the total angular momentum apart from the spin. The interaction between K and S gives the total angular momentum J (see Fig. 8). Figure 9 gives an example of an energy level diagram for $^2\Sigma$ states. The doublet (same value of K) is drawn to a much larger scale than the separation of levels with different K.

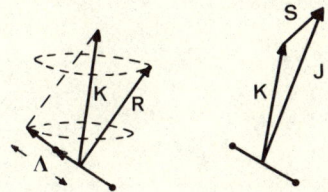

Fig. 8. Vector diagram for Hund's case (b).

ROTATIONAL TEMPERATURE DETERMINATION

From the First Negative System of N_2^+ (1 S$^-$)

The first negative system corresponds to the transition

$$B^2\Sigma^+ \rightarrow X^2\Sigma^+ \text{ (see spectrum of Fig. 10)}. \qquad (4)$$

Such a transition $^2\Sigma \rightarrow {}^2\Sigma$ always belongs to Hund's case (b); consequently selection rule $\Delta K = \pm 1$ holds, the $\Delta K = 0$ transition being forbidden.

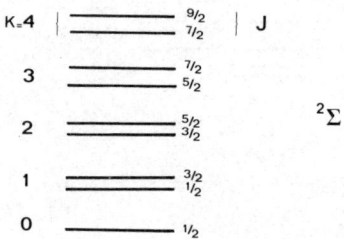

Fig. 9. Energy levels diagram for $^2\Sigma$ levels.

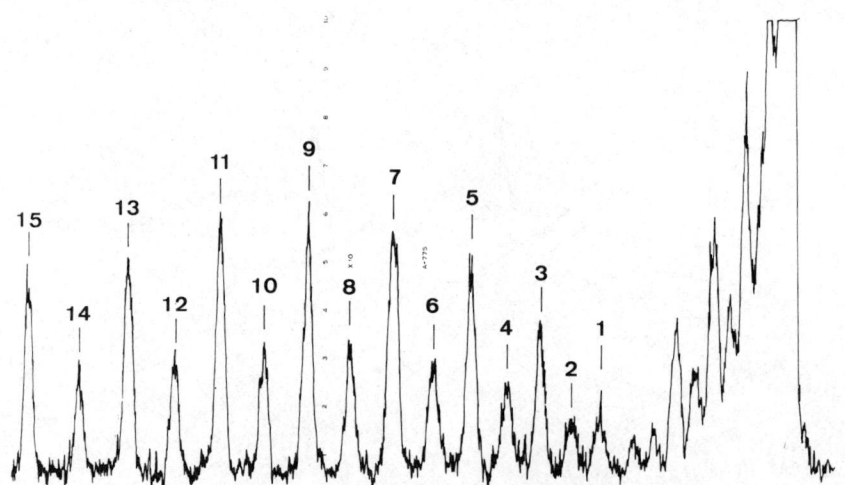

Fig. 10. Emission spectrum corresponding to (0,0) vibrational transition in $B^2\Sigma_u^+ \rightarrow X^2\Sigma_g^+$ electronic transition. For R branch, number corresponds to K' (rotational quantum number of upper state $B^2\Sigma^+$) (after Hartmann, 1977).

As seen previously, the separation of two sublevels with J = K + 1/2 and J = K - 1/2 for a given K is very small. With not too great resolution, we have exactly the same band structure as for $^1\Sigma \to {}^1\Sigma$ transition except quantum number K must be used instead of J.

P and R branches can be observed. Using the R branch, which is better resolved, we have the relation

$$N_{K'}^R = C \, g \, K' \, \exp[-B_v(K' + 1)K' \, hc/kT_r] \qquad (5)$$

where

$N_{K'}^R$ = relative intensity of an R branch line

C = a constant (The frequency is assumed to be almost constant within a branch.)

g = weight factor corresponding to nuclear spin degeneracy

K' = quantum number for the upper level (here $B^2\Sigma_u^+$)

B_v = rotational constant (here 1.9898 cm^{-1})

h,k,c = respectively Plank, Boltzmann constants, and speed of light

T_r = rotational temperature. Rotational temperature of the upper level is often assumed to be equal to the rota- temperature of the lower level.

It is obvious that plotting $\ln(N_K^R/gK')$ as a function of K'(K' + 1) we get a straight line, the slope giving T_r (see example in Fig. 11).

From Second Positive System of N_2 (2 S$^+$)

This system corresponds to the $C^3\Pi_u \to B^3\Pi_g$ transition. Such $^3\Pi \to {}^3\Pi$ transitions present three P branches (P_1, P_2, P_3) and three R branches (R_1, R_2, R_3). Q branches are very weak with negligible intensity.

P_1 and R_1 branches correspond to $^3\Pi_o \to {}^3\Pi_o$ with K = J - 1 (Fig. 12). For the upper as well as the lower state a very rapid transition from case (a) to case (b) is assumed. K values are not defined in case (a), consequently K is not indicated for lower levels.

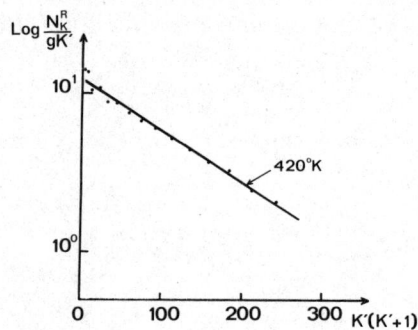

Fig. 11. Rotational temperature determination (· experimental points) (after Hartmann, 1977).

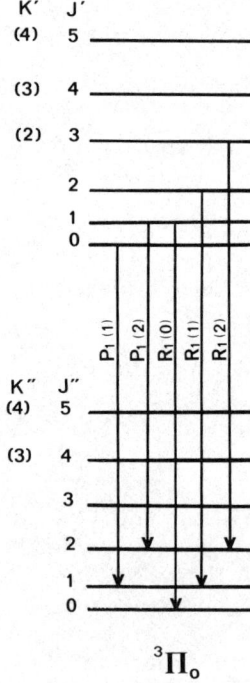

Fig. 12. Transitions for $^3\Pi_o$ level. Some examples of lines in P_1 and R_1 branches are given. Quantum number J corresponds to the total angular momentum. Quantum number K corresponds to the total angular momentum apart from spin (it is defined only in Hund's case (a)).

In the case of the R_2 branch ($^3\Pi_1$), for example, we have the relation

$$N_{K'}^{R_2} = C \frac{K''(K''+2)}{K'+1} \exp[-B_v(K''+1)(K''+2) \, hc/kT_r] \qquad (6)$$

$N_{K''}^{R_2}$ = relative intensity of an R_2 branch line

K'' = quantum number corresponding to $B^3\Pi_g$ level

C = a constant. (The frequency is assumed almost constant within a branch.)

This relation can be used to get T_R. We can also use the R_1 branch to avoid overlapping for a large value of T_R.

Overlapping Problem

Let us now examine some problems of overlapping. If high rotational temperatures are expected, the P branch overlaps the R branch. For example, in the (0,0) band of 1 S$^-$ of N_2^+ $(K+27)^{th}$, P line overlaps K^{th} R line (see Fig. 13). Nevertheless, three different methods can be used in order to get T_R, namely:

a) line abstraction (Bleekrode and Van Benthen, 1969). This method uses a selection of lines from the R and P branches where overlapping is zero or negligible.

b) Moore and Doering method (Moore, 1968). Uses the computation of the ratio of R branch line to total line intensity as a function of T_R for 1 S$^-$ of N_2^+.

c) Wink et al. method (Wink et al., 1971). This method, valid only for the (0,0) band, uses the 2/1 intensity alternation of lines due to nuclear spin degeneracy

(a) $\quad I(K'') = R(k'') + 2 P(K''+27) \quad$ for K'' odd $\qquad (7)$

(b) $\quad I(K'') = 2R(K'') + P(K''+27) \quad$ for K'' even $\qquad (8)$

where $R(K'')$ is the intensity of an R-branch line for quantum number K'' (related to lower states).

$P(K''+27)$ is the intensity of P-branch for quantum number $K''+27$, apart from a factor due to nuclear spin degeneracy. From relation (a) it is possible by interpolation (smoothed curve method) to define $I(K'')_{odd}$ for any K.

In the same way, from relation (b) you get $I(K'')_{even}$ defined for any K. Consequently, for any K you get

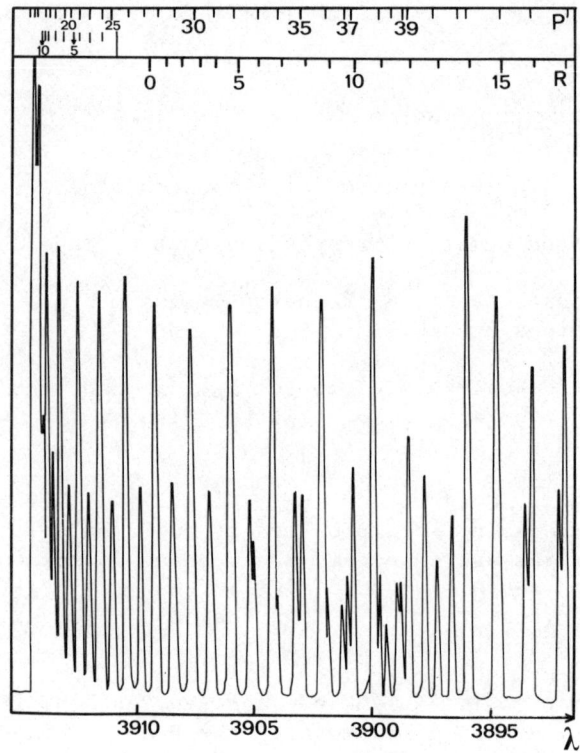

Fig. 13. (0,0) band of the $B^2\Sigma_u^+ \to X^2\Sigma_g^+$ electronic transition of N_2^+ (after Cramarossa et al., 1974).

$$R(K") = \frac{1}{3}\left[2\,I(K")_{even} - I(I")_{odd}\right] \tag{9}$$

$$P(K" + 27) = I(K")_{even} - 2\,R(K") \tag{10}$$

These relations solve the problem.

VIBRATIONAL TEMPERATURE

The vibrational excitation can be obtained from the relative values of the band intensity

$$I_{v'v"} = \text{const}\; q_{v'v"}\, \nu^4_{v'v"}\, N_{v'}\;, \tag{11}$$

where $I_{v'v"}$ is the intensity for a transition from v' to v" vibrational level, and

SPECTROSCOPIC DIAGNOSTICS

$q_{v'v''}$ = Franck-Condon factor

$\nu_{v'v''}$ = transition frequency

$N_{v'}$ = v' level population.

Here the electronic transition moment R_e is considered as a constant. If this is not possible, we have to write $I_{v'v''} \alpha\ q_{v'v''}\ \nu^4_{v'v''}\ N_{v'} R^2_{v'v''}$. If a Boltzmann distribution is assumed, it is possible to define T_v from

$$I_{v'v''} = \text{const}\ q_{v'v''}\ \nu^4_{v'v''}\ \exp\left[-G_o(v')hc/kT_v\right] \qquad (12)$$

where $G_o(v')hc$ is the energy of the v' level.

Now, we shall examine two questions:

(a) How to measure band intensity in spite of the overlapping?

(b) What is the meaning of T_v?

Measure of Band Intensity

In order to get the relative band intensity, we can use the relative values of the <u>band head</u> intensities. This procedure can be applied if:

(1) the rotational energy distributions in each level are equal.

(2) the rotational structure of each vibrational band is identical.

Within the same vibrational sequence, (2) is often fulfilled. Often condition (1) is assumed to be satisfied.

Cramarossa et al. (1974) use, for example (1,0), (2,1) and (3,2) transitions of the 2^d positive system (see Fig. 14). Band head intensity is proportional to the dashed area.

We can also introduce a "slit function" (Hartmann, 1977 or Hartmann and Johnson, 1978). The slit function is defined by the relations

$$\phi^T_{v'v''} = W_{v'v''}\ \phi^P_{v'v''} \qquad (13)$$

where $\phi^T_{v'v''}$ is total band intensity, and $\phi^P_{v'v''}$ is the peak intensity at the band head. If $W_{v'v''}$ has been computed, we can get $\phi^T_{v'v''}$ from $\phi^P_{v'v''}$. We shall examine briefly the method of $W_{v'v''}$ computation.

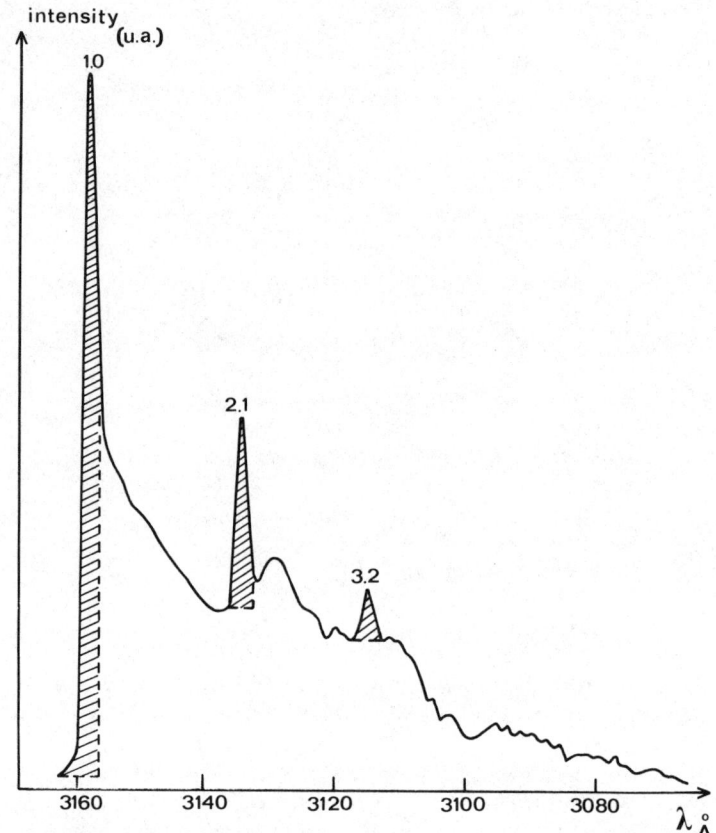

Fig. 14. Band intensities observed with low resolution in the 1,0 sequence of N_2 ($C^3\Pi_u \rightarrow B^3\Pi_g$). Dotted areas correspond to integrated intensity between the band origin and the band head. These areas represent the same number of unresolved P branch rotational lines (after Cramarossa et al., 1974).

A known slit field is scanned over the <u>computed generated band</u> structure.

The computed generated band is obtained from the following principles:

(a) line structure. The line structure is obtained from the semiempirical formulae of Budo. The importance of coupling for energy level computation must be underlined.

(b) intensity distribution. Several identical formulae are used. We give here only an example: $\phi_{\Omega,J}^P$, is the intensity for a line of the P branch with J' the total angular momentum quantum number, and Ω the electronic angular momentum quantum number

$$\phi_{\Omega,J'}^{P} = \text{const } S_{\Omega,J'}^{P} (2J' + 1) \exp\left(-J'(J' + 1) \frac{hcB_v}{kT_r}\right) . \tag{14}$$

$S_{\Omega,J'}^{P}$ is the line strength (Hönl-London factor)

$$S_{\Omega,J'}^{P} = \frac{(J' + 1 + \Omega)(J' + 1 - \Omega)}{J' + 1} . \tag{15}$$

Hund's case (a) is assumed (2 S^+ N_2) and it is assumed that the rotational distribution is <u>unchanged</u> in the excitation process.

Meaning of T_v

The vibrational temperature T_v associated with excited electronic states is not, in general, a "true temperature" because level populations do not follow the Boltzmann law; nevertheless T_v is an interesting parameter.

In some circumstances, the value of the ground state vibrational temperature existing before excitation of the upper electronic level can be deducted from the vibrational temperature of the excited levels. This happens when the molecular electronic level is rapidly excited by a fast electron as in a streamer. The excited levels being vertically excited, the following relation applies to a first approximation (with $R_e = 1$, R_e being the electronic transition moment)

$$N_0^r = P_0 q_{00} + P_1 q_{10} + P_2 q_{20} + \ldots \tag{16}$$

$$N_1^r = P_0 q_{01} + P_1 q_{11} + P_2 q_{21} + \ldots \tag{17}$$

$$N_2^r = P_0 q_{02} + P_1 q_{12} + P_2 q_{22} + \ldots, \tag{18}$$

where N_i^r is the relative population of upper level $v = i$,

P_i = population of lower level $v = i$

q_{ij} = Franck-Condon factors.

If the experiment gives N_i^r, assuming that the terms not written on the right are negligible, and getting q_{ij} from the literature, the population P_i of lower level can be easily computed and so therefore can T_v.

ROTATION TEMPERATURE WITHOUT BAND RESOLUTION

Previous considerations can be used to measure rotational temperature without band resolution.

a) Computational methods used for "slit functions" can also be used for generating the band head profile. This computed band head profile by comparison with experiment can give T_r without rotational structure separation. An analogous method is used by Bauer (1979).

b) It is also possible to use directly the "slit function" (Hartmann, private communication). Two spectrometers are adjusted in order to give for the first the maximum intensity $I_{v'v''}$ of the transition $v'v''$ and for the second the maximum intensity $I_{v'v''_1}$ of another transition $v'v''$ from the same vibrational level v' to another vibrational level v''_1. $I_{v'v''}$ and $I_{v'v''_1}$ can be related to T_r by

$$I_{v'v''} = \text{const } \nu_{v'v''} N_{v'} A_{v'v''} W_{v'v''}(T_r) \tag{19}$$

$$I_{v'v''_1} = \text{const } \nu_{v'v''_1} N_{v'} A_{v'v''_1} W_{v'v''_1}(T_r) ; \tag{20}$$

$\nu_{v'v''}$, $\nu_{v'v''_1}$ are transition frequencies,

$A_{v'v''}$ is the transition $v'v''$ probability, and

$W_{v'v''}(T_r)$ is the slit function for the transition $v'v''$ which is a function of rotational temperature T_r

$$\frac{\nu_{v'v''_1} I_{v'v''} A_{v'v''_1}}{\nu_{v'v''} I_{v'v''_1} A_{v'v''}} = \frac{W_{v'v''}(T_r)}{W_{v'v'_1}(T_r)} . \tag{21}$$

I is given by experiment and A from the literature and the knowledge of the slit function allows one to compute T_r.

To give an idea of the method's sensitivity, Table 2 shows the slit function for (2,0) and (2,5) transition in 2 S^+ N_2 as a function of T_r. The values correspond to a spectrometer with an exit slit width of 2 Å.

SPECTROSCOPIC DIAGNOSTICS

Table 2. "Slit function" for various rotational temperatures and an exit slit width $\Delta\lambda = 2$ Å. (2,0) and (2,5) vibrational transitions of the second positive system of nitrogen (after Hartmann, 1977).

T_r	300 K	800 K	1600 K
(2,0)	1.843	2.168	2.910
(2,5)	2.505	3.819	6.508

STARK BROADENING

Theory of Stark Broadening and External Field

Usually the line Stark broadening is used to determine the electron density n_e. Atomic hydrogen is generally used because the broadening is important and the broadening theory well established. Even if the discharge is not studied in hydrogen, it is often possible to introduce minute quantities of hydrogen which acts as a "test gas".

In a plasma dominated by Coulomb collisions, the high value of conductivity prevents the existence of a static field (produced externally or by space charge) and the "classical" broadening theory applies (Griem, 1964 and 1974). Let us examine the situation in the presence of a static field. For this purpose we shall examine the principle of the theory. Consider the situation of a hydrogen atom, if we do not take electrons into account. For an atom subjected to a given field, the emission depends on the field and the direction of observation. For example, if the H_β line is observed perpendicular to the field, the initial line is split into different π components (see Fig. 15).

If the effect of electrons is taken into account, the electron impacts produce a broadening of each component, and the line aspect is given in Fig. 16.

With the impact approximation and Griem - Kolb - Shen (G.K.S.) formalism (see Griem et al., 1959) the line profile J_e as a function $\omega = 2\pi c/\lambda$ and F (total field) is

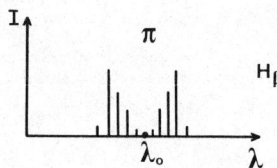

Fig. 15. Π components of H_β (observation perpendicular to constant field). Scheme of line intensities as a function of wave length. λ_o is the initial position of the line.

$$dJ_e(\omega, F) = \pi^{-1} \sum_{\alpha'\alpha''m\beta} R_e <\beta|\vec{e}\cdot\vec{r}|\alpha'm><\alpha''m|\vec{e}\cdot\vec{r}|\beta>$$

$$<\alpha'm|[\imath\Delta\omega_{\alpha''\beta}(F) - \phi a]^{-1}|\alpha'm> d\Omega_{obs} \qquad (22)$$

where the radiation is polarized in $\vec{e}$ direction, and

$\vec{r}$ — vectorial position of the atomic electron

$d\Omega_{obs}$ — solid angle of observation

ϕa — the collision operator,

and the transition $|a> \to |b>$ is split by the Stark effect into substates $|\alpha>$ and $|\beta>$ respectively.

The emitted line ω_{ab} will be shifted towards different $\omega_{\alpha\beta}$ where $\Delta\omega_{\alpha''\beta}(F) = \omega - \omega_{\alpha\beta}$ is the frequency variable.

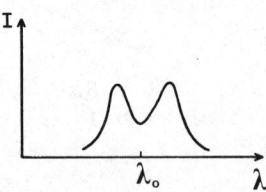

Fig. 16. Profile of H_β taking into account the effect of electrons.

We have to average over field F, and to do that we need to know the field F distribution. When the static field is negligible, it is generally accepted that the field distribution is isotropic.

A characteristic field for this distribution is the so-called Holtzmark field $F_o = 2.6\, e\, N_e^{2/3}$ (C.G.S. units). For example, with an electron density $N_e = 10^{15}$ cm^{-3} we obtain:

$$F_o = 11.3 \text{ statvolt/cm} = 3.4 \text{ kV/cm}. \tag{23}$$

We have to compare this field F_o with the static field F_c (produced externally or by space charge). If the static field F_c is not negligible compared with F_o, the field distribution is no longer isotropic and the broadening theory must be reconsidered. This work has been done by Bastien and Marode (1977). They use G.K.S. formalism with a field distribution computed from composition of the static field F_c and the ionic microfield according to Hooper (1968). As a result, the line profile is not only a function of n_e and T_e, but also of F_c and the direction of observation.

Before examining the useful results, we shall point out the possibility of achieving a realistic situation corresponding to the previous computation. In other words, the following assumptions can be fulfilled when F_c and F_o have the same order of magnitude. The electron distribution is almost isotropic for E/P < 24 Vcm^{-1} torr^{-1} (see Bastien and Marode, 1977 or Bastien, 1977). Consequently, the collision operator ϕ_a can be computed assuming an isotropic distribution. "E.T.L." applies for the sublevel $|\alpha'>$ of the upper level of hydrogen excited electronic states in a large range of T_e and n_e values, for example, with $T_e = 10^4$ K, $n_e = 10^{15}$ cm^3 for H$_\beta$ and with a total pressure equal to atmospheric pressure. The collision frequency between electrons and sublevels $|\alpha'>$ is about $8 \cdot 10^{10}$ s^{-1} greater than $1/\tau$ (τ = radiation time $\simeq 10^{-8}$ s).

Furthermore $P_c = F_c/F_o$ (anisotropic parameter) can easily reach a value of 3 in a discharge-like corona not too far from transition to arc.

Let us now examine the results. The profile $L(\omega)$ is a complicated function of n_e, T_e, F_c and θ_{obs} (angle between the direction of observation and a direction perpendicular to F_c). The difference between these results and results with an isotropic distribution becomes important when $P_c > 2$ (Fig. 17). In order to show that this result is not surprising, we have drawn (Fig. 18) a schematic representation of the addition of the constant field F_c and the ionic microfield F_i (figured here as an isotropically distributed vector with an average length F_o). In Fig. 18a, $F_c \simeq F_o$ ($P_c \simeq 1$), the resulting vector (dotted arrow) has an almost isotropic distribution, and the classical theory can be applied. In Fig. 18b, $F_c \simeq 3\, F_o$ ($P_c \simeq 3$) and the resulting vector has a very anisotropic distribution. Classical theory does not apply.

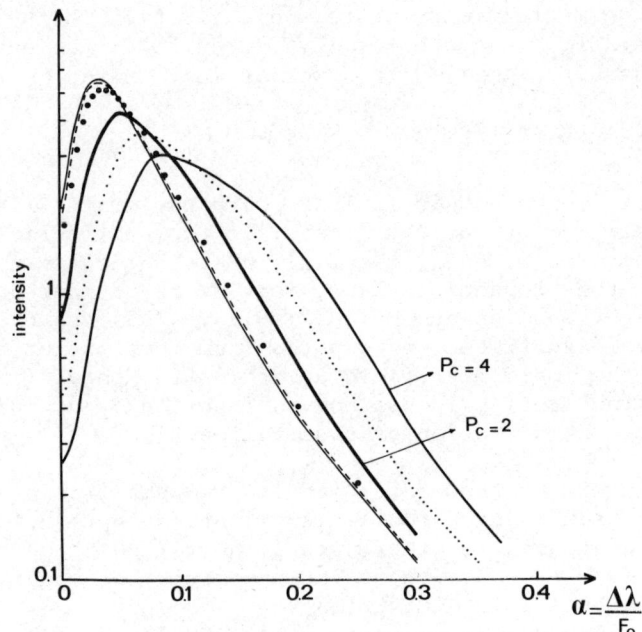

Fig. 17. H broadening for $N_e = 10^{15}$ cm^{-3}.
T_e = 20,000 K and various values of the anisotropy factor
$P_c = F_c/F_o$
$P_c = 0$ ─────
$P_c = 0.5$ ─ ─ ─ ─
$P_c = 1$ ••••••
$P_c = 2$ ─────
$P_c = 3$ ••••••••
$P_c = 4$ ─────

$$\alpha = \frac{\Delta\lambda(A)}{F_o(C.G.S.)}$$

Furthermore, for a given direction of observation and if T_e is approximately known (the profile is only a weak function of T_e), we can write the profile $L(\omega)$ as a function of N_e and F_c.

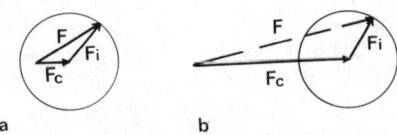

Fig. 18. Schematic representation of the field composition for $P_c < 1$ (a) and $P_c > 1$ (b).

Practical Consequences

If you can make observations in different directions, for example, $L\perp(\Theta_{obs} = 0°)$ and $L//(\Theta_{obs} = 90°)$, it is possible to try to obtain information on both n_e and F_c. The difference, which is not very large, between $L\perp$ and $L//$ is present in Fig. 19. If we have only one direction of observation, we measure L, for example, we cannot get n_e and F_c separately. But it has been demonstrated (Bastien and Marode, 1977, 1979) that if $1 \lesssim P_c \lesssim 4$ at first approximation the half-max intensity width $\overline{\Delta\lambda}$ is not only a function of n_e and F_c, but also a function of the product $n_e F_c$; see $\overline{\Delta\lambda} = f(n_e F_c)$ in Fig. 20.

Consequently, the measure of $\overline{L}$ can give the product $n_e F_c$ and, therefore, the current density j is given by

$$j = e (\mu_e + \mu_i) n_e F_c \simeq e\mu_e n_e F_c \tag{24}$$

where μ_e and μ_i are the mobilities of electrons and ions respectively taken from the literature with F/P approximately evaluated. This last method has been applied to the oxygen corona discharge in a short gap ($\simeq 1$ cm) to measure current density j. Furthermore, from

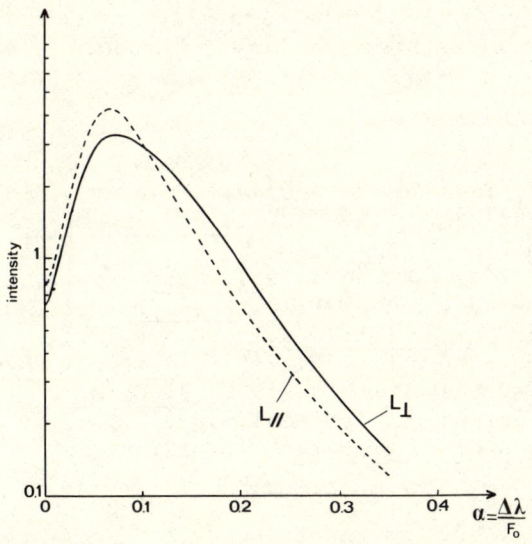

Fig. 19. Comparison of L and L profiles for H_β with $N_e = 10^{16}$ cm^{-3}, $T_e = 20{,}000$ K, $P_c = 3$.
($L\perp$ ——— ; $L//$ - - - -)

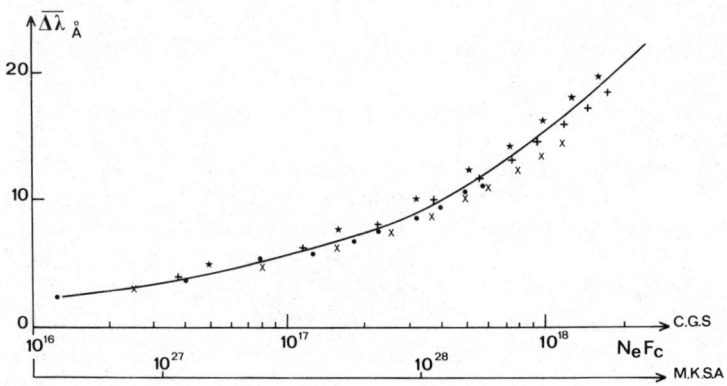

Fig. 20. Half-max intensity width for H_β as a function of the product $N_e F_c$ and with $P_c = 1(\cdot)$, $2(X)$, $3(+)$, $4(\star)$.

j and i (total current), the average radius of the discharge filament has been obtained.

CONCLUSION

We have just examined some emission spectroscopy methods. It appears that these methods can be rather powerful, but the interpretation of the results must be carefully adapted in each particular case. General methods can hardly be used.

REFERENCES

Bastien, F., 1977, Thèse d'Etat, Orsay, France, n. 1871.
Bastien, F. and Marode, E., 1977, J. Quant. Spectrosc. Radiat. Transfer, 17:453.
Bastien, F. and Marode, E., 1977, Rev. Phys. Appl., 12:1121.
Bastien, F. and Marode, E., 1979, J. Phys. D, 12:249.
Bauer, G. H., 1979, J. de Phys. Collo. C7, 40:295.
Bleekrode, R. and Van Benthen, W., 1969, J. Appl. Phys., 40:5270.
Campbell, D. H. and Muntz, E. P., 1980, J. Chem. Phys., 72:1487.
Cramarossa, F., Ferraro, G., and Molinari, E., 1975, J. Quant. Spectrosc. Radiat. Transfer, 14:419.
Drawin, H. W., 1970, High Pres. - High Temp., 2:359.
Griem, H. R., Kolb, A. C. and Shen, K. Y., 1959, Phys. Rev., 116:4.
Griem, H. R., 1964, "Plasma Spectroscopy," McGraw-Hill, New York.
Griem, H. R., 1974, "Spectral Lines Broadening by Plasmas," Academic, New York.
Hartmann, G., 1977, Thèse d'Etat, Orsay, France, n. 1783.
Hartmann, G. and Johnson, P. C., 1978, J. Phys. B, 11:1597.

Herzberg, G., 1950, "Spectra of Diatomic Molecules," Vol. I of "Molecular Spectra and Molecular Structure," Nostrand, Princeton.
Hooper, Jr., C. F., 1968, Phys. Rev., 165:215.
Hooper, Jr., C. F., 1968, Phys. Rev., 169:193.
Marr, G. V., 1968, "Plasma Spectroscopy," Elsevier, Amsterdam, New York.
Moore, J. H. and Doering, J. P., 1968, Phys. Rev., 174:178.
Pecker-Wimel, Ch., 1968, "Introduction à la Spectroscopie des Plasmas," Dunod, Paris.
Wink, R. D., Fergusson, H. I. S. and Lowe, R. P., 1971, J. Quant. Spectrosc. Radiat. Transfer, 14:419.

PLASMA CHEMISTRY

A. Goldman* and J. Amouroux**

*Laboratoire de Physique des Décharges (CNRS)
Ecole Supérieure d'Electricité
91190 Gif-sur-Yvette, France

**Laboratoire de Génie Chimique
Ecole Nationale Supérieure de Chimie de Paris
75231 Paris, France

GENERALITIES

"The passage of electricity through a gas, as well as through an electrolyte . . . is accompanied and affected by chemical changes; also that chemical decomposition is not to be considered merely as an accidental attendant on the electrical discharges, but as an essential feature of the discharge without which it could not occur." If this citation of Thomson (1893) does not give the most suitable definition of plasma chemistry, it explains quite well why a chapter on plasma chemistry has its place in a course on electrical discharges. When Thomson wrote this, pioneering and extensive work on electrically induced chemical reactions in gases was already being performed by Berthelot (1869, 1898, 1899, etc.) who can be regarded as a forerunner of plasma chemistry, although, about one hundred years before, Henry and Dalton had realized the first laboratory reaction of plasma chemistry, succeeding in acetylene synthesis with a capacitive discharge in methane.

The study of electrolytic behavior to which Thomson refers has always prevailed over the study of their gaseous homologues. Thus, the concept of Debye length, currently used as a relevant definition parameter of gaseous plasma, arose from experiments performed by Debye and Hückel (1923) with strong electrolytes, which behave as liquid plasmas when they are almost fully dissociated into ions. However, plasma chemistry is becoming a discipline with its own prerogatives.

Since 1973, it has had its own congress series sponsored by the International Union of Pure and Applied Chemistry, the biennial International Symposium on Plasma Chemistry; it gives rise to various working meetings, e.g., the NSF Workshop on Plasma Chemistry and Arc Technology which took place in Minneapolis in 1980 immediately after a Gordon Conference on Plasma Chemistry. A new specialized scientific journal Plasma Chemistry and Plasma Processing is being published by Plenum Publishing Company. The reader may refer to all of them for up-to-date data and technological development. On the subject, he may also consult the basic treatise of Venugopalan (1971) or more applied books, such as those of Baddour and Timmins (1967), McTaggart (1967), Hollahan and Bell (1974), Veprek and Venugopalan (1980). Finally, he may also find useful information in reviews such as those made by Suhr (1973) and Fauchais and Rakowitz (1979), and in the volumes devoted to the subject which appear from time to time in the Advances in Chemistry Series published by the American Chemical Society and in the Advances in High Temperature Chemistry published by Academic Press.

Plasmas are usually defined in terms of electron density n_e and temperature T_e (or energy kT_e). Further, one knows that their electrical neutrality is insured only if their dimensions are distinctly larger than the Debye length, related to n_e and kT_e by the relation:

$$\lambda_D = \left(\frac{\varepsilon_o kT_e}{n_e e^2}\right)^{1/2} \quad (1)$$

where ε_o is the permittivity of free space; k, Boltzmann constant; and e, the electron charge. The classical diagram of Fig. 1 characterizes a number of natural and laboratory plasmas in terms of the three basic parameters n_e, T_e, λ_D. The plasmas produced by electrical discharges in gases fit into a region of small λ_D ($10^{-8} - 10^{-3}$ m), corresponding to relatively low T_e ($\leq 5 \times 10^4$ °K) and high n_e ($\geq 10^{16}$ el. m^{-3}); this means that the plasma producing regions of the discharge, i.e., the multiplication regions, basically involve rather small volumes.

Another relevant parameter for plasma chemistry is the temperature, T_g, of the background gas. According to T_g, these plasmas are usually divided into two groups, the high-temperature and low-temperature discharge plasmas:

(1) The high-temperature discharge plasmas, often referred to as hot or thermal plasmas, are in local thermodynamic equilibrium (LTE plasmas). In these plasmas, typically produced by arcs and plasmas jets, the gas temperatures are nearly identical to the electron temperatures:

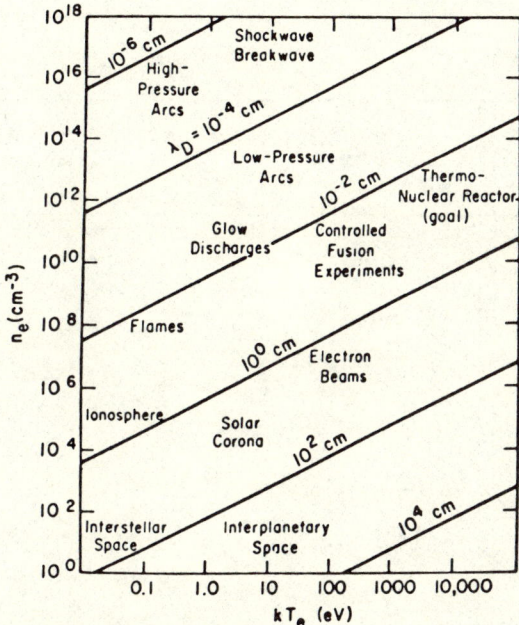

Fig. 1. Typical plasmas characterized by the electron energy kT_e, the electron density n_e, and the Debye length λ_D (Hollahan and Bell, 1974, Chap. 1).

$$T_g \sim T_e .$$

(2) In contrast, the low-temperature discharge plasmas, often referred to as cold plasmas, are characterized by a lack of equilibrium between T_e and T_g (non-LTE plasmas):

$$T_g \ll T_e .$$

T_g may be as low as room temperature. Grouped in Fig. 1 under the name "glow discharges," the discharges concerned include glow and corona discharges as well as high-frequency (HF), radio-frequency (RF) and microwave discharges.

Table 1 summarizes the discharge characteristics, important in plasma chemistry, which differentiate the two groups; for further data, see the detailed papers devoted in these proceedings to the different kinds of discharges.

Figure 2 indicates the main fields of application for each of the two groups. Which type of discharge to use may sometimes be quite evident: for instance, high-temperature plasmas should

Table 1. Characteristic features of the discharges used in plasma chemistry.

Discharge type → Characteristics ↓	Arcs Plasma jets	Glows and coronas HF, RF and microwave discharges
plasma state	LTE ("hot")	non-LTE ("cold")
gas pressure	≥ 1 atm	≤ 1 atm
reduced field	$< 1.4 \times 10^{-20}$ Vm^2	$0.2 - 8.5 \times 10^{-19}$ Vm^2
electron temp.	$0.5 - 5 \times 10^4$ °K	$10^3 - 10^5$ °K
gas temp.	- " -	300 - 1500 °K
current	100 - 100 A	≤ 1 A
electrical power	up to 5-10 MW	≤ 1 kW

LTE: Local Thermodynamic Equilibrium.

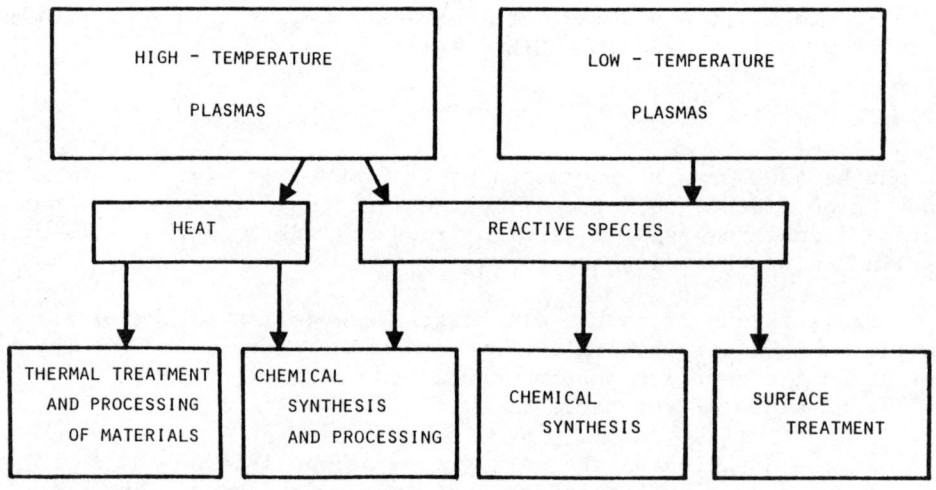

Fig. 2. Main active processes of plasma chemistry and related applications.

be employed in applications needing high heat transfer, while low-temperature plasmas are better suited to the processing of temperature sensitive materials. However, the choice is often less obvious; it is one of the present purposes of plasma chemistry research to develop optimized working conditions not only for new, challenging applications, but also for older ones.

The present paper is written as an introduction to plasma chemistry. Two main sections are devoted to the fundamental aspects: the specific features related to the use of high-temperature plasmas and nonequilibrium processes which play an increasing role in the present development of plasma chemistry. These processes, which indeed dominate in the low-temperature plasmas, may also play an important role in high-temperature plasma chemistry. For instance, encouraging results have been obtained in the field of NO synthesis by high-temperature plasmas, showing that nonequilibrium effects may increase the yield to substantially higher values than predicted by equilibrium calculations. Laboratory scientific approaches to selected applications are described in the final section.

SPECIFIC FEATURES OF HIGH-TEMPERATURE PLASMA CHEMISTRY

Constricted arc plasmas provide the best means of obtaining very high gas temperatures ($5 \times 10^3 < T_g < 5 \times 10^4$ °K), well beyond those which can be reached by conventional means, combustion flames, for instance. Before going further, as underlined by Reed (1967), it must be emphasized that solids or liquids no longer exist above 5,000 °K at 1 atm. pressure, and that above 10,000 °K the ionization rates are so high that no more molecules and only very few atoms exist. Hot plasmas will be well suited:

(1) as a source for heating either materials directly (melting, alloying, welding, cutting), or the gaseous medium which will transfer its energy to injected reactants (i.e., spheroidization),

(2) as a hot medium to accelerate chemical reactions (i.e. coal gasification or cracking fuels),

(3) to produce new chemical species or reactive intermediates needing high temperatures for their formation (i.e. NO synthesis).

As stated the main specific feature of these plasmas is their thermodynamic equilibrium which is achieved locally for sufficiently high gaseous pressure and discharge current (Fig. 3). The heating of the gas resulting from the collisions increases with pressure. On the other hand, high currents imply high electrical powers, in the range of 100 kW or more, while in low-temperature plasma chemistry the powers lie under some kW. Depending on the energy involved, the mass treated is generally also more important in the case of high-temperature plasma chemistry.

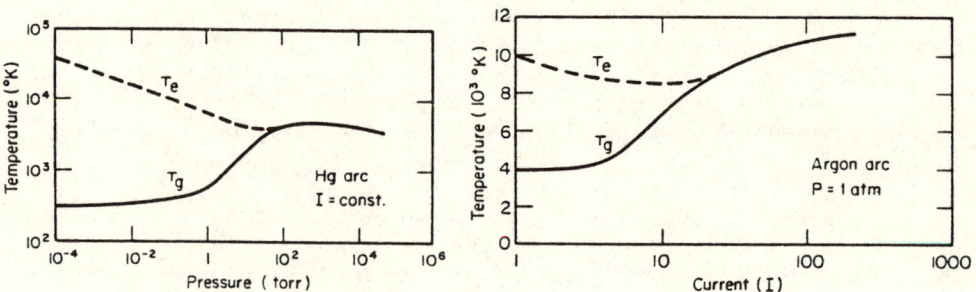

Fig. 3. Variation of electron and gas temperature with pressure and discharge current (Baddour and Timmins, 1967, Chap. 1).

Another most important parameter is the thermal conductivity of the gaseous medium, which is the determining parameter for the heat transfer efficiency of the plasma reactor. Gases usually have a higher thermal conductivity in the hot plasma temperature range than in the cold plasma range (Fig. 4). This is particularly true for the diatomic gases in the temperature range where dissociation processes greatly contribute to energy transfer, namely in the 2,500 - 5,000 °K range for H_2 and 5,000 - 8,000 °K for N_2, as illustrated by the first peak observed at increasing temperature on the curves of Fig. 4; the further peak, observed on the Ar curve as well as on the H_2 and N_2 curves, is due to the high reaction conductivity associated with the ionization processes (Gorse, 1975). Thus the generator yield reaches respectively 70 to 80 % with H_2 and 50 to 70 % while N_2 only reaches 30 to 40 % with Ar (Tallandier, 1979). However, Ar remains quite interesting as a source of metastables, used, for instance, for selective excitation of N_2, and as a gas needing quite low voltages for arc formation. A good solution is obtained by mixing Ar with H_2 or N_2. Figure 4 shows how the thermal conductivity of such Ar - H_2 and Ar - N_2 mixtures increases with H_2 and N_2 concentration.

In general, a typical hot plasma chemical reaction proceeds by the steps sketched in Fig. 5. Hereafter we shall briefly analyze the major problems related to these steps:

- plasma generation,

- reactant injection and mixing with the plasma,

- product quenching and recuperation.

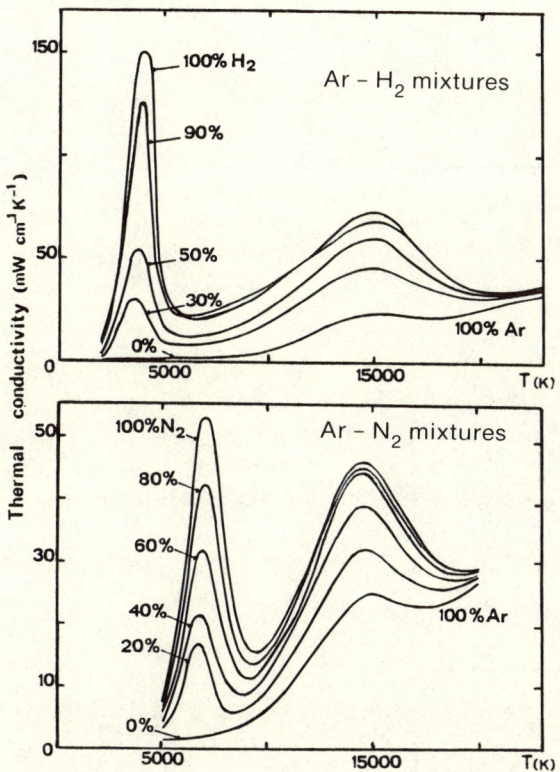

Fig. 4. Thermal conductivity of Ar-H_2 and Ar-N_2 mixtures at 1 atm. as a function of gas temperature (Gorse, 1975).

Plasma Generation

In the arc devices designed for plasma chemistry, the electric arc is generally constricted into a smaller circular cross section than would ordinarily exist in a classical open arc-type device, and thus it achieves higher temperatures.

Let us consider for instance a DC arc generator in which the discharge is fed by a rapid gas flow to generate plasma jets. According to the position of the electrodes with respect to each other and to the arc plasma, two different types of devices can be distinguished: transferred and nontransferred arc plasma generators.

(1) In the transferred arc plasma generator, the anode is some distance away from the cathode, and the arc is constricted between both electrodes. Such a device is represented in Fig. 6a. The nozzle around the cathode acts as an accelerating guide for the plasma which takes a shape close to a cylindrical column.

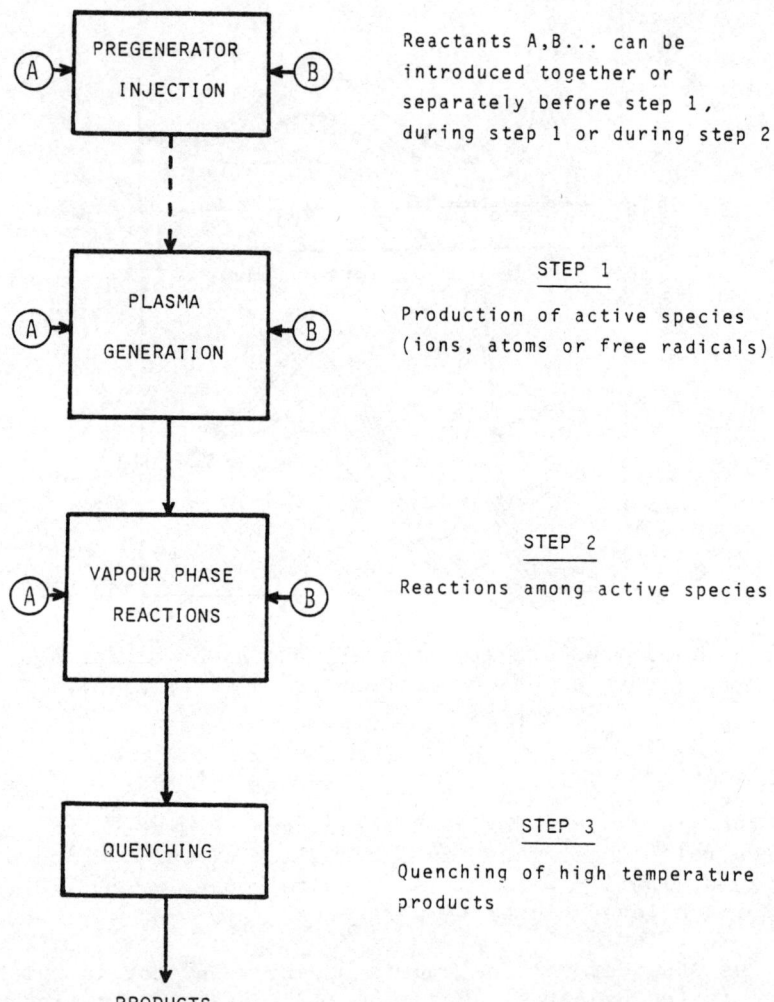

Fig. 5. General scheme of the different stages often involved in hot plasma chemical processing (Venugopalan, 1971, Chap. 11).

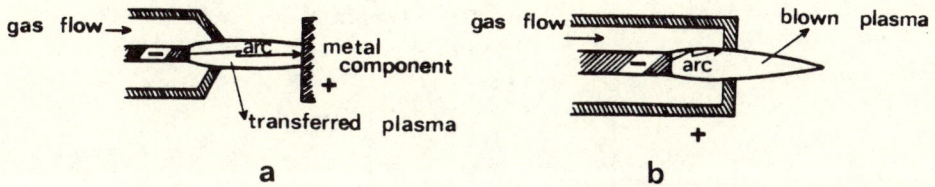

Fig. 6. Scheme of DC arc generators for the production of transferred (a) and nontransferred (b) plasmas.

This type of generator is especially well suited to the machining of metal components because these components can directly serve as anode.

(2) In generators of nontransferred arc plasma (also referred to as "blown" arc plasma), the arc is struck between the electrodes, represented by a cylindrical cathode and a tubular anode in Fig. 6b; the gas-fed plasma is ejected from the anode through an opening made in it for this purpose, and it emerges in the shape of a plume.

In the design of an arc plasma generator, much attention has to be paid to the electrodes. To limit their destruction by the plasma jet which carries high energies, they must be strongly cooled, but, unfortunately, this cannot be done without important energy losses. This problem can be avoided with the use of HF torches. When reactive gases are to be used or when a high degree of purity is desired, this is an undeniable advantage. However, since energy transfer is lower in HF torches, higher power input is needed to obtain equivalent results. Better yields can be obtained with gliding discharges from RF microtorches (Leprince, P. 1981, private communication).

Reactant Injection and Mixing with the Plasma

Thermal plasma processing routes are often based on the injection of matter into the plasma. As can be seen from Fig. 5, this injection can be made before, during, or after plasma generation. The mixing of the reactants with the plasma is more or less difficult according to their nature. For instance, in the case of solid particles, the results will strongly depend on the location, the speed, and the orientation chosen for the injection. Figure 7 gives a practical example of a DC arc reactor designed for the treatment of such particles which may be, for instance, a powder feed for spraying or UO_2 particles to be spheroidized for dispersed nuclear fuels. The particles are injected near the cathode tip, and the gas flow, which already serves here to feed the arc discharge, to cool the cathode, and to stabilize the arc aerodynamically, also

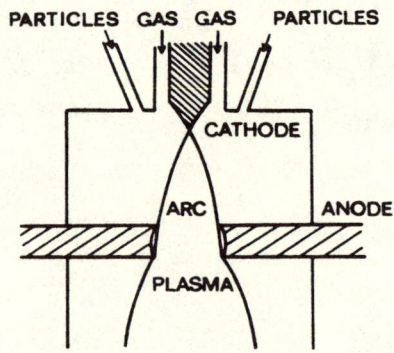

Fig. 7. Arc plasma reactor designed by IONARC-TAFA (U.S.A.) for injection of particles ahead of the plasma.

serves to drag the particles inside the plasma; it is helped in this last function by the Maecker effect (1953), which generates a magneto-hydrodynamic pumping in the vicinity of the cathode.

The major problem for the mixing of the reactants with the plasma comes from the high viscosity of the plasma. When possible, the best way for heat and mass transfer is to inject the reactants before plasma generation. In this case, the reactants participate in the reactions in the plasma, but sometimes reactions inside the plasma, or with the electrodes, may be harmful. With the electrodes, this is the case for corrosive reactants. Within the plasma, it is the case, for instance, for nonselective processes reducing the desired reactions; thus it is preferable in NO synthesis to dissociate N_2 first and to inject O_2 afterwards, in order to favor the reactions of O_2 with N rather than with N_2. On the other hand, the injection of cold reactants after plasma generation strongly modifies the temperatures, leading to a medium quite out of equilibrium in which excitation transfer often becomes the main reaction process.

Calculations assuming thermodynamic equilibrium can be made to predetermine the products of a given chemical reaction (Fauchais and Rakowitz, 1979), provided that the reaction is expected to occur more rapidly than the physical phenomena, such as relaxation of energy levels or maxwellization. This can be expressed by means of the Damköhler parameter (Barrère and Prud'homme, 1973)

$$D_i = t_c/t_m,$$

giving the ratio of the chemical reaction time t_c to the mechanical residence time t_m which strongly depends on the injection mode and parameters. Equilibrium conditions are satisfied when $D_i \ll 1$, i.e., in practice for $T_g \gtrsim 5,000$ °K.

Product Quenching and Recuperation

In plasma chemical synthesis, a quenching step is necessary to stabilize the final products. Its roles are, for times shorter than those involved by the reaction kinetics:

• On the one hand, to remove the excess kinetic energy from the new species formed;

• On the other hand, to deactivate the species which might destroy the final products by inverse reactions.

The importance of quenching is illustrated by the curves of Fig. 8b which are to be compared with those of Fig. 8a. In particular, those concerning NO production show that the maximum yield, reached for 1 atm. air at about 3,500 °K, can nearly be kept at lower temperatures by rapid quenching.

The most important quenching parameters for a reaction yield are the quenching delay before temperature begins to decrease and the quenching speed. Orders of magnitude are given by the following examples taken from Polak (Venugopalan, 1971, Chap. 13): a 2 ms delay should diminish methane conversion in acetylene from 15.5% to 10%, and a decrease of the quenching speed from 10^8 to 10^7 °K sec^{-1} should reduce NO production from 9.6 to 6.4 %.

The range of quenching speeds useful in hot plasma chemistry lies between 10^5 and 10^8 °K sec^{-1}. They can be obtained by different techniques already in use, such as fast expansion of the jet in a nozzle, contact with cooled walls, liquid pulverization, and cold-gas injection, but, due to the importance of quenching on the production of chemicals, there is still interest in the development of new improved techniques as well as in further theoretical studies.

The step after quenching, which consists of the products recuperation, is closely connected to it. Thus, for instance, the quenching of gaseous products by dilution with a colder gas leads to very large energy losses, more than 25%. A part of the energy could be recovered in a heat exchanger, but until now this has only been done on a laboratory scale.

NONEQUILIBRIUM PLASMA CHEMISTRY

First, recall that, if the nonequilibrium processes dominate in the low-temperature plasmas, they may also play an important

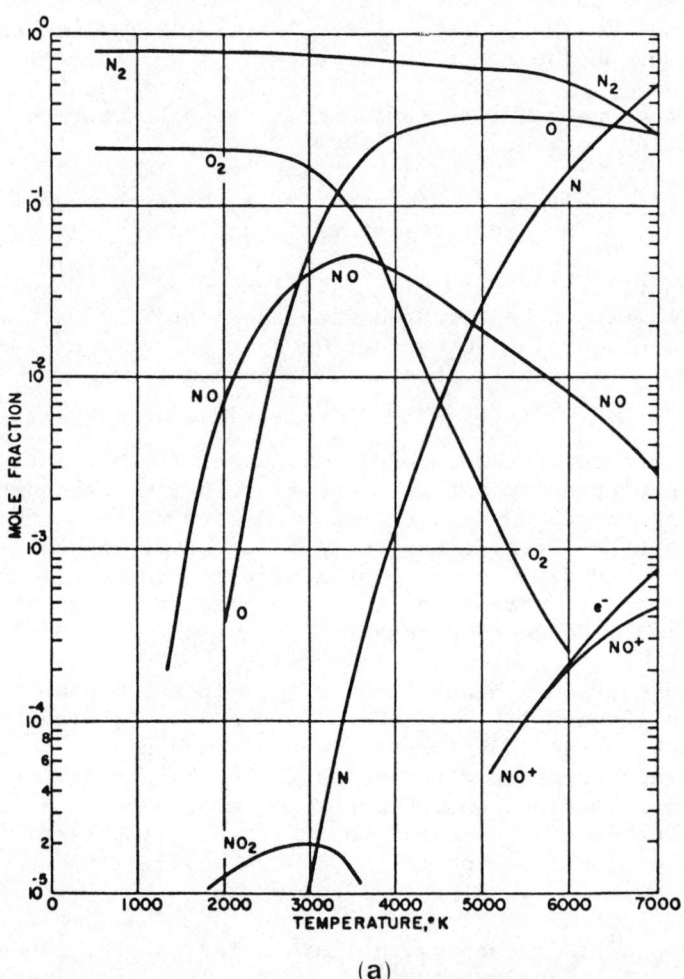

(a)

Fig. 8. Calculated curves of the composition of air at 1 atm. pressure (Baddour and Timmins, 1967, pp. 100-111). (a) at chemical equilibrium, (b) with quenching from 7000 °K at an initial rate of 3.5×10^7 °K sec^{-1}.

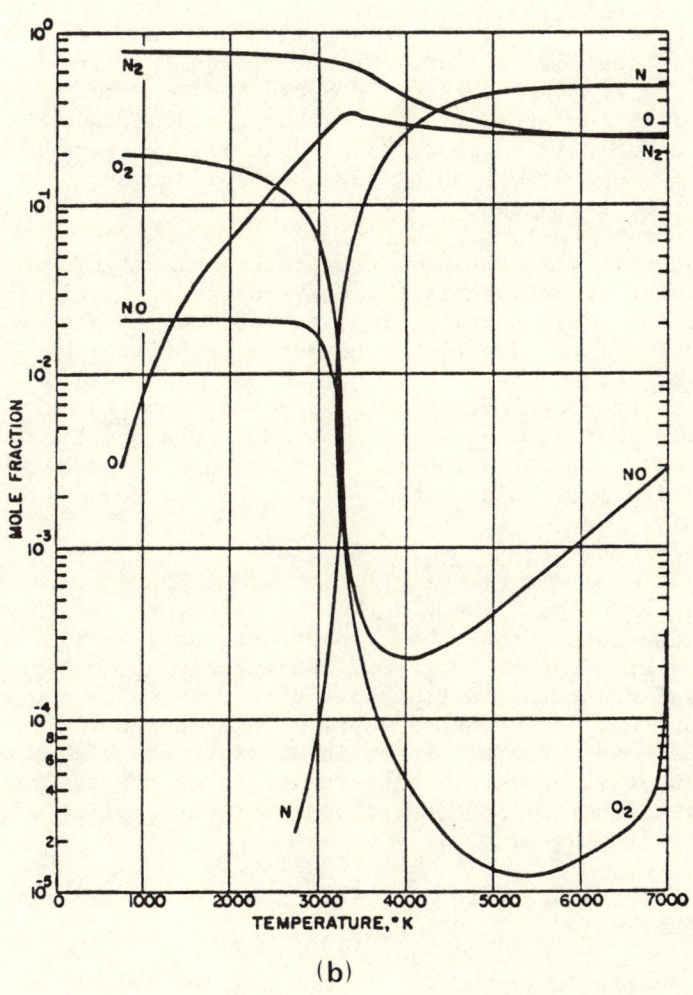

(b)

Fig. 8. Continued.

role in the high-temperature plasma chemistry, as in the case of NO synthesis cited.

Cold plasma discharges generally operate at lower pressures than hot plasma discharges (see Table 1), which makes energy exchanges more difficult and leads to a temperature spreading with a large gap between the electron temperature T_e and the gas temperature T_g (see the low pressure part of T_e and T_g curves in Fig. 3). This suggests that pressure could be varied to enhance or attenuate the nonequilibrium state of the discharge plasma; however, the fact that corona discharges, which typically generate nonequilibrium plasmas, operate at relatively high pressures, shows that this action must be taken cautiously.

In contrast with the high-temperature plasma chemistry, the low-temperature plasma chemistry only involves low electrical powers and concerns processes which are used more for "adding value" treatments than for large quantity treatments. Their <u>high electron temperatures</u> permit one to obtain chemically active species which could not be produced outside discharges at any feasible temperatures, while their <u>low gas temperatures</u> ensure practically instantaneous cooling of the products and allow one to process temperature-sensitive materials.

Production of Chemically Active Species

The interaction between a flux of electrons and the particles of a gaseous fluid leads to a transfer of energy which is responsible for the creation of a plasma. Chemically active species are produced and constitute a reservoir of energy which can be used for the applications of the so-called "plasma chemistry" which, besides the heat transfer aspect involved in the high-temperature plasma chemistry, concerns the interaction of the activated chemical species with other plasma constituents or with solids or liquids in contact with the plasma.

Usually, the activated chemical species are separated into three categories:

(1) species carrying electrical charges (electrons, ions)

(2) neutral fragments of molecules (atoms, radicals)

(3) excited molecules (vibrational, rotational, electronic excitation).

As will be seen a little later, *the reactivity of these different species depends not only on the energy transported by them, but also on other different factors related to their structure.*

A chemical reaction proceeds by modification of the molecular structure by means of changes in the repartition of the internal energy. It often depends on an electron exchange. Electrons and ions often initiate chemical reactions, for instance, oxydoreduction reactions, which result in displacement of electrical charges in the molecular architecture. This type of chemical reaction initiation becomes rapidly very complex, since ions polarize molecules and create molecular agregates called "clusters," the reactivity of which increases with the strength of the "ligand's" electrostatic field created by coordinated radials around a central ion.

Radicals and atoms are formed from molecules by decomposition reactions. This kind of reaction is more efficient when the energy is provided in a vibrational and/or an electronic form rather than in a kinetic form. Cross sections for vibrational and electronic (optically allowed or forbidden) excitation are usually very high in the low temperature discharge plasmas. The extremely high chemical reactivity of the so-formed species leads to a multitude of fast reactions which find many applications, in particular, in surface treatments (grafting, etching, plasma polymerization, etc.).

To give an idea of the diversity of reactions which take place in a discharge plasma we can look in some detail at what already occurs when only one parent gas is present. To this end, let us consider oxygen discharges which we shall meet again in the next section as a source of "singlet oxygen," a most useful metastable molecule, and later as a source for ozone production.

We may begin by considering the energy balance in steady-state conditions, when the loss of energy due to electron-molecule collisions and other processes is just balanced by the energy gain of the electrons from the electric field. Bell (Hollahan and Bell, 1974, Chap. 1) writes this energy balance by means of the following relation which takes into account the loss of energy by elastic and inelastic collisions, respectively, in the first two terms; the energy needed to bring each newly formed electron up to the average electron energy in the third term; and the loss of energy associated with ambipolar diffusion in the fourth term:

$$\bar{P} = \frac{2m}{M} <\varepsilon> k_m \, nN + \Sigma_j \, \varepsilon_j \, k_j \, nN + <\varepsilon> k_i \, nN + <\varepsilon> k_d \, n \, . \quad (2)$$

In this expression, $<\varepsilon>$ is the average electron energy, and ε_j the energy loss for the jth inelastic process between the electrons, of mass m and density n, and the molecules, of mass M and density N. The fractional losses of energy are here expressed as functions of rate constants: $k_m = \nu_m/N$ for momentum transfer characterizing the collisions of frequency ν_m, k_j for the jth inelastic process, k_i for the ionizing collisions and $k_d = D_a/\Lambda^2$ where D_a is the ambipolar diffusivity and Λ the diffusion length. Used in plasma chemistry, in preference to the cross section, to express the probability of occurrence of a particular kind of collision process, the reaction rate coefficient k is related to the cross section $\sigma(v_r)$ for a two-body process between two colliding particles evolving with a relative velocity v_r, by the expression (Hasted, 1972):

$$k = \int_0^\infty v_r \, \sigma(v_r) \, f(v_r) \, dv_r \quad . \tag{3}$$

where $f(v_r)$ is the fraction of the collisions in which v_r lies between v_r and $v_r + dv_r$. This relation should apply to different kinds of reactions provided one makes appropriate changes. Because k is determined by an evolution equation in the form

$$\frac{\partial N_1}{\partial t} = k \, N_1 \, N_2 \quad \text{or} \quad k \, N_1 \, N_2 \, N_3 \quad , \tag{4}$$

it will have units of $cm^3 sec^{-1}$ (molecule^{-1}) in the case of two-body processes between particles of concentrations N_1 and N_2 and units of $cm^6 sec^{-1}$ (molecule^{-2}) in the case of three-body processes between particles of concentrations N_1, N_2 and N_3.

Since the rate constants for the collisional processes between the electrons and the molecules are functions of the reduced field E/N, the fractional losses of energy may also be expressed in terms of this parameter. Figure 9 shows the losses for a variety of such processes occurring in oxygen. Concerning first the elastic collisions, we see that no more than about 1.5% of the total power is consumed by them; this explains why the gas temperature here remains low, since it is the kinetic energy (2 m/M) ε gained by the molecule from the energy ε of the colliding electron which leads in the hot discharge plasma to an important increase of temperature, due in this case to the higher values of both electron and molecule densities. In contrast, the inelastic collisions dominate in the cold discharge plasmas. At low fields, they mainly occur through the excitation of vibrational states. As the field is increased, various electronic excitation and dissociation processes occur, consuming more and more energy until the ionization processes take the most important part of the power input.

The positive and negative ions produced by the electron-molecule collisions can react further with the molecules, with the electrons, or between them. Table 2 shows how long the list of elementary reactions can be, even in the case when only one parent gas is present. Whether the reactions are exothermic or endothermic, they proceed in the first case at a rate essentially independent of the energy or in the second case only if the energy is greater than the activation energy of the reaction.

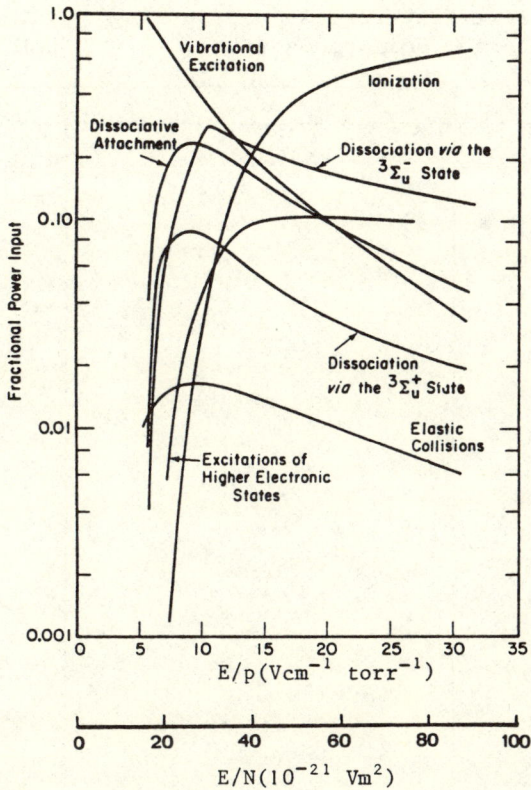

Fig. 9. Fractional electrical power transferred by electron-molecule collisions for a variety of processes in oxygen, as a function of the reduced field (Hollahan and Bell, 1974, Chap. 1).

Reactivity of the Chemical Species

In the preceding section, the chemical species have been essentially considered from an energy point of view. Now, we shall see that their properties are also influenced by their molecular structure and thus depend at the same time on:

(1) the electronic level of the molecular oribtals,

(2) the vibrational level,

(3) the symmetry of the atomic arrangement, i.e., the bonding angles,

(4) the hybridation of the structure.

Table 2. Elementary reactions occurring in an oxygen discharge (Hollahan and Bell 1974, Chap. 1).

Reaction	k	σ_{max} (cm²)	Reference
Ionization			
1. $e + O_2 \to O_2^+ + 2e$		2.72×10^{-16}	[9]
2. $e + O \to O^+ + 2e$		1.54×10^{-18}	[10]
Dissociative ionization			
3. $e + O_2 \to O^+ + O + 2e$		1.0×10^{-16}	[11]
Dissociative attachment			
4. $e + O_2 \to O^- + O$		1.41×10^{-18}	[12]
5. $e + O_2 \to O^- + O^+ + e$		4.85×10^{-19}	[12]
Dissociation			
6. $e + O_2 \to 2O + e$		2.25×10^{-18}	[13]
Metastable formation			
7. $e + O_2 \to O_2(^1\Delta_g) + e$		3.0×10^{-20}	[13]
Charge transfer			
8. $O^+ + O_2 \to O_2^+ + O$	2×10^{-11} cm³/sec		[14]
9. $O_2^+ + O \to O^+ + O_2$		8×10^{-16}	[15]
10. $O_2^+ + O_2 \to O_3^+ + O$		1×10^{-16}	[16]
11. $O_2^+ + 2O_2 \to O_4^+ + O_2$	2.8×10^{-30} cm⁶/sec		[17]
	2.5×10^{-14} cm³/sec at $E/p = 20$ V/cm torr		[18]
12. $O^- + O_2 \to O_2^- + O$	3.4×10^{-12} cm³/sec at $E/p = 45$ V/cm torr		
13. $O^- + O_3 \to O_3^- + O$	5.3×10^{-10} cm³/sec		[19]
14. $O^- + 2O_2 \to O_3^- + O_2$	$1.0 \pm 0.2 \times 10^{-30}$ cm⁶/sec		[18]
15. $O_2^- + O \to O^- + O_2$	5×10^{-10} cm³/sec		[20]
16. $O_2^- + O_2 \to O_3^- + O$		$<10^{-18}$	[21]
17. $O_2^- + O_3 \to O_3^- + O_2$	4.0×10^{-10} cm³/sec		[19]
18. $O_2^- + 2O_2 \to O_4^- + O_2$	3×10^{-31} cm⁶/sec		[22]
19. $O_3^- + O_2 \to O_2^- + O_3$		4×10^{-17}	[21]
20. $O_4^- + O \to O_3^- + O_2$	4×10^{-10} cm³/sec		[23]
21. $O_4^- + O_2 \to O_2^- + 2O_2$	6×10^{-15} cm³/sec		[22]
Detachment			
22. $O^- + O \to O_2 + e$	3.0×10^{-10} cm³/sec		[20]
23. $O^- + O_2 \to O + O_2 + e$		7×10^{-16}	[24]
24. $O^- + O_2(^1\Delta_g) \to O_3 + e$	$\sim 3 \times 10^{-10}$ cm³/sec		[25]
25. $O_2^- + O \to O_3 + e$	5.0×10^{-10} cm³/sec		[20]
26. $O_2^- + O_2 \to 2O_2 + e$		7×10^{-16}	[24]
27. $O_2^- + O_2(^1\Delta_g) \to 2O_2 + e$	$\sim 2 \times 10^{-10}$ cm³/sec		[25]
Electron–ion recombination			
28. $e + \begin{Bmatrix} O^+ \\ O_2^+ \\ O_3^+ \\ O_4^+ \end{Bmatrix} \to \begin{Bmatrix} O \\ 2O \\ O + O_2 \\ 2O_2 \end{Bmatrix}$	$\leqslant 10^{-7}$ cm³/sec		[26]
Ion–ion recombination			
29. $\begin{Bmatrix} O^- \\ O_2^- \\ O_3^- \\ O_4^- \end{Bmatrix} + \begin{Bmatrix} O^+ \\ O_2^+ \\ O_3^+ \\ O_4^+ \end{Bmatrix} \to \begin{Bmatrix} O \\ O_2 \end{Bmatrix}$	$\sim 10^{-7}$ cm³/sec		[27]
Atom recombination			
30. $2O + O_2 \to 2O_2$	2.3×10^{-33} cm⁶/sec		[28]
31. $3O \to O + O_2$	1.5×10^{-34} cm⁶/sec		[28]
32. $O + 2O_2 \to O_2 + O_2$	$1.9 \times 10^{-35} \exp(2100/RT)$ cm⁶/sec		[29]
33. $O + O_3 \to 2O_2$	$2.0 \times 10^{-11} \exp(-4790/RT)$ cm³/sec		[29]
34. $O \xrightarrow{wall} \tfrac{1}{2} O_2$	$\gamma = 1.6 \times 10^{-4}$ to 1.4×10^{-2} ($T = 20$–$600°C$)		[30]

For instance, consider the case of a diatomic molecule such as the molecule N_2. An increase of its internal energy may be stored in the $X^1\Sigma_g^+$ state, the ground state, by an increase of the vibrational excitation only or also in a higher electronic state, for instance the $A^3\Sigma_u^+$ excited state, with a part of the energy in a vibrational level (Fig. 10). In so far as the total energy stored in the molecule is the same, the result is the same as long as only the energy transfer is considered (Gray, 1973). But if now we look at the chemical reactivity of the molecule in the two cases, it is quite different because, as mentioned above, the chemical properties depend on the molecular oribtals:

(a) For the molecule N_2 in the ground state ($X^1\Sigma_g^+$), an increase of the vibrational level does not modify the electronic configuration of the σ and π bonds between the two atoms N; the chemical

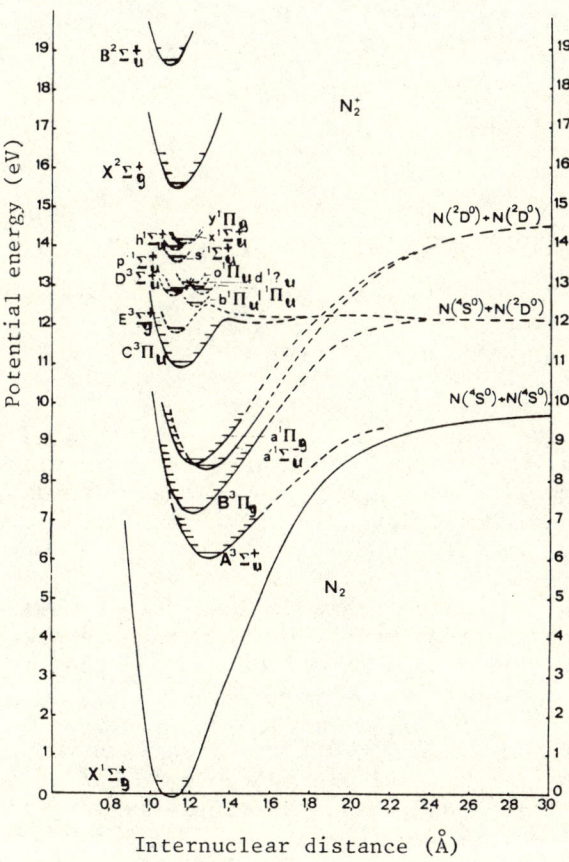

Fig. 10. Potential energy curves for N_2.

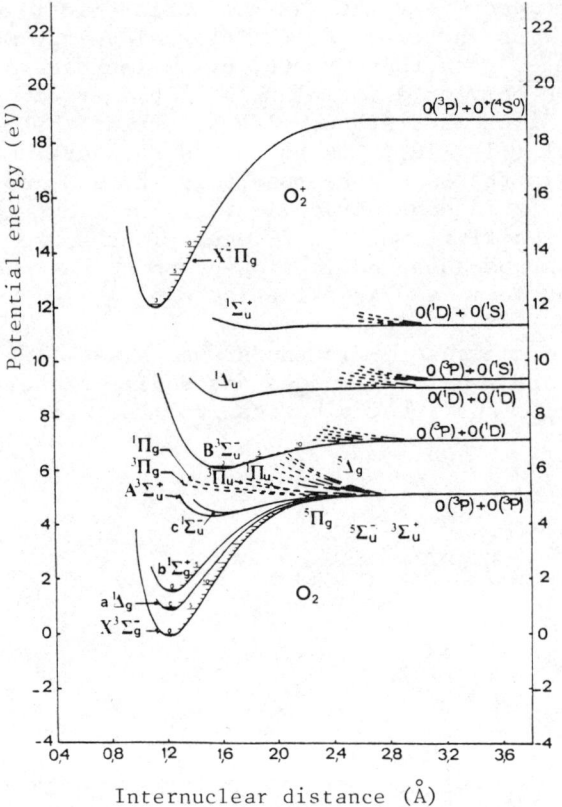

Fig. 12. Potential energy curves for O_2.

are not exactly the same as in the case considered before, with the ground state and an excited state of the molecule N_2 both vibrationally excited. But again, the two species exhibit quite different chemical reactivity, depending on the electron repartition in the π^* orbital (see Table 3):

(a) Though optically forbidden like the $^1\Delta_g$ state, which should imply long radiative lifetimes, the $^1\Sigma_g^+$ state has a relatively short lifetime because of its unpaired electrons and thus is not very interesting for applications.

(b) In contrast, the $^1\Delta_g$ state (already introduced in Table 2), with an electron pair in a single molecular orbital leaving one of the highest orbitals empty, has a particularly long lifetime (45 min in vacuum, 5×10^{-2} sec at 1 atm, $10^{-6} - 10^{0.3}$ sec in different kinds of solutions) and can diffuse relatively deeply into thin films ($\gtrsim 100$ Å). Its chemical properties are those of a double properties are poor because the molecule in the ground state cannot easily exchange bonds with other chemical compounds.

(b) The molecule on the $A^3\Sigma_u^+$ state has two unpaired electrons, one on the π_{2p} orbital and the other on the antibonding orbital π^*_{2p} (Fig. 11). So, independently of the fact that the strength of the bond between the two atoms decreases from 9.7 eV ($X^1\Sigma_g^+$) to 3.6 eV ($A^3\Sigma_u^+$), the electronic configuration and the chemical properties of the triplet state of N_2 are similar to those of dinitrogen and amine functions.

Another interesting case is the "singlet oxygen." Under appropriate excitation conditions, two low electronically excited singlet states of oxygen can be obtained from the ground state ($X^3\Sigma_g^-$): the $a^1\Delta_g$ state at 0.98 eV and the $b^1\Sigma_g^+$ at 1.63 eV (Fig. 12). Though quite close to each other, the energies of the two excited molecules

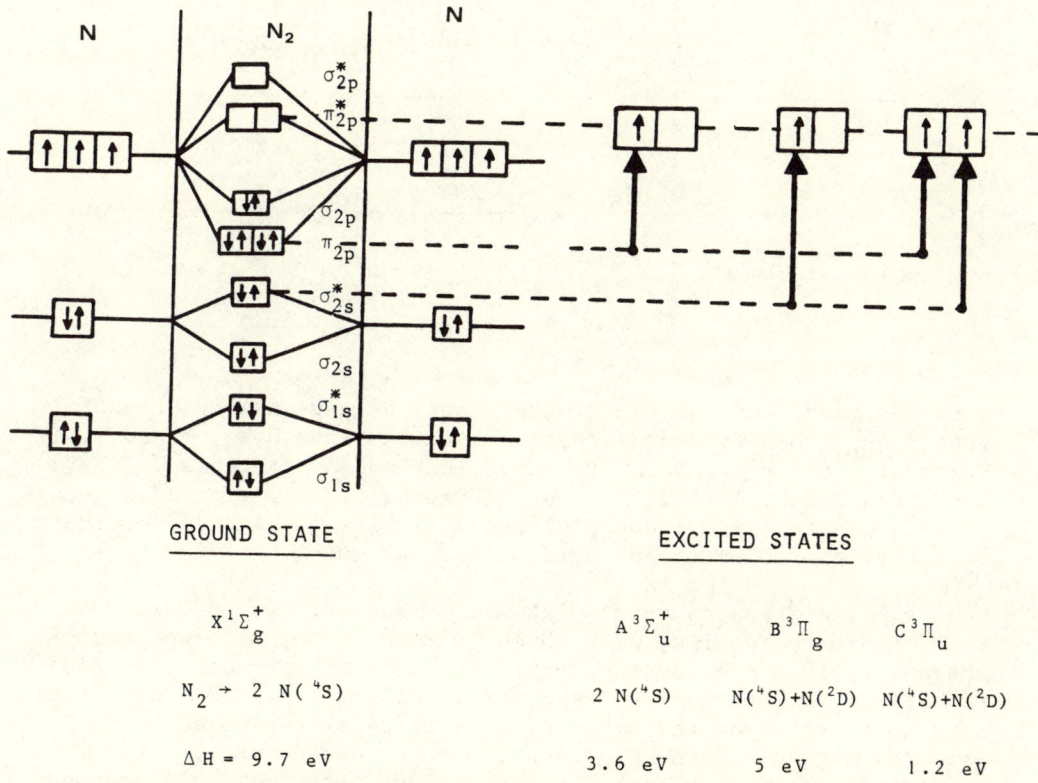

Fig. 11. Main features of the molecule N_2 in the ground state and in the excited triplet states $A^3\Sigma_u^+$, $B^3\Pi_g$ and $C^3\Pi_u$: electronic configuration and bonding strength between the two atoms of the molecule.

Table 3. Configuration of the external 2 p^4 electron shell of O_2 in the ground state and in the first excited singlet states $a^1\Delta_g$ and $b^1\Sigma_g^+$.

spectroscopic formulation of the state	$X^3\Sigma_g^-$	$a^1\Delta_g$	$b^1\Sigma_g^+$
energy level above ground state	0	0.98 eV	1.63 eV
chemical figure			
orbital angular momentum	opposite	identical	opposite
occupancy scheme in the π^* orbital	↑ ↑	↑↓ —	↓ ↑
radiative life-time in vacuum		45 min	7 - 12 sec

ethylene bond, and, at the present time, a large number of useful reactions with different organic substances have been or are still being studied on a laboratory scale (Kaplan, 1971; Schaap, 1976; Rånby and Rabek, 1978). The peculiarities of this excited molecule should lead to large industrial use, practically limited for the moment to pharmaceutic and perfume industries.

All the above concerns diatomic molecules. In the case of more complex molecules, with three or more atoms, a large part of the energy is stored by the modification of the structure hybridation, i.e., of the molecule symmetry. Gillepsie (1967) rules indicate an increase of the symmetry properties of the molecule for an exothermic reaction and a decrease for an endothermic reaction. These rules point out that the reactivity of the molecule depends on the spatial repartition of the energy in the electron molecular orbitals; generally a decrease of the symmetry permits an easier chemical reaction, in connection with the energy stored by the molecule.

Chemical Reactions in the Gas Role of Heterogeneous Catalytic Processes

The general concept of temperature as classically used for systems in thermodynamic equilibrium must be reconsidered in the case of a plasma out of equilibrium, where the energy is unequally shared between different processes. One needs in each case to take into account the different forms of energy involved as well as the different temperatures associated with them.

The concept of nonequilibrium reactions in gas phase systems was first developed by Polanyi's group (Polanyi, 1972) on the endothermic reaction:

$$H + HF \rightarrow H_2 + F \quad (\Delta H = 1.37 \text{ eV})$$

induced by a molecular beam of hydrogen. Their experiments showed that the reaction rate of the chemical system depends on the repartition of the energy on the different degrees of freedom. The diagram in Fig. 13 illustrates the evolution of this reaction rate as a function of the energy distribution between the vibrational, rotational, and translational levels with which are respectively associated the ordinates V axis, the abscissa R axis, and the diagonal T axis. While a vibrational excitation of the molecule HF allows an increase in the reaction rate by a factor on the order of 10^3, an increase of energy on the rotational or the translational level

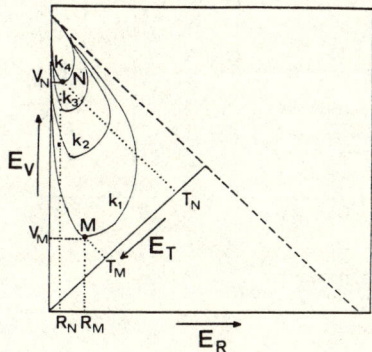

Fig. 13. Polanyi's schematic representation of the reaction rate k as a function of the vibrational energy E_V, the rotational energy E_R, and the translational energy E_T. The contour lines represent curves of constant reaction rate, increasing from k_1 up to k_4.

does not modify the reaction rate. This led Polanyi to interpret the endothermic and exothermic reactions by the diagrams respectively sketched in Fig. 14.I and 14.II for a colinear triatomic reaction:

$$A + BC \rightarrow AB + C$$

It is the only one which allows such a graphic representation, but the same treatment could be applied to follow the dynamical evolution of more complicated systems (Ashmore, 1977). The diagrams of Fig. 14 represent isoenergetic lines of a potential energy surface section; r_1 is the distance between the incident A atom and the atom B of the reactant molecule BC (reactants' side) and r_2, the distance between the atom B of the product molecule AB and the leaving C atom (products' side). The energy barrier in the valley corresponds to the formation of the complex chemical species [A-B-C]. The reaction proceeds by a displacement on the potential energy surface from the right to the left, and the excitation mode depends on the position of the energy barrier:

• In case I, which corresponds to an exothermic reaction, the energy barrier sits in the entrance valley, and the most difficult step is to bring A close to B. This could be achieved by concentrating the energy on the translation of A relative to BC.

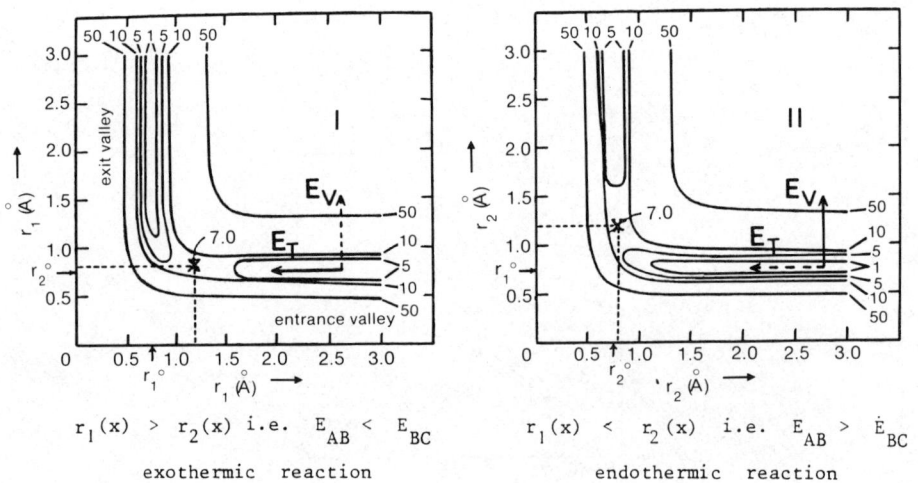

Fig. 14. Potential energy diagram as a function of interatomic distances for the reaction: $A + BC \rightarrow AB + C$, (after Polanyi). $r_1 = r_{A-B}$; $r_2 = r_{B-C}$. Curves represent isoenergetic lines of the potential energy surface, in kcal/mol (1 kcal/mol ≈ 0.043 eV). The point (x) denotes the lowest energy path (here 7 kcal/mol).

- On the other hand, in case II, corresponding to an endothermic reaction, the energy barrier sits in the exit valley, and the most difficult step concerns the separation between B and C. The reaction will be favored by preferential vibrational excitation of the molecule BC.

Thus, for the reaction under consideration:

$$A + BC(v,j) \rightleftarrows AB(v',j') + C,$$

the reaction rate coefficient will be written, according to Eq. (3), as:

$$k = \int_0^\infty v_r \sigma(v,j,v',j',E_r) \, f(v_r,T) \, dv_r, \qquad (5)$$

which takes into account the different forms of energy which control the transformation, i.e., the vibrational and rotational energies of the molecules BC and AB and the relative kinetic energy E_r of the species.

Gas Phase Reactions in a Plasma Chemical Reactor. The plasma system is more complex than the preceding one: the electronic excitation participates directly in the chemical reaction, which means that the reaction rate also depends on this parameter not involved in the case studied above. Polanyi's rules may apply to the plasmas out of equilibrium produced by electrical discharges, but care must be taken as will now be shown.

Let us consider the two endothermic reactions controlling NO synthesis in a $N_2 + O_2$ plasma, on the one hand and HCN synthesis in a $N_2 + CH_4$ plasma on the other. As shown by the results of the Amouroux' team grouped in Table 4, the nitrogen fixation yield which controls both processes increases with pressure in NO synthesis and varies inversely in HCN synthesis. According to the effect of pressure variations on the excitation processes indicates in the table by the spectroscopic results, this means that the former reaction is favored by a high vibrational temperature and the latter by a high translational temperature, as can also be seen from the curves of Fig. 15a and 15b (gas phase curve). These results seem to contradict Polanyi's rules stated above, but the difficulty has been solved by considering the chemical reactional system not as a whole but as a series of three successive steps:

(1) excitation of the reactants by the discharge,

(2) interaction of the excited species in the bulk of the plasma,

Table 4. Active energy processes controlling NO synthesis in $N_2 + O_2$ plasma, according to Amouroux et al. (1979), and HCN synthesis in a $N_2 + CH_4$ plasma, according to Rapakoulias and Amouroux (1979).

Increasing parameter	Spectroscopic results			% N_2 fixed	
	T_{vib}	T_{rot}	T_{el}	$N_2 - O_2$ reaction	$N_2 - CH_4$ reaction
↗ energy	↗	↗	↗	↗	↗
↗ pressure	↘	↗	⇗	↘	↗
↗ N_2 flow	≈	⇗	↗	↘	↘

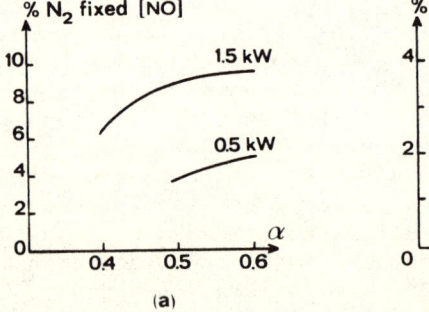
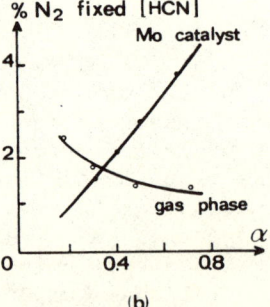

Fig. 15. Conversion yield as a function of the parameter $\alpha = (T_{vib} - T_{transl})/(T_{vib} + T_{transl})$ characterizing the nonequilibrium degree of the plasma (Rapakoulias et al., 1981) (a) in a $N_2 + O_2$ plasma for NO synthesis, (b) in a $N_2 + CH_4$ plasma for HCN synthesis.

(3) quenching of all the excited species at the walls of the discharge vessel or on a target.

Each step constitutes a separate "reaction" with a reactant and a rather long lifetime product. Polanyi's rules should be applied to these individual reaction steps. For example, in NO synthesis, the global reaction is controlled by the endothermic step of N_2 bond dissociation (excitation) and thus is favored by high vibrational temperature. In the case of HCN, the global synthesis is controlled by the exothermic step of the release of the excess energy from the product molecule and thus is favored by a high translational temperature (quenching).

Another particular point of the reactivity in electrical discharges is the electronic excitation of the molecules. As far as not very low pressure discharges are concerned, this electronic excitation concerns only a minority of the molecules. But electronically excited states are often very reactive, and Polanyi's rules are not valid for them.

Role of Heterogeneous Catalytic Processes. Different chemical and physical phenomena may occur when the plasma comes into contact with solid surfaces. These phenomena will be studied in the next section. However, one class of processes, heterogeneous catalytic, is the concern of the present section since, in this case, the role of the solid is only to "lend" its surface for a surface reaction to occur (Eremin and Rubatsova, 1973; Meubus, 1975; Botchway and Venugopalan, 1980).

The only theoretical treatment of heterogeneous catalysis in the case of nonequilibrium gas phase systems has been carried out by Wolken and co-workers (Purvis et al., 1979). For their calculations, which have not yet been experimentally confirmed, they used simple reacting systems (H_2 dissociation, O_2 adsorption) and did not consider electronic excitation of the gas. As in the case of Polanyi's studies of gas phase reactivity, Wolken's conclusions can be extended to discharge plasmas but need an adaptation, because of the presence of excited electronic states and the multiform excitation of molecules in the plasma.

The heterogeneous catalytic processes involve three successive steps (Rapakoulias et al., 1981).

(a) *Chemisorption of the reactants on the surface of the catalyst*. There are three types of adsorption: activated adsorption (with an energy barrier), adsorption without an energy barrier and spontaneous adsorption (attractive energy potential). Wolken's calculations show that in the first two cases preferential vibrational excitation of the reactants will favor the adsorption process. Experiments in plasmas confirm this conclusion: as shown in Fig. 15b the efficiency of the catalytic process is enhanced when the nonequilibrium degree α, i.e., vibrational excitation, is increased. Furthermore, the electronic excitation of the molecules in the plasma is higher when nonequilibrium is increased. These states chemisorb very easily because of their different orbital (space) symmetry. So, catalytic reactions from electronic excitation are also favored by nonequilibrium.

(b) *Reactions of the absorbed species on the surface.* The influence of vibrational excitation of the gaseous phase on the surface reactions is not well established. It seems, however, that the high vibrational excitation of the gas is transferred to the solic-adsorbate bond, facilitating the diffusion on the surface (i.e., the reaction probability) and the desorption of the products. The favorable effect on the catalytic reaction is, however, limited by the possibility of activating at the same time the desorption of the adsorbed reactants.

(c) *Desorption of the products.* At the end of the heterogeneous reaction, the energy balance between the bond energy of the products and their desorption energy is often in excess. The desorbed gaseous molecule conserves a great part of this excess energy, and sometimes all of it (Halpern and Rosner, 1978). The theoretical calculations of Wolken indicate that this excess will be channelled into the vibration of the molecule in the case of quasi-symmetric molecules and that for highly asymmetric molecules, the vibrational and translational excitation of the desorbing molecule will be equal. Amouroux and co-workers' experiments have confirmed these theoretical conclusions. For example, in the case of N_2 catalytic dissociation:

$$N_2 + O_2 \xrightarrow[\text{or } WO_3]{MoO_3} 2 NO,$$

the spectroscopic analysis of the gas in the close vicinity of the catalyst surface has shown the existence of a boundary film in which the vibrational excitation of N_2 is significantly higher than that in the bulk of the plasma (Rapakoulias et al., 1981).

Energy Yield of the Plasma Reactor. The energy yield of plasma reactors is generally low. Effectively, as excitation in the discharge is not selective, an important part of the induced energy is consumed by the excitation of nonreacting low energy states. For example, in the case of N_2 plasmas, most of the energy is consumed by the population of the metastable N_2 $A^3\Sigma_u^+$ state or by the vibrational excitation of the ground state $N_2 X^1\Sigma_g^+$, which do not fix nitrogen by gas-phase reaction.

Preferential excitation of high energy electronic states will increase the chemical yield of the reactor but not its energy yield which will be all the lower as the main reaction will proceed by a higher excitation state. For example, if we look again at the reaction of NO synthesis:

$$N_2 + O_2 \rightarrow 2 \text{ NO}$$

which gives two NO molecules with a reaction enthalpy of 1 eV for each of them, it is easy to see that the energy yield can be at best (i.e. assuming 100% chemical yield), as sketched in the diagram of Fig. 16:

- 13% only by preferential use of the N_2^+ ion which is very reactive but needs 15.5 eV to be produced (see Fig. 10 for the potential levels of N_2);

- 20% by excitation of the dissociative N_2 $C^3\pi_u$ state which needs about 11 eV;

- 31% by catalytic means using for instance the excitation of the low energy metastables N_2 $A^3\Sigma_u^+$ which consumes about 6 eV only.

Physical and Chemical Effects on the Surfaces

In this section, we shall analyze the general aspects of the modifications of the surfaces by the discharge plasma in contact

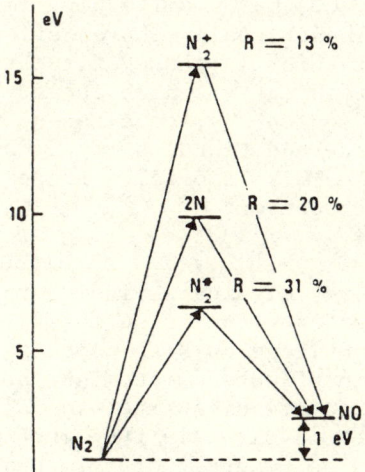

Fig. 16. Theoretical energy yield of NO synthesis for three main ways of dissociating N_2 (Rapakoulias et al., 1981).

with them. One can distinguish three classes of processes occurring between them:

(1) chemical processes (e.g., oxidation, nitriding, etc.)

(2) mechanical processes (e.g., sputtering)

(3) electrostatic processes (charging of insulating materials or of poor conducting spots or layers on conducting materials).

The effects are not independent and consequently a prolonged chemical surface treatment will necessarily lead to morphological changes of the surface. So such treatments will have to be stopped in any case before degradation effects become significant.

In contrast to the case treated in the preceding section of heterogeneous catalysis, where the essential function of the surface is to help a gas phase reaction to occur and where the solid can remain practically unchanged at the end of the reaction, the surface can, in the present case, act as a reaction partner, either as a reactant (etching) or as a product (coating). Furthermore, the solid can react at its surface, either completely (metal nitriding) or only at definite sites (polymer treatment).

The effects are obviously quite different according to the kinetic energy transported by the impinging particles, to the conducting or insulating nature of the surface involved, and to its polarity in the case of a conducting surface. Note also that the gas, the bulk material of solid components, and certain constitments or impurities like H_2O, CO, O_2 or oil vapor, join together to form surfaces which are no longer really characteristic of the bulk material. This accounts, for instance, for changes of the surface secondary electron coefficient γ, which is very sensitive at the same time to impinging ion energy, to clean-up effects produced by ions and heat, and to the additional surface constituents. Further, for potentials below 1 - 10 V, a metal surface very often acts as an insulator or a semiconductor.

We shall not look in great detail at either the basic processes by which the energy dissipates inside the solid or all the elementary subsequent effects on its surface (there is in any case an important lack of knowledge on both aspects), but we shall first briefly examine the main effects which might be expected from the action of the low temperature discharge plasmas and then consider separately the processes involved in the impact of the species activated by the discharge, whether or not they transport kinetic energy.

PLASMA CHEMISTRY

By direct chemical reactions with the surface material, the discharge plasma can produce very different kinds of modifications, such as:

• With oxygen, strong adhesive bonds, increased wettability by formation of polar groups, sites for chemical attachment of other molecules, clean removal of impurities, protective coatings.

• With nitrogen, scavenging of free radicals, destruction of organic compounds, formation of carbon-nitrogen bonds, increased hardness of metals.

• With fluorine, decreased wettability and formation of a chemically non reactive layer on polymer surfaces, changes of friction coefficients of metals.

• With hydrogen, regeneration of catalyst active sites.

The discharge plasma can also operate by the formation of free radials on the surface, which in turn react with the material or with other media subsequently brought into contact with the surface. Free radicals, formed not only at the surface, but also up to the depth of penetration of UV radiation from the discharge, are responsible in particular for graft polymerization producing films with quite interesting properties for a broad range of applications from dielectrics for capacitors to semipermeable membranes for water purification, gas separation, or medical applications.

Phenomena Induced by the Impact of Particles with Negligible Kinetic Energy. In this case, only potential energy supplied by the impinging particles and the discharge photons is available for surface reactions. This energy can be supplied by different exothermic processes including heterogeneous recombination of atoms, ion neutralization, and deexcitation of species, all processes able to cause, for instance bound-breaking of polymer chains thus creating active sites for further chemical reactions, ashing of organic molecules, decomposition of chemical compounds, migration and coalescence of surface atoms.

All these phenomena can be easily observed either on insulating materials or on conductive materials, except at cathode surfaces where kinetic impact effects involving field accelerated positive ions often dominate and hide the other effects.

While the charged species are transported by the electric field, the neutral activated species proceed by diffusion owing in particular to the kinetic energy kept after charge-transfer processes or to the energy acquired from ion drift through momentum transfer. The effect of the latter process is most easily seen in coronas, as an electric wind, directed from the high-field to the low-field electrode. Figure 17 shows a negative point-to-plane corona system in ambient air where the gas flow induced by the discharge has been visualized by means of the dust particles present in the gas; one can see that the corona acts as a pump attracting the surrounding gas and projecting it toward the low-field electrode. Up to now, the electric wind has primarily been used to blow out candles and drive windmills in demonstration experiments. However, it carries a large population of excited and chemically very active species, which may account for an important energy transfer and give to even "passive" electrodes quite an active role in electrochemical processes. Calorimetric measurements made on negative point-to-plane coronas in air (Goldman, 1974, Goldman, 1979) have demonstrated that more than 50% of the energy input may be transferred to the low-field electrode in short air gaps by the neutral activated species transported by the electric wind (Fig. 18a and b), while the contribution of the ions remains very low. As shown in Fig. 18a, this percentage decreases rapidly with increasing gap distance because of progressive degradation of the energy of the activated species population from their production region located in the vicinity of the point. The very important chemical activity of the neutral species in negative point-to-plane corona discharges has also been demonstrated in oxygen by means of a chemical method employing liquid surfaces, specially chosen to react with the specific

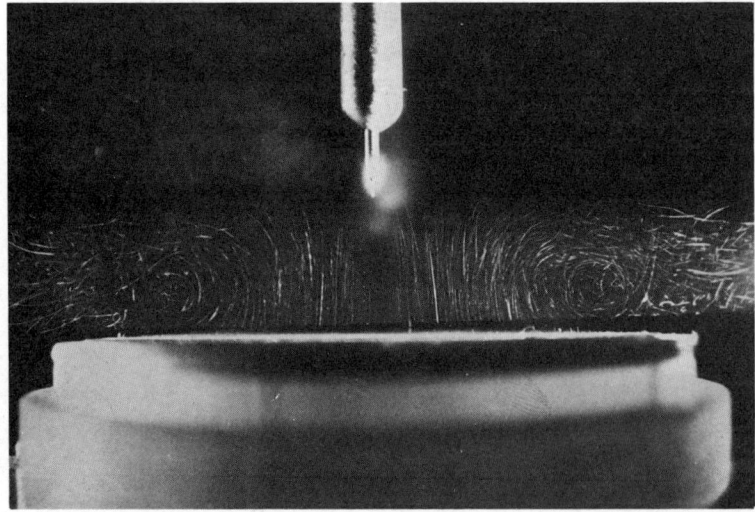

Fig. 17. Visualization of the gas flow induced by DC corona discharge in a negative point-to-plane air gap (d = 31 mm; I = 20 µA) which acts as a transport means for the neutral chemically active species formed and activated in the gap (Teisseyre et al., 1982).

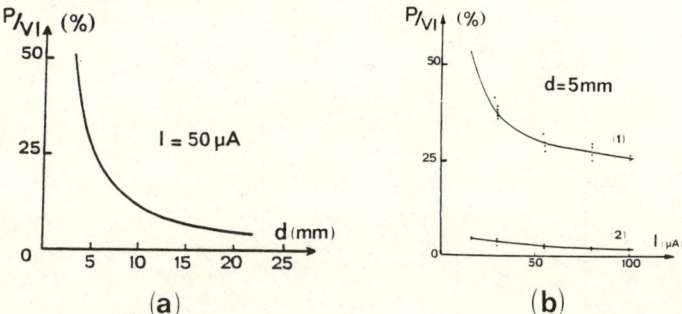

Fig. 18. Fractional power transferred to the low-field electrode by a DC negative point-to-plane corona in ambient air (Goldman, 1974; Goldman, 1979). (a) Variation of the power, all processes included, as a function of gap length. The curve indicates that the active species disappear with a mean macroscopic time constant on the order of 1 msec. (b) Variation of the power transferred, neutrals contribution included (curve 1) and excluded (curve 2), as a function of the average discharge current. The curves show that most of the power transferred to the low-field electrode is due to the neutrals action.

corona products, instead of a solid anode (Lecuiller et al., 1972); the results, illustrated by Fig. 19, show that the neutral activated species population, denoted [x], clearly exceeds the ion population.

Though the charged and neutral activated species move in sometimes quite separate reaction channels, important conversion and synergistic effects exist which give the corona reactor properties of peculiar interest for certain applications, for instance, polymer treatment. The combined effects of both types of species are well illustrated, for, instance by observations which have been made on corona pitting corrosion of aluminum (Sigmond et al., 1980): anode pitting ceases if the ion current is interrupted, and decreases substantially if the electric wind which transports the neutral activated species is diverted. Bringing us back to the parallelism evoked at the very beginning, Fig. 20 proposes equivalent schemes for the ion activity in electrochemical and in corona anode pitting corrosion which proceed in a similar way. While anodic passivation of aluminum in electrolytes proceeds by reactions between atomic oxygen ions of the electrolyte and aluminum ions which diffuse through the oxidized layer covering the bulk material, it is known that traces of negative ions such as Cl^-, BO_2^- or NO_3^- ions are sufficient to generate a counter process, leading on preferential sites to pitting corrosion which can even be observed outside the electrolytic cell, simply in a corrosive atmosphere. In the same manner, aluminum

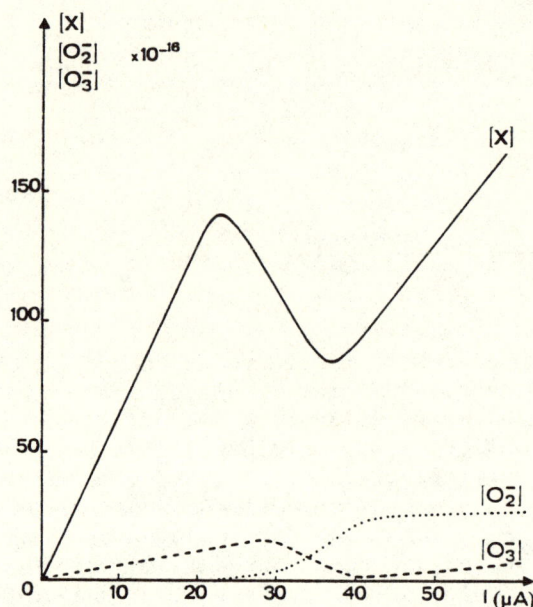

Fig. 19. Chemically active species produced by a DC negative point-to-plane corona discharge (d = 20 mm) in oxygen at 1 atm. pressure as a function of the average discharge current (Lécuiller et al., 1972). The curves give the number of particles formed for a duration of 15 min: $[O_2^-]$: O_2^- ions and derivatives of comparable chemical activity (O_4^-, CO_4^-, $N_2O_2^-$, ...); $[O_3^-]$: O_3^- ions and derivatives (CO_3^-, NO^-, ...); [X] : total of neutral oxidizing species.

plasma etching used in the fabrication of semiconductor devices is obtained in RF discharges with only tiny amounts of chlorine; this can be explained by reactions leading to the regeneration of the Cl^- ions.

Some remarks concerning the use of discharge plasma for surface processing are appropriate here:

• It is sometimes difficult to distinguish useful and harmful processes; for instance oxidation may be regarded as a corrosive as well as a protective process.

• Both types of effect can sometimes act simultaneously and one or the other dominate, depending on the operating conditions; for instance, corona treatment lasting some minutes improves the insulating properties of dielectric films while a longer exposure to the discharge acts in the opposite way.

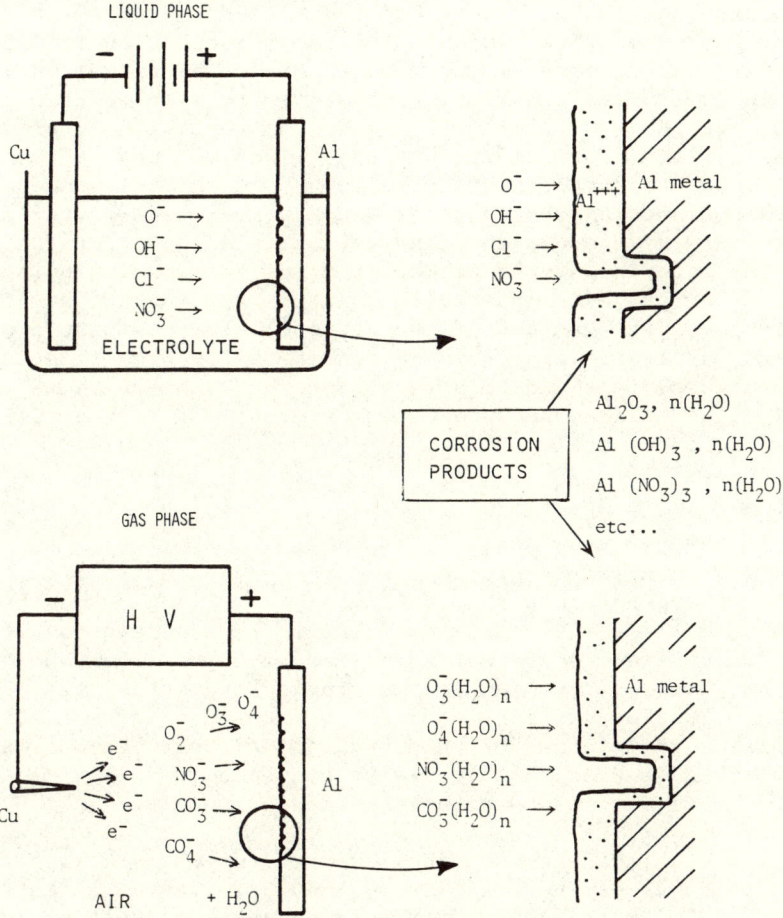

Fig. 20. Schematic representation of the activity of "corrosive" negative ions, leading to comparable effects on aluminum anodes in an electrolyte and in a gaseous plasma (Sigmond et al., 1980).

• Ablation and addition effects may act together in a competitive way as could be the case for hydrogen in metal nitriding with N_2-H_2 plasmas (see next section).

<u>Impact Phenomena Which May Imply Kinetic Effects.</u> Energetic ions, or neutrals, allow removal of matter from surfaces in a more or less selective manner. With ions of nonreactive species, the process, known as "sputtering," represents a purely physical type of plasma etching. Sputtering is typically obtained at the cathode of DC glow discharges where positive ions may acquire relatively

high kinetic energy, up to some hundreds or eV, by crossing the cathode fall region with a minimum of collisions. The ions impart their kinetic energy and momentum to atoms of the target on which they impinge, and the result is that atoms are ejected from the surface. The process, still subject to controversial interpretations, is largely described in Kaminsky (1965). The sputtering rate depends on numerous factors related in particular to parameters of the discharge such as the value of the abnormal cathode fall, the discharge current and current density, but also to other factors such as the material of the target, the mass of the impinging positive ions, the parameter pd (gas pressure x gap length) and the temperatures of the gas and of the cathode. In the first approximation, the sputtering rate can be considered as proportional to the discharge current and to a factor $(V_c - V_o)$ where V_c is the abnormal cathode fall and V_o a constant greater than the value of the normal cathode fall:

$$m = k I (V_c - V_o) \tag{6}$$

where k is a coefficient depending on the gas nature and the target material. The phenomenon is not limited to metals and can be observed as well on insulating materials. It has been used for a long time to clean surfaces, to prepare metal samples and other inorganic specimens for microscopic studies, to metallize surfaces placed in suitable positions to receive the sputtered metal, or to prepare films as thin as 50 A. More sophisticated methods, such as ion plating (Mattox, 1963), are presently being developed to achieve the two last items with still better results.

Plasma etching becomes a chemical process when chemically reactive gases are used, such as O_2, CO_2, H_2, halogens; the reactive species react with the surface to form volatile products which sublime. Their action alone, without cathode sputtering assistance, can be sufficient for different ashing and etching applications (Hollahan and Bell, 1974, Chap. 8). Thus ashing applications for specific removal of certain constituents mostly utilize a gas flow (often O_2) through a RF electrodeless discharge at low temperature, and various etching applications also employ the same sort of discharge or other different types, such as DC glow discharges with the specimen on the anode side as well as on the cathode side. For instance, in "plasma pyrolysis" of coal by a RF discharge, the impinging ions can work their etching effect simply by gasifying hydrocarbon polymer components of the material to methane, ethane, etc. and leave the elemental carbon and graphite components unaltered (Fu and Blaustein, 1969). But physical sputtering might bring a useful contribution to the etching. When chemical and physical processes act simultaneously, the relative importance of the effects depends in particular on the specimen, and the gas and is sometimes difficult to assess. So let us consider the case of iron hardening obtained by ion-nitriding of samples working as the cathode in a glow discharge (Hollahan and Bell, 1974, Chap. 3). Figure 21 shows the beneficial action of hydrogen in the gas composition, which can be ascribed to its cleaning properties, and at the same time to its chemical action since higher efficiency is obtained with $N_2 + H_2$ plasmas which produce essentially molecular ions such as NH^+, NH_2^+, $N_2H_2^+$, ..., than with N_2 + Ar plasmas which generate large amounts of nitrogen ions. The figure reproduces results obtained with additional heating favoring both the reaction rates and the diffusion of the reactive species into the bulk material. In a similar way, since the abnormal regime of the glow discharge operates with maximum current density, i.e., maximum heating, it provides optimized discharge conditions for this application. However, it must be emphasized that the penetration of species into the bulk can have harmful effects. For instance, if they accumulate on preferential sites just under a thin surface layer, they might lead to the formation of gaseous pockets in over-pressure, likely to explode. This phenomenon, called "blistering" because of the blistered aspect it gives to the surface, is known to occur in accelerator environments with ion energies in the range of keV; a similar aspect is observed on materials which have been subjected to discharge plasmas for a long time, with evidence that both phenomena should involve the same mechanism.

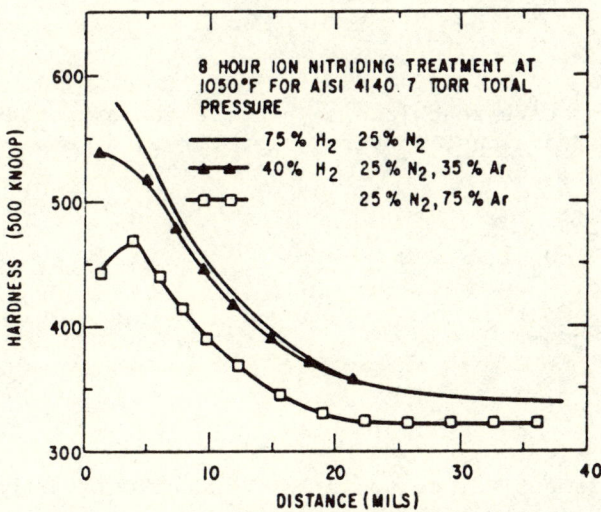

Fig. 21. Effect of gas composition on iron hardening by ion-nitriding (Hollahan and Bell, 1974, Chap. 3) (10 mils = 0.254 mm).

LABORATORY STUDIES ON SOME APPLICATIONS

Ozone Synthesis

Ozone production has great importance because of its broad spectrum of applications (disinfection, bleaching, chemistry, etc.), among which water purification takes a larger and larger place, as a substitute for chlorine which produces a bad smell and taste in water. Even if they still give low ozone yield, electrical discharges now indisputably provide the best way for ozone synthesis. It is a technique which has been investigated for a long time and is now largely industrialized, but much progress has to be and can be made.

The global equation describing ozone formation

$$3 O_2 \rightarrow 2 O_3$$

implies an energy ΔH = 2.95 eV (assuming that the molecules O_2 and O_3 are in their ground state, respectively $3 \Sigma_g^-$ and ^1A) and a maximum yield of 1200 g/kWh.

We have already seen (Table 2) many of the chemical reactions which can occur in oxygen discharges. Concerning ozone formation in such discharges (Goldman and Lécuiller, 1980), it is now usually admitted that the production is mainly governed by the <u>creation reaction</u>

$$O + O_2 + O_2 \xrightarrow{k_1} O_3 + M$$

implying dissociative reactions to provide the oxygen atoms, and by <u>destruction</u> <u>reactions</u>, in particular for O_3 by the reaction of thermal decomposition:

$$O_3 + O_2 \xrightarrow{k_2} O_2 + O_2 + O ,$$

and the reaction

$$O_3 + O \xrightarrow{k_3} 2 O_2$$

which, at the same time, destroys the product O_3 and the useful intermediate species O.

The occurrence of these basic reactions emphasizes two points remarkably important for ozone production. First, let us consider the expressions giving the different reaction rate coefficients involved in the above reactions:

$$k_1 = 8.17 \times 10^{-35} \exp(447/T) \quad cm^6 \, sec^{-1}$$

$$k_2 = 3.35 \times 10^{-9} \exp(-12060/T) \quad cm^3 \, sec^{-1}$$

$$k_3 = 5.59 \times 10^{-11} \exp(-2865/T) \quad cm^3 \, sec^{-1}$$

One sees that temperature plays a most important role in all these reactions, slowing down slightly the production reaction and rapidly accelerating the destruction reactions, especially the one governed by k_2, the so-called thermal decomposition reaction.

Secondly, a sufficient production of oxygen atoms implies dissociative reactions with large cross sections σ_d; it should be obtained via the $A^3\Sigma_u^+$ state (max. $\sigma_d = 2.10^{-17}$ cm^2) and the $B^3\Sigma_u^-$ state (max. $\sigma_d = 8.10^{-17}$ cm^2) of the molecule O_2 which have relatively high onset energies, respectively 4.5 and 8 eV (see Fig. 12 and 9), i.e., which need relatively high electric fields to occur.

Both requirements needed for ozone production, low temperature to reduce the destruction reactions and high-field to enhance O_2 dissociation, are best fulfilled in the streamer channels of corona discharges. Figure 22 shows a typical example of an industrial plant presently operating in France. Each ozonizer of this plant is composed of 550 elemental tubes in parallel. One single tube works with an axial air flow circulating between two coaxial metal cylinders. Corona discharge is established between these cylinders, and the insulating material covering the inner cylinder prevents spark formation; the presence of this insulator implies an alternative voltage supply. With this type of device, no more than 70 g/kWh (i.e., energy yield hardly higher than 5%) is obtained, and the production can be approximately doubled by operating with oxygen instead of air.

The influence of the electric field on ozone yield has been studied theoretically by Davidson and Lécuiller (1979) who considered the ozone production in the body of a single electron avalanche developing from time 0 to time τ under uniform field conditions. They built a simplified model, considering only some reactions in pure oxygen and assuming that all the oxygen atoms react to form ozone, i.e. that

$$\frac{d[O_3]}{dt} = \frac{d[O]}{dt} = \chi n_e v_e , \qquad (7)$$

where $[O_3]$ and $[O]$ are the concentrations of ozone molecules and oxygen atoms respectively, n_e the concentration of electrons, v_e

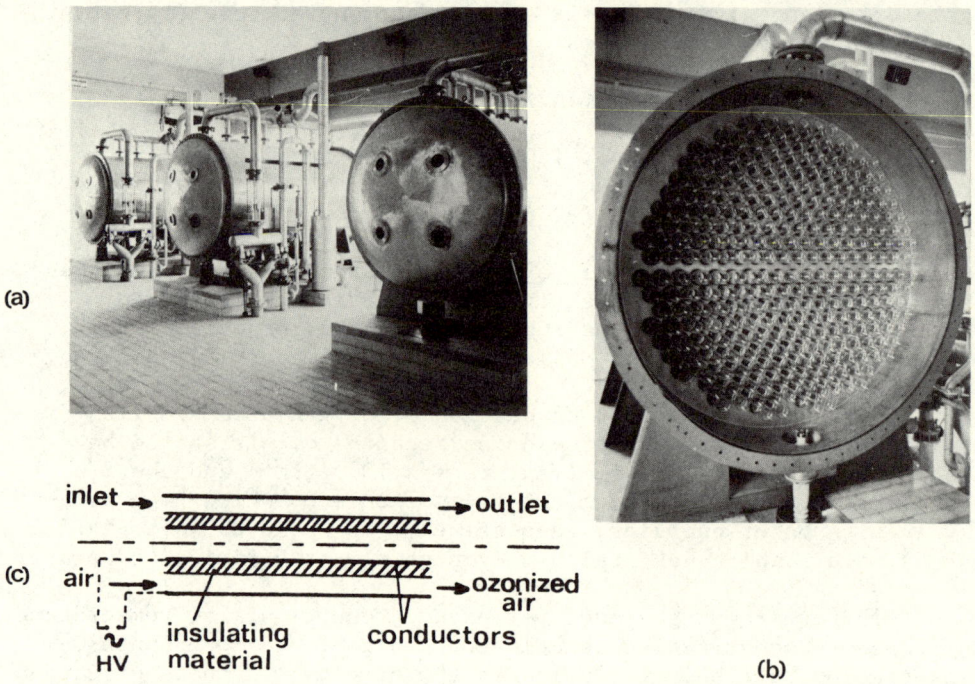

Fig. 22. Ozonizers of a water treatment plant working at Nantes, France. (a) general view; (b) sectional view of one ozonizer, composed of 550 elemental tubes; (c) longitudinal section of one elemental tube.

the electron velocity, and χ the number of oxygen atoms formed per electron and per unit length parallel to the electric field. Considering, furthermore, that the energy expense for this production is the energy dissipated by the electrons formed in the avalanche

$$W = V \int_0^\tau i_e(t) \, dt \qquad (8)$$

where V is the voltage across the avalanche length $v_e \tau$ and i_e the electron current at time t, they obtained the expression:

$$\eta_{(g/kWh)} = \frac{[O_3]}{W} = 1500 \frac{\chi/N}{E/N} \qquad (9)$$

relating the maximum possible yield of ozone to the reduced electric field E/N by means of the mean dissociation cross section χ/N. This term χ/N was then estimated as a function of E/N, using simplifying assumptions with data given by the literature. Figure 23 shows the

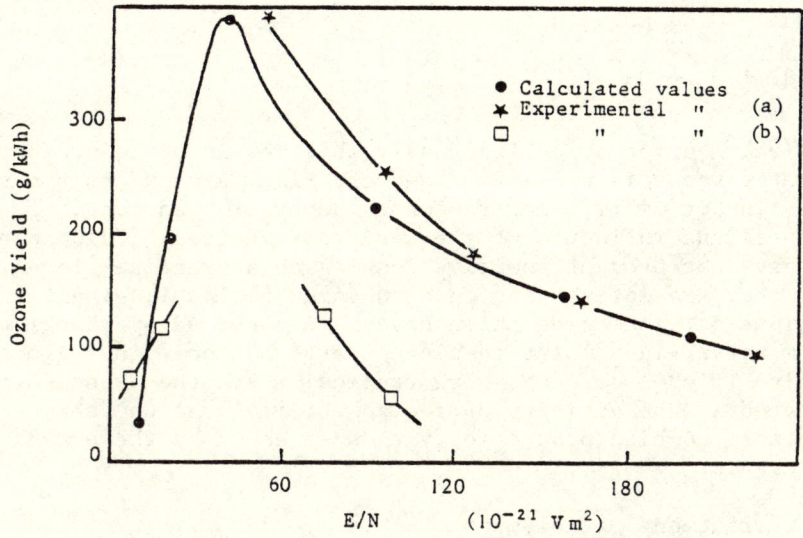

Fig. 23. Variation in ozone yield with the reduced electric field (Davidson and Lécuiller, 1979) (a) values from experiments at low current density $j \sim 8$ µA cm^{-2} (Warburg and Rump, 1925); (b) values from experiments at higher current density $j \sim 1$ mA cm^{-2} (Brewer and Westhaver, 1930).

agreement obtained between the theoretical curve of η as a function of E/N traced out from Eq. (9) and old experimental curves taken from the literature (Warburg and Rump, 1925; Brewer and Westhaver, 1930). A maximum yield of 400 g/kWh (i.e., energy yield of ~ 30%) should be reached for E/N values around 40×10^{-21} Vm^2. The results of Brewer and Westhaver might be lower because they were obtained at higher current densities (~ 1 mA cm^{-2}), likely leading to a gas heating incompatible with the initial assumption whereby each oxygen atom forms one ozone molecule.

On a technological level, the greatest progress presently expected concerns the voltage supply in short pulses, so as to increase the number of streamers and to limit the duration of the individual discharges to the useful propagation phase, the following phase being only a source of heating, i.e., of destruction reactions for the ozone formed. Further, for the ozone production, there is an interest in using dried gas, oxygen, or air in order to avoid, in particular, the creation of the vibrationally excited radical OH^v which can be formed in collisions of H_2O molecules with electrons or $O(^1D)$ atoms, and then inducing destructive reactions of ozone by the following steps:

$$OH^V + O_3 \rightarrow HO_2 + O_2$$

$$HO_2 + O_3 \rightarrow OH^V + 2\,O_2 \quad , \text{ etc.}$$

Effectively, a reduction is usually observed in ozone production when water vapor is present. However, Fig. 24 shows that negative coronas in air exhibit contradictory behavior. In contrast, recent studies on the influence of the electrode material (Goldman et al., 1982) have not brought positive results on a practical level, although they are interesting on a fundamental level; significant variations of ozone production have been observed by changing the point material in a point-to-plane system but only in regimes (negative Trichel coronas, glow regimes) where the ozone formation is located at the vicinity of the point, i.e., at the same time unfavorable to high production yields as shown by the experimental measurements.

Surface Treatment of Polymers

Plasma surface treatment of polymers and assimilated materials is now most widely used in various industries, such as packaging, clothing, electronics, photographic and magnetic support manufacture, to improve the adhesive properties of the materials, allowing a better adherence on them of dyes, inks, adhesives, etc. The adhesive properties of a surface are related to various physical and chemical properties of the surface and can be improved by plasma

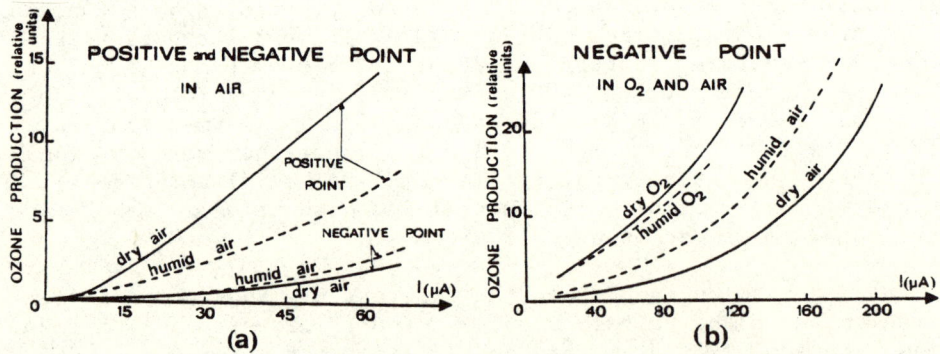

Fig. 24. Effect of humidity on ozone production by point-to-plane corona discharge (Goldman and Lécuiller, 1977). The curves show that humidity increases ozone production in air with negative corona and acts in the opposite way when polarity (Fig. a) or gas (Fig. b) is changed. Note the increase of production at the beginning of the positive corona streamer regime (Fig. a).

treatment through different combined processes. Table 5 gives a short survey of these processes and of the related effects on the surface properties involved. All processes allowing an increase of the surface energy of the material will lead to improved adhesive properties.

Corona discharges give the best means to achieve such treatments. Compared to the classical chemical methods which they are replacing, corona treatments are relatively fast (seconds), easier to control, less polluting, and often give better results leaving, for instance, the bulk properties unchanged. A corona reactor for polymer surface treatment typically consists of a rotating roll, which supports the film to be treated and constitutes the grounded electrode, and of a high-voltage electrode, in the shape of a long knife-blade, parallel to the roll axis (Fig. 25).

Table 5. Plasma-surface interaction processes and related effects on the surface properties.

Plasma-surface interaction processes	Surface effects
Physical processes · impact of electrons, ions and neutral particles · thermal effect of the deactivation impact · charge trapping (ions or electrons) in the lattice	increase of the surface roughness decrease of the crystallinity of the material increase of the surface potential
Chemical processes · action of the activated species · diffusion of the activated species into the material	cleaning of the surface oxydation of the molecules by radical mechanisms decrease of the molecular weight carboxyl groups formation cracking or cross-linking (according to the plasma gas composition) and related increase or decrease of the porosity

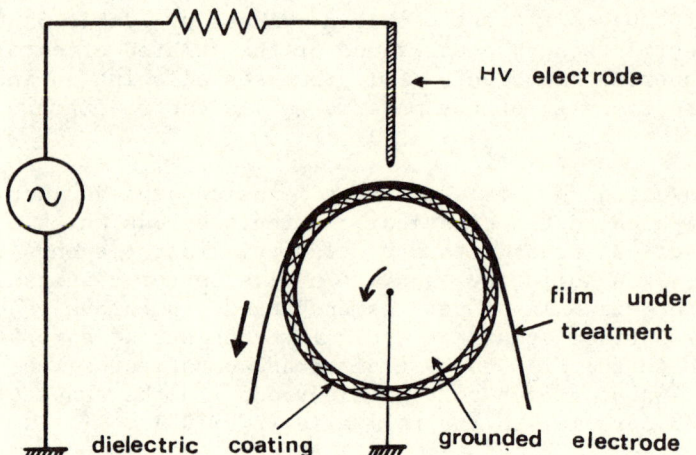

Fig. 25. Simplified scheme of an apparatus for plasma surface treatment of polymer films and assimilated materials.

An overview of the fundamental aspects involved in the surface treatment of polymers can be made from the results of an extended study which has been carried out in the authors' laboratories. They concern two kinds of films of about 50 μm thickness:

(1) Films of polyethylene (PE), pure material of low density, 50-60% crystallinity and without surface pollutants, formed by $[CH_2]_n$ groups (Amouroux et al., 1978; Tran, 1980);

(2) Films of polyethylene terephtalate (PET), higher density material which contains oxygen in its matrix:

$$[- O - CH_2 - CH_2 - O - \overset{O}{\overset{\|}{C}} - \quad - \overset{O}{\overset{\|}{C}} -]$$

and which has a higher crystallinity at the surface (90%) than in the bulk (60%) (Amouroux et al., 1981, 1982).

The films were treated by AC coronas using an electrode system similar to that of Fig. 25, but with the sample fixed on the roll turning at the rate of 4 revolutions/min = 80 cm/min; this led to a discontinuous treatment, a given part of the sample passing several times again under the discharge.

The main data concern the evolution of the chemical species on the surface, as determined by ESCA analysis (for details on the

method, see, for instance, Clark, 1978). They also concern the
correlated variations of the surface energy, which are obtained
by measurement of the contact angle of a drop of an appropriate
liquid laid on the surface and which are indicative of the surface
wettability. For instance, Fig. 26 shows the increase of oxygen
fixation obtained on PE films by increasing the current intensity
of the discharge established in air, while Fig. 27a shows that,
in the range of values considered, the increase of oxygen fixation
is correlated with a linear increase to the polar component γ_S^P of
the surface tension γ_S. In the case illustrated by Fig. 27b, the
purpose was, on the contrary, to graft halogens on the surface of
PE films with the objective this time, to reduce the surface energy
and subsequently the wettability of the surface. Variations in the
concentration of the halogen grafted on the surface (fluorine,
chlorine, or bromine) were obtained by adding $C Br F_3$ or $C Cl F_3$

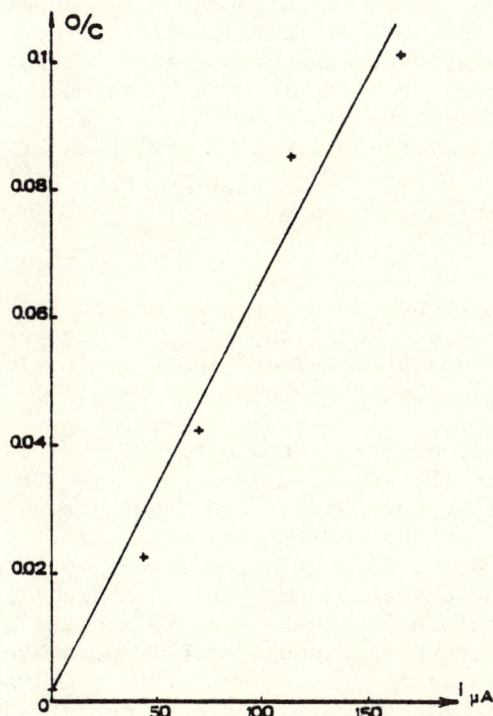

Fig. 26. Variation, as a function of the discharge current, of
the rate of oxygen fixation on PE films relative to the
carbon fixed (Amouroux et al., 1978). Treatment conditions: 7 min total treatment duration by AC corona
in air at 1 atm. pressure, with 5 mm gap length.

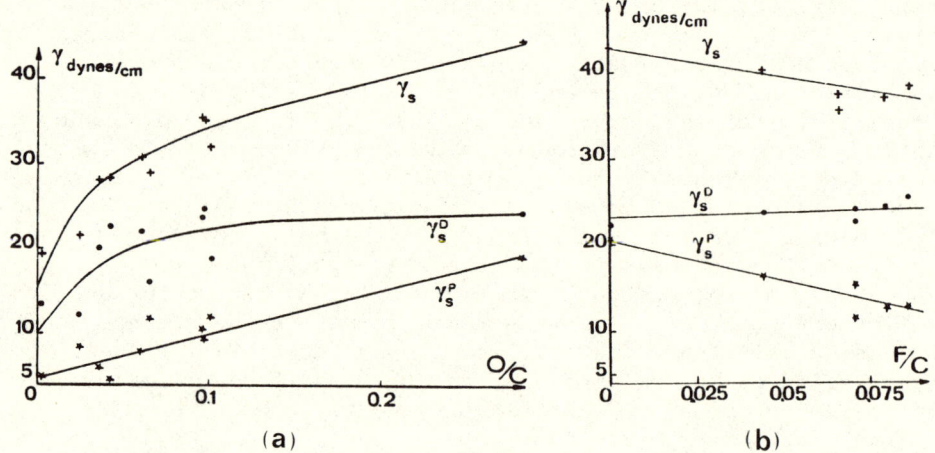

Fig. 27. Variation of the surface tension γ_S, of its polar component γ_S^P and of its dispersion component γ_S^D, as a function of the rate of fixation X/C of the element X on PE films relative to the carbon fixed (Tran, 1980). Treatment conditions: 8 min total treatment duration by AC corona, with 100 µA current and 2 mm gap length: (a) oxygen fixation in air at 1 atm. pressure; (b) fluorine fixation in $CF_2 = CH_2$ + air mixtures at 1 atm. pressure rates, corresponding to O/C = 0.3.

in air, in varying proportions. Figure 28 shows that C Br F_3 allows a high level of Br and F fixation, probably because of the relatively low energy of the bond C-Br (~3 eV) in the molecule considered (Tran, 1980).

We have already seen that the efficiency of the surface treatment increases with the discharge current. The treatment duration plays a similar role, both parameters showing a saturation effect for large values. Another most important factor for the treatment efficiency, which has to be optimized in the building phase of the reactor, is the electrode gap length controlling the residence time of the chemical and activated species in the discharge. In Fig. 29, one can see the influence of this parameter on the fixation of oxygen on PET films and at the same time the influence of the material of which the high-voltage electrode is made.

Concerning the first point, the curves show that an optimum value exists for the gap length, probably related to an equilibrium state between the mechanisms of oxidation and destruction of carboxyl groups in competition in the process (see later). The decay of the curves for larger gap lengths may be due to the progressive degra-

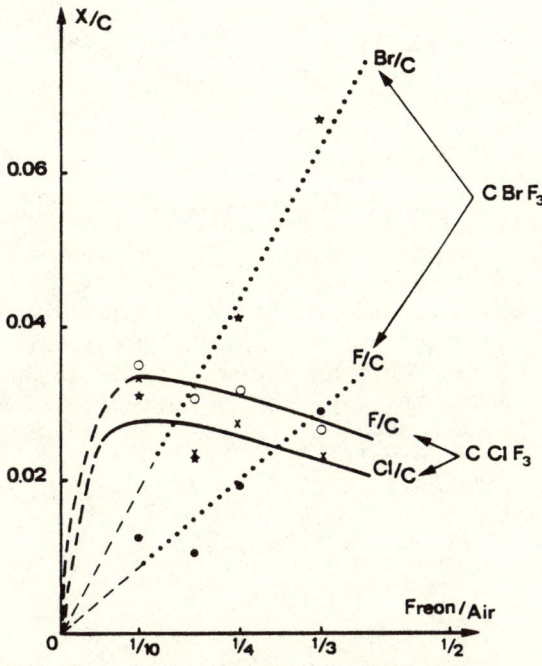

Fig. 28. Variations of the rate of fixation X/C of the halogen X(F, Cl, Br) fixed on PE films relative to the carbon fixed, for C Br F$_3$ + air and C Cl F$_3$ + air mixtures in varying proportions at 1 atm. pressure (Tran, 1980). Treatment conditions: the same as in Fig. 27.

dation of the energy of the neutral active species along their trajectory from their activation region in the vicinity of the point (see Fig. 18a).

Concerning the second point, the results indicate the existence of a catalytic effect when transition metals such as vanadium, molybdenum or titanium are used for the high-voltage electrode (Amouroux et al., 1979); this observation is reminiscent of that made in connection with NO synthesis (see "Role of Heterogeneous Catalytic Processes").

As indicated in Table 5, the effects of the plasma-surface interaction also involve changes in the physical properties of the surface subjected to the discharge. For instance:

• X-ray diffraction measurements on PET films which have been subjected to positive DC coronas have indicated a loss of crystallinity.

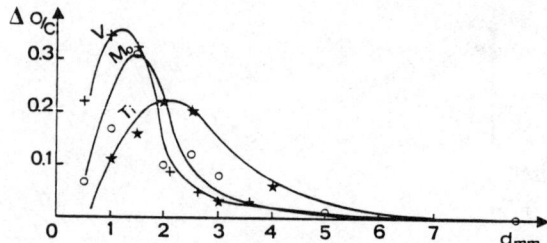

Fig. 29. Variation, as a function of the gap length, of the rate of oxygen fixation on PET films relative to the carbon fixed (Amouroux et al., 1981). Treatment conditions: 7 min total treatment duration by AC corona in air at 1 atm. pressure, with 130 μA current. Note that the samples nontreated have an O/C content of 0.75 taken here as a zero reference for the ordinates axis. The curves related to different materials (V, Mo, Tr) for the high-voltage electrode show the great influence of this parameter on the surface treatment.

• Porosity measurements on PE films have shown that a treatment in NH_3 atmosphere produces a decrease in the porosity while in air atmosphere acts in the opposite way; this result has been interpreted as implying a longer diffusion path into the material for the species activated in NH_3 than for those activated in air.

Coming back to the chemical aspect of the plasma-surface interaction, the following mechanisms for the grafting of oxygen and nitrogen on the polymer surface can be proposed. According to the competitive aspects of oxidation and decarboxylation observed during the surface treatment, the fixation of oxygen should proceed in the four following steps suggested by Rånby and Rabek (1978):

(1) Transfer of energy to the surface and activation of the absorbed species with formation of radicals

$$RH \xrightarrow{O_2^*, \, O_2, \, O_2^+, \, O} R\cdot + OH\cdot$$

(2) Surface reactions between radicals and O_2^*

$$R\cdot \xrightarrow{O_2^*} ROO\cdot \quad \text{(peroxide group)}.$$

(3) Evolution of peroxide to carboxyl groups

ROO· $\longrightarrow$ RCOOH

(4) Destruction of the carboxyl groups

$$RCO\ OH \xrightarrow{h,\ e^-,\ O_2^*} R'\cdot + CO_2\ .$$

As for the fixation of nitrogen, the ESCA data point out competitive mechanisms between NH formation and NH oxidation in NO function. Figure 30a shows the different forms in which nitrogen is fixed and Fig. 30b that the amines NH are reaction intermediates for they rapidly reach a steady state while the nitrogen oxides regularly increase with time. The following steps are proposed to describe the process:

1. $R\cdot \xrightarrow{N_2^*,\ N} R-NH\cdot$

2. $R\cdot NH\cdot \xrightarrow{O_2^*} R-N=O + OH\ .$

Grafting of new chemical functions on polymer surfaces will give material for further development in surface treatment studies. The choice of the gaseous atmosphere will determine the nature and the concentration of these new chemical functions, but in any case, it must not be forgotten that the gases dissolved inside the polymer will bring their own contribution to the chemical evolution of the surface.

Fabrication of Microelectronic Devices

The semiconductor industry is a most attractive field for the application of plasma chemical processes since, in particular, it can (Hollahan and Bell, 1974, Chap. 9-10) greatly reduce the number of steps required by the conventional chemical methods and can be easily used with automatic control or programming of the operating variables. The importance of present development in this field is proved by the number of papers which have been presented on the subject at the last International Symposium on Plasma Chemistry (Edinburgh, 1981).

Low temperature, low pressure plasmas find a usefulness in three directions which we shall only briefly introduce.

(1) The first step in the fabrication of integrated circuits is the transfer of the circuit pattern onto the surface of a silicon wafer. The medium into which the image is projected is usually an organic polymer, called a photo-resist, sensitive to U V, X-ray, and electron beam irradiation. Electron and X-ray lithography are

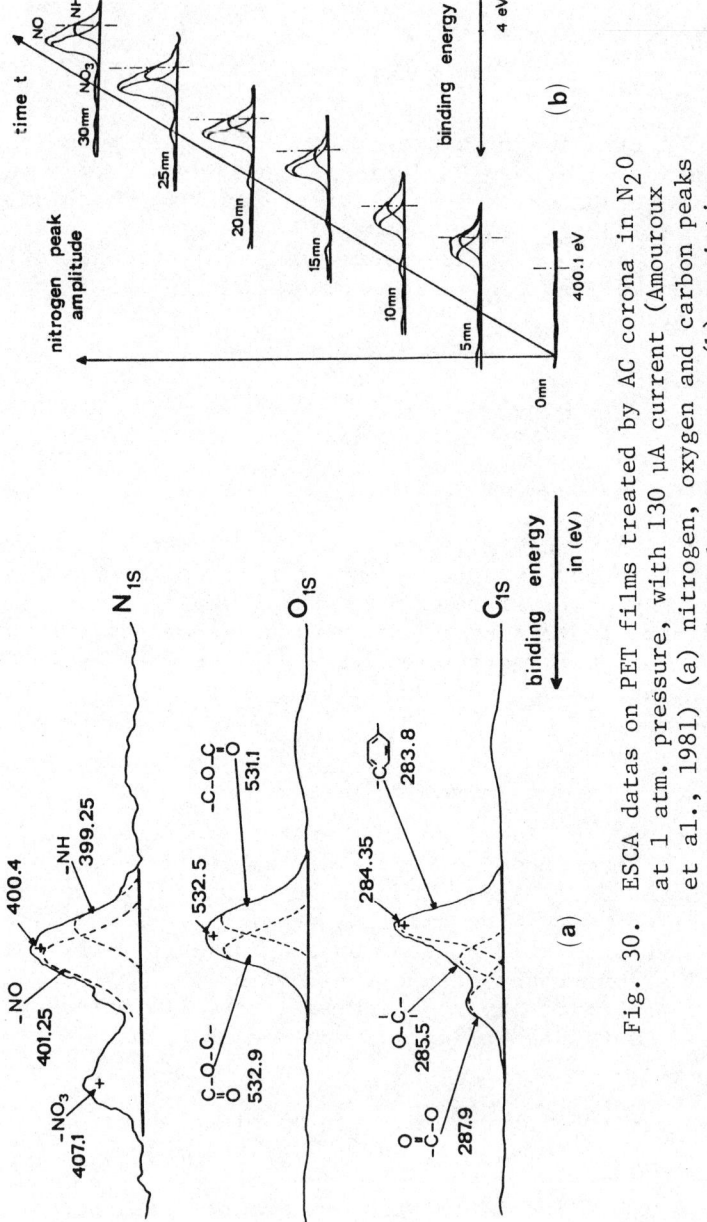

Fig. 30. ESCA datas on PET films treated by AC corona in N_2O at 1 atm. pressure, with 130 µA current (Amouroux et al., 1981) (a) nitrogen, oxygen and carbon peaks with their deconvoluated components; (b) variation of the nitrogen components as a function of time.

both promising techniques for the production of patterns with narrow line widths (~0.25 µm), and with ultra-high molecular weight polymethyl methacrylate (UHMW-PMMA), produced by plasma-initiated polymerization, line widths of 0.5 µm can be achieved, giving good quality figures, i.e., line resolution and edge definition.

(2) The etching of silicon, SiO_2, Si_3N_4, Al, Ti is realized by plasmas of RF discharges at low pressure with the main object of obtaining an etching speed of 1500 - 2000 Å/min. The choice of the gaseous mixture (CF_4+O_2, CF_4, $CBrF_3+O_2$, $CF_4BF_3+O_2$, +He, +Ar,...), the flow rate, the electrical field and the ionic density are the main parameters studied on a laboratory scale to obtain a good anisotropic etching. The main industrial objectives of the manufacturing process are the size and the thickness of the figures to be obtained and the speed of the treatment; the control of the speed is today developed by using the chemiluminescence of the chemical compounds produced from the surface.

(3) Deposition of thin layers of amorphous material are today a new way for the preparation of new electronic and optical materials, and, in this field, plasma processes can be used for the deposition as well as for its control. The main results are now obtained by using RF discharge plasmas at low pressure (0.1-1 torr) with gaseous mixtures such as SiH_4, CH_4, H_3P, BF_3, and AsH_3, and parameters such as pressure, electrical field, ion concentration, and radical mechanisms to control to a certain extent the deposition speed and the layer characteristics such as the thickness and the nature. One can, for instance, use such techniques for the deposition of thin amorphous layers of Si, SiC, C, AsGa, SiGe, on substrates. Using this process, the fabrication of photovoltaic cells of amorphous Si on SiC, working with an efficiency of 7%, is being developed in Japan.

Finally, it must be said that sputtering, ion plating, and interaction between laser and plasma discharge are just beginning to produce new solutions and new routes for the elaboration of electronic materials.

CONCLUSION

According to Oskam and Pfender, co-chairs of the NSF Workshop on Plasma Chemistry and Arc Technology (Minneapolis, 1980):

> "Plasma Chemistry (and Arc Technology) are now approaching a most exciting period. The challenges of improving the efficiency of energy utilization and conversion and of innovative methods for extraction, processing and recycling of materials point out serious gaps

in our fundamental understanding of the plasma state and
its interaction with materials. Developments over the
past years and the enhanced understanding gained by present
research point the way to new technologies which
were beyond our imagination a few decades ago."

Let us conclude with the same optimistic view of the future.

REFERENCES

Amouroux, J., Goldman, M. and Tran, M., 1978, J. Chem. Phys., 75: 662.
Amouroux, J., Cavvadias, S. and Rapakoulias, D., 1979, Rev. Phys. Appl., 14: 969.
Amouroux, J., Goldman, M. and Revoil, M. F., 1979, Patent EdF - France N° 79.17.050.
Amouroux, J., Revoil, M. F. and Goldman, M., 1981, in: "Proceedings, 5th International Symposium on Plasma Chemistry, Edinburgh," Heriot-Watt University, Edinburgh, p. 388.
Amouroux, J., Goldman, M. and Revoil, M. F., 1982, J. Polym. Sc.- Part A1, to be published.
Ashmore, P. G., ed., 1977, "Gas Kinetics and Energy Transfer," The Chemical Society, London.
Baddour, R. F. and Timmins, R. S., eds., 1967, "The Application of Plasmas to Chemical Processing," Pergamon, New York.
Barrère, M. and Prud'homme, R., 1973, "Equations Fondamentales de l'Aerothermochimie, Masson, Paris.
Berthelot, M., 1869, C.R. Acad. Sci., 67: 233.
Berthelot, M., 1898, C.R. Acad. Sci., 126: 566.
Berthelot, M., 1899, Ann. Chim. Phys., 16: 32.
Botchway, G. Y. and Venugopalan, M., 1980, Z. Phys. Chemie Neue Folge 120: 103.
Brewer, A. K. and Westhaver, J. W., 1930, J. Phys. Chem., 34: 1280.
Clark, D. T., 1978, in: "Polymer Surfaces," D. T. Clark and W. J. Feast, eds., Wiley Intersci., Chichester, Chap. 16.
Davidson, R. and Lécuiller, M., 1979, in: "Proceedings, 4th International Symposium on Plasma Chemistry," Zurich Univ., Zurich, p. 723.
Debye, P. and Hückel, W., 1923, Phys. Z., 24: 305.
Eremin, E. N. and Rubatsova, V., 1973, Russ. J. Phys. Chem., 47: 356.
Faraday, M., 1839, Res. Elec., p. 663.
Fauchais, P. and Rakowitz, J., 1979, J. Phys., 40: C7-289.
Fu, Y. C. and Blaustein, B. D., 1969, Ind. Eng. Chem. Process Des. Dev., 8: 257.
Gillepsie, L., 1967, in: "The Electronic Repulsion Theory of Chemical Bond," W. L. Luder, ed., Dunod, Paris.
Goldman, A., 1974, in: "Proceedings, 3rd International Conference on Gas Discharges, London," IEE Conf. Publ. No. 118, p. 275.

Goldman, A., 1979, in: "Proceedings, 3rd International Symposium on High Voltage Engineering, Milan," Paper 53: 12.
Goldman, M., Lécuiller, M., and Palierne, M., 1982, in: "Proceedings, 3rd International Symposium on Gaseous Dielectrics, Knoxville, Tennessee." Pergamon, New York.
Gorse, C., 1975, Doctorate de 3e cycle Thesis, University of Limoges, France.
Gray, H. B., 1973, "Les Électrons et la Liaison Chimique, "Ediscience-McGraw, New York.
Halpern, B. and Rosner, D., 1978, J. Chem. Soc., Farad. Trans., 8: 1883.
Hasted, J. B., 1972, "Physics of Atomic Collisions," Butterworth, London.
Hollahan, J. R. and Bell, A. T., eds., 1974, "Techniques and Applications of Plasma Chemistry," Wiley - Interscience, New York.
Kaminsky, M., 1965, "Atomic and Ionic Impact Phenomena on Metal Surfaces," Academic Press, New York.
Kaplan, M. L., 1971, Chem. Tech., p. 621.
Lécuiller, M., Julien, R. and Pucheault, J., 1972, J. Chim. Phys., 9: 1353.
McTaggart, F. K., 1967, "Plasma Chemistry in Electrical Discharges," Elsevier, Amsterdam.
Maecker, H., 1953, Z. Physik, 136: 119.
Mattox, D. M., 1963, J. Appl. Phys., 34: 2493.
Meubus, P. 1975, J. Electroch. Soc. 122: 298.
Oskam, H. J. and Pfender, E., eds., 1980, in: "Proceedings, NSF Workshop on Plasma Chemistry and Arc. Technology, Minneapolis," Univ. Minn.
Polanyi, J. C., 1972, Acc. Chem. Res., 5: 161.
Purvis, G. D., Redmon, M. J. and Wolken, G. J., 1979, J. Phys. Chem., 83: 1027.
Rånby, B. and Rabek, J. F., eds., 1978, "Singlet Oxygen, Reactions with Organic Compounds and Polymers," Wiley-Intersci., Chichester.
Rapakoulias, D. and Amouroux, J., 1979, Rev. Phys. Appl., 14: 961.
Rapakoulias, D., Gicquel, A. and Amouroux, J., 1981, in: "Proceedings, 5th International Symposium on Plasma Chemistry, Edinburgh," Heriot-Watt Univ., Edinburgh, p. 688.
Reed, T. B., 1967, in: "The Application of Plasma to Chemical Processing," R. F. Baddoud and R. S. Timmins, eds., Pergamon, New York, Chap. 3.
Schaap, A. P., 1976, "Singlet Molecule Oxygen," Halsted, New York.
Sigmond, R. S., Goldman, A. and Brenna, D., 1980, in: "Proceedings, 6th International Conference on Gas Discharges, IEE Conf. Publ. No. 189, Vol. 1, p. 82.
Suhr, H., 1973, in: "Proceedings, 11th International Conference on Phenomena in Ionized Gases, Prague," Invited Papers, Czech. Acad. Sci., Inst. Phys., p. 413.
Tallandier, F., 1979, "Rapport Direction Etudes et Rech. EdF.-Renardiéres, France," HP 40/79/307 (a).

Teisseyre, Y., Haug, R. and Ballereau, P., 1982, J. Phys. D: Appl. Phys., to be published.
Thomson, J. J., 1893, "Recent Researches in Electricity and Magnetism," Oxford, London. p. 189.
Tran, M., 1980, Docteur-Ingénieur Thésis, University of Paris 6, France.
Venugopalan, M., ed., 1971, "Reactions under Plasma Conditions," Vol. II, Wiley - Intersci., New York.
Veprek, S. and Venugopalan, M., eds., 1980, "Plasma Chemistry," Vol. I - II, "Topics in Current Chemistry," 89-90, Springer, Berlin.
Warburg, E. and Rump, W., 1925, Z. Phys. 32: 245.

MICROWAVE DISCHARGES

J. Marec, E. Bloyet, M. Chaker, P. Leprince and P. Nghiem

Laboratoire de Physique des Gaz et des Plasmas
Orsay, France

INTRODUCTION

Microwave discharges first appeared as unwanted and disturbing effects. However, beginning about the end of World War II, Professors Allis and Brown at the Massachusetts Institute of Technology started to investigate the physics of these discharges. During the next few years, many experimental and theoretical studies were undertaken. However, in the early 60's and for about 15 years, there were few studies of such discharges because of the theoretical difficulties encountered. Effectively, the impossibility of modeling microwave discharges prevented a good understanding of their behavior, and their future use did not appear promising. Recently there has been new interest in these discharges. The plasmas produced by microwave discharges find applications in areas such as: 1) spectroscopy (because of their low contamination), and 2) plasma chemistry. Another advantage of these discharges as compared to d.c. discharges is their ease of operation.

The present paper is a brief survey of the field of microwave discharges, considering both theoretical and experimental investigations. We consider two types of physical phenomena: first, those associated with producing breakdown, and second, those related to the conditions for discharge maintenance.

In the first part, we describe the breakdown conditions and the difficulties of predicting them theoretically. Then, we give some classical experimental results interpreted by the diffusion theory. The limits and physical meaning of this theory are also considered.

In the second part, we describe the steady-state discharges. First, we try to give a classification of microwave coupling structures and then, discuss the different types of plasmas. The particular case of plasma columns sustained by a traveling surface wave, a technique developed by our group, is presented with some detail. Finally, we review briefly some actual applications of microwave discharges.

MICROWAVE BREAKDOWN

Introduction

In this part, we describe the main features of microwave breakdown, beginning with the difficulties of the problem. Some experimental results obtained by the cavity method are qualitatively interpreted through the simplified model of the average electron. Then, we explain how the distribution function can be determined in some particular cases. Finally, we give a comparison between experimental and theoretical results.

Breakdown Problems

To establish the breakdown conditions, we must determine the values of both the electric field initiating the discharge and the corresponding parameters of the discharge. The breakdown conditions can be determined experimentally and theoretically. Measurements of electrical breakdown field and relevant parameters of the discharge can be made by classical microwave methods. From the theoretical point of view, a modeling of breakdown is possible if we know the processes of electron production and loss and the distribution function. At this step, we find the most important difficulties of the breakdown problem. Effectively, as we shall see later, the determination of the distribution function is a very difficult problem. Moreover, the choice of parameters is difficult because the phenomena involved in microwave breakdown depend on a very large number of them. From that point of view, the electrical discharges initiated by microwave fields differ significantly from d.c. discharges. Let's now look at the parameters that the physical phenomena depend on.

Breakdown Parameters

Three types of interaction lead to electron losses. First, the interaction between electrons and container walls leads to a free diffusion process. This process, which is the most important in microwave breakdown, depends on the pressure (mean free path ℓ) and the container dimensions (diffusion length Λ). Second, the interaction between electrons and positive ions leads to the recombination process which, in our case, is negligible. Third, the interaction between electrons and neutral atoms or molecules leads to the attachment process which is also negligible except in air and oxygen gas. Further details on both cases can be found in Mac-Donald (1979). Thus, we may conclude that microwave breakdown is controlled by free-diffusion.

The interaction of electrons with both the electric field and the gas produce energy transfer and ionization. The energy transfer depends on: electric field amplitude, excitation frequency of the microwave field, nature of the gas, gas pressure (collision frequency ν) and energy distribution. The ionization frequency depends on: nature of the gas, gas pressure, electric field distribution, and the distribution function. The second difficulty in the analysis of breakdown is that we must know the distribution function to write the breakdown condition that is a particle balance equation: electron losses and production must be equal. We can solve this equation analytically only in some cases.

Simplified Model of the Average Electron

Energy Transfer. Because free electrons in a vacuum under the action of an alternating electric field oscillate out of phase with it, no energy is transferred from the electric field to the electrons. On the other hand, if the electrons are in the presence of gas, their ordered oscillatory motion is changed to a random one and, on the average, they can take energy from the electric field by elastic collisions. This gain of energy occurs as the energy absorbed by electrons is proportional to the square of the electric field and, hence, does not depend on its sign.

The energy gain of an electron is given by

$$P = e E \bar{v} = \frac{e^2 E^2}{m \nu_m} \cdot \frac{\nu_m^2}{\nu_m^2 + \omega^2} \qquad (1)$$

where E is the amplitude of the electric field, $\bar{v}$ the average drift velocity, ν_m the collision frequency for momentum transfer and ω the excitation angular frequency of microwave field. This equation may be written in terms of an effective electric field E_e

$$P = \frac{e^2 E_e^2}{m \nu_m} \qquad (2)$$

where E_e is the steady field which would produce the same energy transfer as E. This notion will be used later in the discussion of steady-state discharges.

Breakdown Condition. If the electric field is large enough, some electrons acquire sufficient kinetic energy that they may ionize the atoms. The primary ionization is considered to be the only production phenomenon that controls the breakdown, while free diffusion is the main loss process.

When the gas breakdown occurs, the electron production by ionization balances the electron loss by diffusion, recombination, or attachment. Assuming that the last two processes are small, we can obtain a theoretical model of diffusion-controlled breakdown. In this case, the continuity equation may be written

$$\frac{\partial n}{\partial t} + \vec{\nabla} \cdot \vec{\Gamma} - \nu_i n = 0 \qquad (3)$$

where n is the electron density, ν_i the ionization frequency and $\vec{\Gamma}$ the current density given by

$$\vec{\Gamma} = - \vec{\nabla}(Dn), \qquad (4)$$

with D as the free diffusion coefficient. At breakdown if the electron production is equal to the electron loss, the continuity equation becomes

$$\nabla^2 (Dn) + \frac{\nu_i}{D}(Dn) = 0. \qquad (5)$$

If we replace the spatial Laplacian operator ∇^2 by $1/\Lambda^2$ where Λ is the diffusion length determined by the container geometry, we obtain the breakdown condition in the form

$$\frac{\nu_i}{D} = \frac{1}{\Lambda^2}. \qquad (6)$$

From relation (6) the breakdown field can be deduced as a function of the ionization frequency.

Example of Experimental Results. Figure 1 shows a block diagram of the experimental cavity method of determining breakdown at microwave frequencies (Brown, 1959). Through the power divider, a variable incident power is fed into the cavity. The directional coupler provides a known fraction of the incident power which is measured by the wattmeter. A slotted line allows the determination of the standing wave ratio. The discharge tube is inserted into the cavity which uses the TM_{010} - mode. Then, from standard microwave measurements, the electric field is determined through the Q factor of the cavity and the known field configuration.

MICROWAVE DISCHARGES

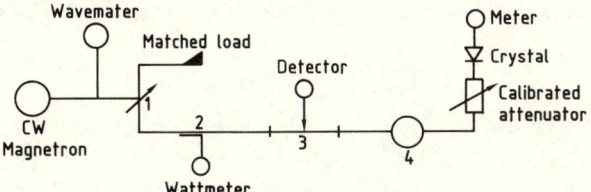

Fig. 1. Block-diagram of an experimental method of determining the breakdown at microwave frequencies.
1 - Power divider
2 - Directional coupler
3 - Slotted section
4 - Resonant cavity using the TM_{010} - mode.

An example is given in Fig. 2 for Heg gas which is a mixture containing mostly helium with a small amount of mercury. The experiment of MacDonald and Brown (1949) was performed at a frequency f = 2.8 GHz in a cylindrical cavity whose length is L.

A theoretical interpretation has been attempted for low and high pressure ranges.

At high pressure, $\nu_m \gg \omega$, the energy transferred from the electric field to the electrons is dissipated into elastic collisions

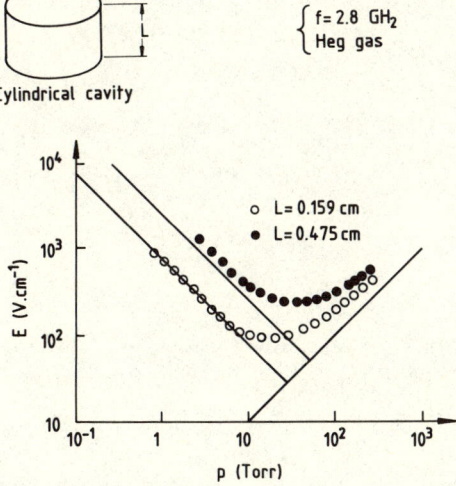

Fig. 2. Breakdown electric field versus pressure (after Brown, 1959).
—— Theoretical curves
o o Experimental points.

between electrons and gas molecules. The energy dissipated per collision is given by

$$\text{Energy/Collision} = \frac{e^2 E^2}{m\nu_m^2} = e \frac{2m}{M} \bar{u} \qquad (7)$$

where $\bar{u}$ is the average energy. From (7), the electric field may be calculated

$$E = 0.94\, p(\bar{u})^{1/2} \qquad (8)$$

for Heg gas in which nearly all the collisions are nonionizing. If one assumes the average energy to be one third of ionization potential u_i ($u_i = 19.8$ v), we find for E the following expression

$$E \simeq 2.4\, p. \qquad (8a)$$

It can be seen in Fig. 2 that at high pressures, breakdown measurements approach this line.

At low pressures, $\nu_m \ll \omega$, the electrons make many oscillations per collision. The number of collisions to ionize is nearly equal to that to diffuse out of the tube. In Heg gas, all inelastic collisions are ionizing ones; hence, all the input power goes into ionization and we may write

$$\nu_i = \frac{P}{eu_i} = \frac{e E^2}{mu_i \nu_m^2} \frac{\nu_m^2}{\nu_m^2 + \omega^2}. \qquad (9)$$

Then the electric field is deduced from the breakdown condition by assuming that ν_m has the same value in helium and Heg gases ($\nu_m \simeq 2.4 \times 10^9\, p$)

$$E \simeq \frac{1284}{p\, \Lambda\, \lambda} \qquad (10)$$

where λ is the wavelength of the microwave field. It can be seen in Fig. 2 that corresponding lines are in good agreement with experimental results.

Of course, this crude model cannot explain quantitatively the phenomena, but it gives a first qualitative approach. The next step must take into account the electron energy distribution.

MICROWAVE DISCHARGES

A Model Taking into Account the Distribution Function

As we have just seen, the model of the average electron cannot describe the real situation. We must use the kinetic theory of gases provided that the distribution of electrons in space and in time is known. To determine the distribution function, we must solve the Boltzmann equation

$$\frac{\partial F}{\partial t} + \vec{v} \cdot \vec{\nabla} F + \vec{a} \cdot \vec{\nabla}_v F = C \qquad (11)$$

where $\vec{v}$ is the velocity vector, $\vec{\nabla}_v$ is the gradient in velocity space, $\vec{a}$ the acceleration, and C the collision term. To solve this equation, we must make some simplifying assumptions. Then, from the distribution function F, we may deduce the ionization frequency ν_i and the breakdown field from the breakdown condition $\nu_i - D/\Lambda^2$.

In this section we shall not give the general method of determining F (the reader will find a complete study in MacDonald (1966), but we shall recall only the physical meaning of the assumptions and some possibilities of calculating analytically the electric breakdown field in particular cases.

Some simplifying assumptions are necessary:

(1) We assume that the number of collisions is large enough to prevent any substantial ordering of the direction of motion of individual electrons. Thus, the velocity distribution may be considered as spherically symmetrical at any point in space configuration. Therefore, we may expand the F function in spherical harmonics and also the collision term.

(2) We assume a rapid convergence of the series.

(3) We assume that most breakdown problems can be adequately approximated by steady-state solutions.

The assumptions 2 and 3 lead us to neglect the higher order terms, and the set of simultaneous equations for components of the distribution function may be reduced to four (MacDonald, 1966).

Then to arrive at the solution of the Boltzmann equation, let us denote by $F_o^°$ the component of the distribution which is zero order in space and in time, i.e., the symmetrical term. We have all the information necessary to determine the electrical properties of the discharge if we can calculate $F_o^°$. To solve the set of equations for $F_o^°$, the classical method uses the possibility of separating the velocity variation from the space variation which appears only in the Laplacian operator. Under these conditions, if we replace the operator ∇^2 by $1/\Lambda^2$ and if we set

$$F_o^° (u,x,y,z) = f(u)g(x,y,z),$$

we obtain an equation for the energy-varying component, f, of the spherically symmetrical term of the distribution function:

$$\frac{2e}{3m} \frac{E^2}{u^{1/2}} \frac{d}{du} \left(\frac{u^{3/2} \nu_m}{\nu_m^2 + \omega^2} \right) \cdot \frac{df}{du} + \frac{2m}{M u^{1/2}} \frac{d}{du} (u^{3/2} \nu_m f)$$

$$= \left(h \nu_c + \frac{2 e u}{3 m \nu_m \Lambda^2} \right) f \quad ,$$

(12)

with $u = \dfrac{m v^2}{2 e}$, (13)

where E is the rms value of the electric field, ν_c the collision frequency and h the efficiency of inelastic collision processes.

If we look at Eq. (12), we find that analytic solutions can be found only in some cases.

(1) The collision frequency ν_m does not depend on energy, the case in helium, hydrogen or Heg gases.

(2) The mean free path does not depend on energy, the case in neon gas. Under this condition, the collision frequency ν_m is proportional to $u^{1/2}$, and analytic solutions are found in the low and high pressure ranges. We give two examples, Heg gas and neon gas.

When the collision frequency is independent of energy as in the case of Heg mixture, $\nu_m(u)$ is a constant. In Fig. 3, we have plotted the results obtained by MacDonald and Brown (1949). It can be seen that theoretical curves and experimental points agree well. The dashed lines describe the results of the average electron model; these were good for low and high pressures.

A typical example in which the mean free path is independent of energy is that of neon gas. Taking into account the proportionality between ν_m and $u^{1/2}$, the differential equation for f may be simplified at low and high pressures. Effectively, if the pressure is low enough that there are many cycles of the field per collision, the ν_m^2 term may be neglected compared with ω^2. At high pressure, the ω^2 term may be neglected compared with ν_m^2. In Fig. 4, we have reported the results obtained by MacDonald and Betts (1952). It can be seen that the agreement between experimental and theoretical results is not nearly as good as for gases whose collision frequency is independent of energy. Further, no information can be obtained on the optimum breakdown region.

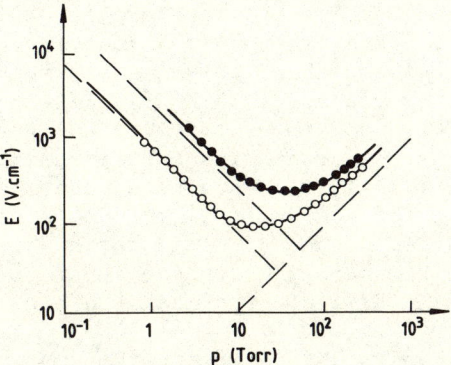

Fig. 3. Breakdown electric field in Heg gas $\nu_m(u)$ = constant.
—— Theoretical curves taking into account the distribution function. — — Theoretical model of the average electron.
● L = 0.159 cm. ○ L = 0.475 cm.

Limits of Diffusion Theory

It is useful to consider the limits of applicability of the diffusion theory, and graphically represent them versus the practical variables. As we have previously seen, the breakdown field may be expressed as a function of the ionization potential u_i, the mean free path ℓ, the wave length λ and the diffusion length Λ:

$$E_b = E(u_i, \Lambda, \lambda, \ell) .$$

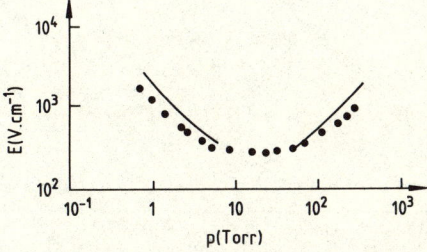

Fig. 4. Breakdown electric field in Neon $\ell(u)$ = constant (after MacDonald, 1966). —— Theoretical curves, ● Experimental points, f = 2.8 GHz - Λ = 0.202 cm.

Thus, from dimensional analysis, there are five variables, but only two are independent. This leads us to treat the problem with three independent dimensionless variables which are functionally related. A possible set of such variables is $E\Lambda/u_i$, λ/ℓ, and Λ/ℓ. However, it is often more convenient to use dimensional variables taking into account the invariance of quantities contained in them. In practice, the ionization potential is a constant for each gas and the mean free path is nearly proportional to $1/p$. So, a convenient set of such variables is $E\Lambda$, $p\lambda$, and $p\Lambda$. This set has the advantage that p, Λ and λ are the experimentally independent parameters which determine the dependent variable E, i.e., the measured breakdown field.

The diffusion limits introduced by the assumptions are expressed in terms of the proper variables $p\Lambda$ and $p\lambda$ and, hence, plotted on the $p\Lambda$ - $p\lambda$ plane (Fig. 5).

The diffusion theory assumes a uniform field. It is clear that at very high frequencies, there exists a limit to the discharge size consistent with this assumption. If Λ becomes longer than λ, we are not able to calculate a meaningful characteristic diffusion length. This is written in terms of the size of a cylindrical container whose length is equal to a single loop of a standing wave of the electric field.

$$\Pi \Lambda = \frac{\lambda}{2}, \tag{14}$$

which may be written

$$p\lambda = 2\Pi(p\Lambda) \tag{15}$$

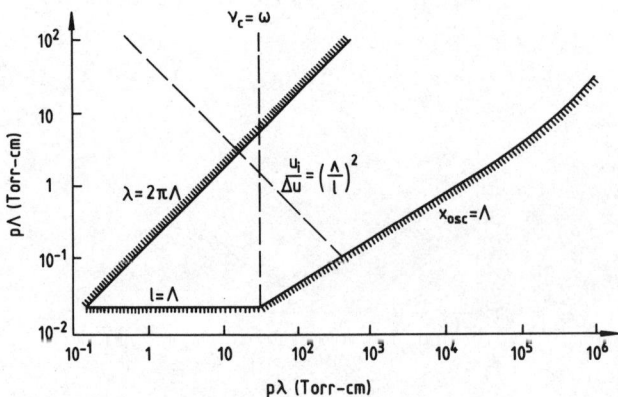

Fig. 5. Limits of diffusion theory for microwave breakdown (after Brown, 1959).

A second limit occurs if the electron mean free path becomes longer or comparable to the vessel size, i.e., if $\ell > \Lambda$, the diffusion concept ceases to have meaning. This limit may be written

$$p\Lambda = p\ell = \frac{1}{P_m} . \tag{16}$$

In the case of hydrogen reported in Fig. 5, we get $P_m = 49$ (cm - torr)$^{-1}$ from the classical values of collision probability.

Finally, there is the oscillation amplitude limit. If the amplitude of the electric field is large enough, the electrons can be swept completely across the tube and collide with the walls during a half-cycle. Thus, the first order, time dependent, in the expansion of the distribution function is no longer negligible, and the series does not converge rapidly. The limit is obtained when the oscillation amplitude is equal to the diffusion length which in hydrogen leads to the following expression

$$p\lambda = 2\pi \cdot 10^5 (\frac{p\Lambda}{E/p}) . \tag{17}$$

The equality between the collision and excitation frequency corresponds to the transition from phenomena in which there are many collisions per oscillations of the field to those in which there are many oscillations per collision. This transition is obviously related to the energy-transfer problem; it may be written

$$p \lambda = 32 \tag{18}$$

in hydrogen.

As we have seen in previous examples, the breakdown field has a minimum as a function of pressure. It varies differently on the two sides of the collision frequency transition.

At high pressure, the breakdown field increases with the pressure. This can be interpreted as follows: The efficiency of energy transfer is high but if the pressure increases, an increasing fraction of the energy is dissipated in elastic collisions.

In hydrogen, the breakdown curves show that, at high pressures, the breakdown electric field is almost independent of frequency and diffusion length and may be represented by

$$E \simeq 10 \, p . \tag{19}$$

At low pressure, the breakdown field increases when the pressure decreases. This can be explained as follows: The electrons oscillate out of phase with the field, and an increasing scarcity of collisions decreases the efficiency of energy transfer. The breakdown condition may be stated: The collision number to reach the ionization potential, u_i, is equal to the collision number to diffuse out of the container. This leads to

$$\frac{u_i}{\Delta u} = \frac{u_i}{(\frac{eE_e^2}{m\nu_m^2})} = \frac{3}{2}(\frac{\Lambda}{\ell})^2 \qquad (20)$$

where Δu is the energy change between two collisions. In hydrogen we obtain for the electric field

$$E \simeq \frac{550}{p\lambda\Lambda} . \qquad (21)$$

The elimination of E between relations (19) and (21) leads to

$$p\lambda = \frac{55}{p\Lambda} . \qquad (22)$$

Conclusion

Summarizing, in some cases we are able to model the phenomena involved in microwave breakdown. Nevertheless, in most cases we cannot describe accurately the breakdown conditions, and it would be very interesting if new theoretical studies were undertaken to model these phenomena. Without a model, it is very difficult to predict the experimental conditions of breakdown of a gas which are of interest for experiments on maintained plasmas and for their potential applications, particularly at high pressure. Despite these difficulties, the theoretical model of Brown (1959) that we have previously reported gives good insight into the phenomena.

STEADY STATE DISCHARGES

Introduction

Under breakdown conditions, electron losses are controlled by free diffusion while in steady-state discharges the electron losses are controlled by ambipolar diffusion because the electron concentration is high enough after breakdown. The free diffusion coefficient is much higher than the ambipolar diffusion coefficient; thus the effective electric field required to maintain the discharge is much smaller than that required to break down the gas.

The difference between the breakdown and maintaining fields depends on the gas and other experimental parameters. An example is shown in Fig. 6, in hydrogen. The figure gives theoretical curves based on the analysis of Rose and Brown (1955) which leads to the solution of the Boltzmann equation by assuming that electron recombination and electron-electron interactions may be neglected.

As in the breakdown problem, the question is, what are the conditions required to maintain the discharge. The problem is now one of energy balance. Indeed, the maintaining condition is that the absorbed energy balances the losses. So, to determine the maintenance conditions and give a good description of the physical phenomena, we must discuss the parameters that play a role in establishing the energy balance.

Problems of Steady-State Discharges

To create a model of the discharge, we must look at the main loss processes involved and estimate them. Energy can be lost by ionization, excitation, gas heating, collisions at the walls, etc. To estimate these losses, we must know

- the electron temperature T_e,

- the collision frequency,

- the electromagnetic field in the discharge (requires solving Maxwell's equations),

- the energy and field distributions.

Solving Maxwell's equations is difficult, and we need some assumptions such as

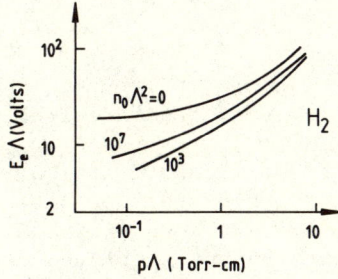

Fig. 6. Steady-state values of electric field versus $p\Lambda$ and electron density (after MacDonald, 1979).

- Maxwell distribution of electrons
- electron temperature equal to that of a positive column.

Having determined the losses, we still need to know the energy transferred from the microwave field to the plasma. Obviously, it is not sufficient to know the amount of absorbed energy; we are also interested in the means by which it is transferred. This leads us to the description of excitation structures. We shall then try to make a classification of microwave structures used for plasma generation.

Classification of Microwave Structures of Plasma Generation

We can consider that we have three types of systems, as shown in Table 1: resonant systems, propagative systems, and absorbing systems. With regard to these systems, the plasma can play several roles. It is either a perturbation of the excitation structure or the excitation structure itself.

We now discuss these different systems and show the corresponding structures with their characteristics and the problems associated with determining the energy balance.

Plasma in a Cavity. An example of a cavity used by Asmussen et al. (1974) is shown in Fig. 7. The plasma tube is centered on the cylindrical cavity axis. One end of the cavity is a fixed short-circuit and on the other end is a moving one which makes it possible to change the cavity's dimensions. The microwave energy is coupled to the cavity plasma system by the antenna.

The plasma may be considered as a perturbation on the resonance cavity whose frequency in the presence of the plasma is

$$f_{res}(\omega_p) = f_{res}(\omega_p = 0) + \Delta f \tag{23}$$

with $\tilde{\Delta f}$ proportional to n, where ω_p is the plasma frequency and n the electron density.

The characteristics of the structure are:

Pressure range: From 0.1. torr up to the atmospheric pressure. However, it must be noted that the dimensions of the created plasma are nearly those of the cavity. Further, at high pressure, the plasma is localized around the antenna, and its volume is about one cubic centimeter.

Stability: The system has an operating point. This particular point will be explained in the next subsection.

MICROWAVE DISCHARGES

Table 1. Classification of Microwave Structures

	PLASMA = PERTURBATION OF EXCITATION STRUCTURE	PLASMA = EXCITATION STRUCTURE
RESONANT SYSTEM	PLASMA in a cavity	eigen resonance
PROPAGATIVE SYSTEM	PLASMA in a waveguide	PLASMA = Propagative structure
ABSORBING SYSTEM	Partially absorbing plasma in a waveguide	PLASMA = load for a line

Matching: To obtain matching, one must change either the frequency or the cavity dimensions.

Energy balance: it is difficult to obtain energy balance as we do not know accurately

- the electric field E
- the quality factor Q
- the operating point on the resonance curve

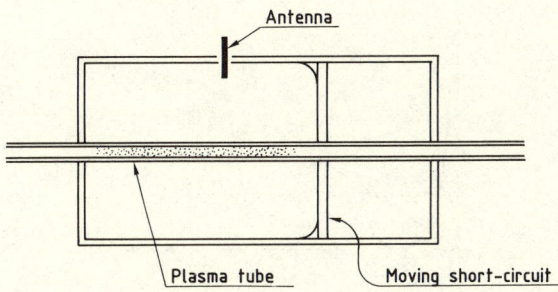

Fig. 7. Typical microwave cavity.

The maintaining condition of such a discharge is simply the equality between the absorbed power and the losses. The absorbed power is

$$P_{abs} = \frac{1}{2} \text{Re} \int_v \sigma |E|^2 \, dv \qquad (24)$$

where σ is the plasma conductivity, E the amplitude of electric field inside the plasma and v the total plasma volume.

At a fixed value of the field, the absorbed power depends on the damping processes in the plasma and the coupling between the fields inside and outside the plasma. Nevertheless, the variations of the absorbed power versus the plasma frequency give a resonance curve like those plotted in Fig. 8. In the same figure, we have plotted the losses, the dashed curve which crosses with resonance curves to give the operating points, after Leprince et al. (1973).

At points like B, the discharge is unstable because decreasing n leads to a decreasing absorbed power, and the discharge disappears. At points like A, the discharge is stable because decreasing n leads to a decreasing absorbed power. Thus, the plasma is maintained with a density slightly higher than that at resonance.

We may further comment on this figure. If we excite the resonance frequency f_o of a cavity, there is breakdown from a power threshold. Usually, the electron densities obtained are weak and a large fraction of the energy is reflected. To obtain plasmas with higher densities, it is possible to increase the excitation frequency f_o, and the maximum density is that of point C where the curves of absorbed power and losses are tangent. Another way of increasing the density is to change the cavity length. Then, a plasma with a density about ten times higher than the cutoff density may be obtained.

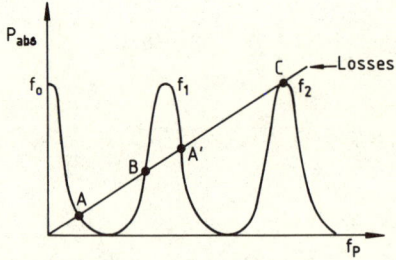

Fig. 8. Determination of operating points of a microwave cavity sustaining a plasma.

MICROWAVE DISCHARGES

This type of microwave discharge is well known to people using microwave spectral lamps like Fehsenfeld et al. (1965). They also know their disadvantages, particularly, the matching difficulties when discharge parameters are changed.

<u>Plasma Created in an Eigen Resonance.</u> A bounded plasma has several resonances; a main resonance and secondary ones. They are connected with two types of modes: tape with even symmetry (e.g., azimuthal, quadrupolar modes) and those with odd symmetry (e.g., dipolar, Tonks-Dattner modes). The frequencies of these resonances may be written

$$\omega_{res} = K \omega_p \qquad (25)$$

with $1<K<3$

These resonances do not occur if the electron density is equal to zero. So, a plasma cannot be initiated by them but only maintained. That means an external source is necessary. Major studies have been reported by Leprince (1968), Messiaen and Vandenplas (1971).

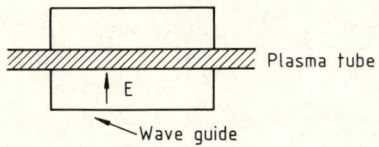

Fig. 9. Structure for plasma creation on an eigen resonance.

An example of such a system is shown in Fig. 9. A plasma tube is inserted in a rectangular waveguide, using TE_{01} mode. The tube axis is set perpendicular to the electric field of the waveguide.

The characteristics of this structure are:

Pressure range: Low and mean pressures

Stability: The problem looks like that of a cavity

Energy balance: It is still a difficult problem as one does not know accurately
- the electric field and its distribution
- the quality factor
- the operating point on the resonance curve

Plasma as a Perturbation of a Propagative Structure. In Fig. 10, we show an example of a structure studied by Kampmann (1979). A microwave source feeds a circular waveguide in the H_{11} mode and a plasma tube whose axis is the same as that of the waveguide is set inside the guide.

The propagation of the waveguide mode is slightly perturbated by the plasma. The main difference between this structure and that of the preceding section is that the guide modes in the previous case may be propagated without plasma.

The plasma that may be created has a density varying along the column which generally, has a value lower than critical $n < n_c$. Unfortunately, this coupling structure has a very low efficiency of energy transfer.

Its characteristics are:

Pressure range: Low and mean pressure

Energy balance: No attempt to determine this has been made.

Plasma Created by a Plasma Eigenmode. Such a structure called surfation is shown in Fig. 11. It has been studied by Glaude and our group (1980). Its principle will be explained in more detail later. However, we now discuss its characteristics. The plasma created is the propagative structure (surface wave, for example). Its density is inhomogeneous along the column and has a value always higher than critical, $n > n_c$.

These coupling structures are wide band structures (from 400 MHz up to 700 MHz). The frequency range which may be used is from about 200 MHz to 2.450 MHz. The limits are those of the practical dimensions. We have obtained energy coupling as high as 90%.

Characteristics are:

Pressure range: Very large, from 10^{-2} torr up to atmospheric pressure or a few bars.

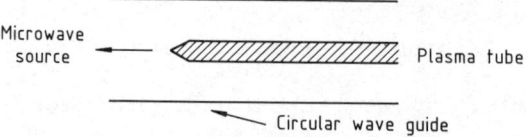

Fig. 10. Geometry of a propagative structure.

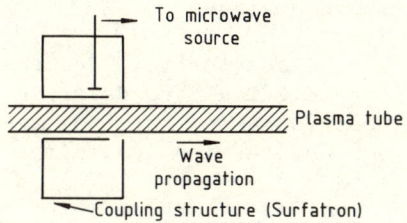

Fig. 11. Surfatron

Energy balance: Some results have been obtained and will be presented in a following section. The collision frequency ν can be measured and from it the electron temperature T_e may be deduced. The electric field E can be calculated from wave properties. Thus, the energy balance can be obtained.

<u>Partially Absorbing Plasma in a Waveguide.</u> The plasma is set in a rectangular waveguide and plays the role of an absorbing obstacle (Fig. 12). Only a part of the microwave energy is absorbed, and the waveguide must have a matched load at its end. So, the plasma is a complex impedance whose resistive term is due to collisions.

Usually, the collision frequency is smaller than the excitation one, $\nu < \omega$, and the density is smaller than the critical one, $n > n_c$. Obviously, in this case, the absorbed power can be measured from the relation

$$P_{abs} = P_{in} - P_{ref} - P_{trans} \tag{26}$$

where P_{in}, P_{ref} and P_{trans} are respectively the incident, reflected, and transmitted power. The only available results on such a structure are those published by Maksimov (1974). His interest is in obtaining the energy balance. The main characteristics are, hence, the following:

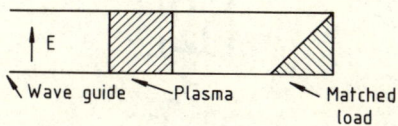

Fig. 12. Partially absorbing plasma.

Pressure range: Mean pressure

Energy balance: The electric field and the collision frequency are difficult to determine. Some results have been obtained but, sometimes, using assumptions of undetermined validity.

Plasma, A Load at the End of a Line. In this case, the plasma behaves as a purely resistive impedance closing a propagation line. Usually, the characteristics of such a plasma are that the collision and plasma frequencies are higher than the excitation frequency, $\nu > \omega$ and $\omega_p > \omega$.

A structure of this type has been studied by our group. It is shown in Fig. 13.

This structure looks like the system we have called surfation but the "capacitive coupling" has been replaced by a short circuit and the dielectric tube containing the gas by a metallic tube, which is simultaneously the inner tube of a coaxial output system. Gas (argon or helium in our experiments) can flow through this inner tube at the end of which a flame is created. The matching of the system is obtained by moving the cylindrical plunger. We obtain good energy transfer, about 90%, at 2.450 MHz with input power about 1 Kw. Care must be taken with regard to the radiated energy. The general characteristics of the structure are:

Pressure range: High pressure, from a few tenths of torrs up to the atmospheric pressure.

Energy balance: It is now impossible to obtain.

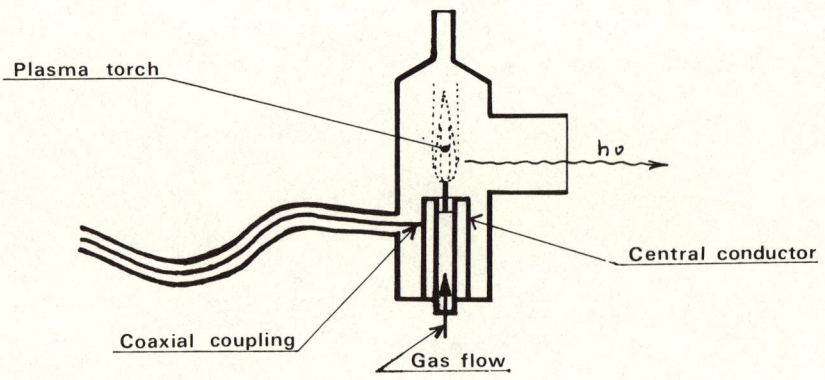

Fig. 13. Plasma torch where the plasma is a microwave load.

Discharges Created by Surface Waves

Characteristics of the Surface Wave. Along a plasma column, waves with the electric field $\vec{E}$ and the wave number $\vec{k}$, both parallel to the column axis, can be propagated. These waves for which the radial dependence of the axial field shows a maximum at the plasma-dielectric container boundary are surface waves. They were first studied by Trivelpiece and Gould. Figure 14 shows a example of dispersion curves when the container is a glass tube of relative permittivity ε_g, f_o being the wave frequency. Only the modes m = 1 (dipolar mode) and m = 0 (axially symmetrical mode) are plotted. The only mode excited by our couplers (shown in Fig. 11) is the m = 0 mode.

The limit marked in Fig. 14 shows that the wave can be propagated if $f_o \leq f_{pe}/\sqrt{1+\varepsilon_g}$, i.e., if the electron density is higher than a critical value n_c. Thus, we can write the propagation condition of this wave as

$$n > \alpha \, n_c \qquad (27)$$

with $\alpha = 1 + \varepsilon_g$.

Because a microwave field can ionize a gas, a plasma is created when enough microwave power is fed into the system. Condition (27) implies that the wave cannot be propagated without a plasma; thus, we have a system where a traveling surface wave sustains a plasma column acting as the propagating medium.

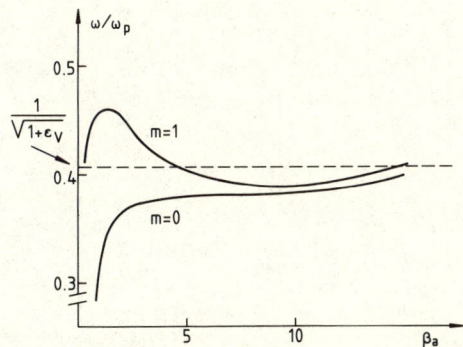

Fig. 14. Dispersion curves of the axially symmetrical mode m = 0 and the dipolar mode m = 1. Inner plasma radius a = 10 mm – outer plasma radius b = 12 mm; ε_g = 5 and $\frac{\omega_p}{c}$ = 1.

Further information on this type of wave is given in Fig. 15 which shows the shape of the radial dependence of the three field components E_r, E_z, and H. We must note that the main component in the plasma is E_z, while outside it is E_r. In practice, the wave frequency is constant and more convenient parameters to use are the dimensionless variables ω/ω_p and βa, where $\beta = \frac{2\Pi}{\lambda}$ is the wave number and a the plasma radius. The corresponding curve is shown in Fig. 16.

From the beginning to the end of the plasma column, energy is taken from the wave to ionize the gas until the wave energy becomes lower than a threshold value. This value is such that the electrons created do not reach the minimum density necessary for wave propagation. Hence, the electron density decreases along the column, and the operating point covers a portion of the curve of Fig. 16, e.g., from region 1 to region 2. That fact has two main consequences. First of all, the field distribution of the wave varies along the column; second, the energy repartition also varies along it. Results from numerical computations have been plotted in Fig. 17. It shows the variations of the field distribution between the end of the column (region 2 of Fig. 16) and its beginning (region 1 of Fig. 16). For convenience, these computations have been performed in the ideal case of two-layer structure (without glass). Figure 18 also shows mean variations of energy repartition along the column. Only a small part of the energy lies in the plasma for low values of the ω/ω_p ratio (the case of region 1 of Fig. 16), i.e., at the end of the column, while for higher values of ω/ω_p, near the column beginning, the amount of energy is practically the same inside and outside the plasma (the case of region 2 of Fig. 16).

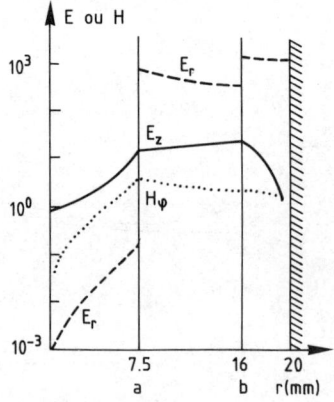

Fig. 15. Radial distribution of fields for $P_{in} = 1$ w. Values of E_z and E_r are expressed in V/m, those of H_ϕ in ampere-turn/m $\varepsilon_g = 4.$ - $\omega = 2\pi \times 200$ MHz. - $n = 10^{19} m^{-3}$.

MICROWAVE DISCHARGES

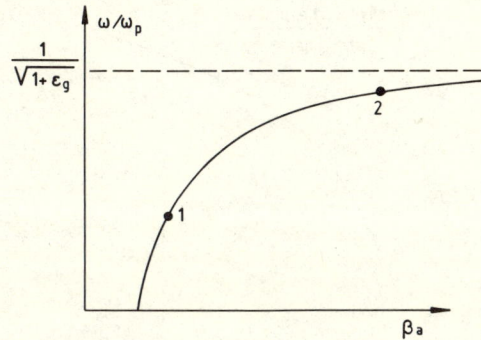

Fig. 16. Curve deduced on the dispersion one when ω = constant.

Discharge Modeling. As explained earlier, a good understanding of a microwave discharge can be obtained if we are able to write the energy balance. Our theoretical model is based on that idea, and we assume that the energy lost by the wave, i.e., the absorbed power, over a small length Δ_z is proportional to the number N of electrons produced over the same length. Thus, we may write

$$P_{abs} = \Theta N \tag{28}$$

where Θ is a proportionality constant depending on gas, pressure, tube radius, and wave frequency, but being independent of the

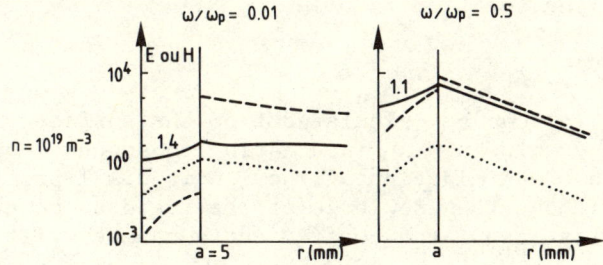

Fig. 17. Influence of ω/ω_p on the radial distribution of fields for $P_{in} = 1w$ and $\varepsilon_g = 1$.
——— E_z component (V/m)
— — E_r or $-E_r$ component (V/m)
····· H_ϕ component (ampere-turn/m)

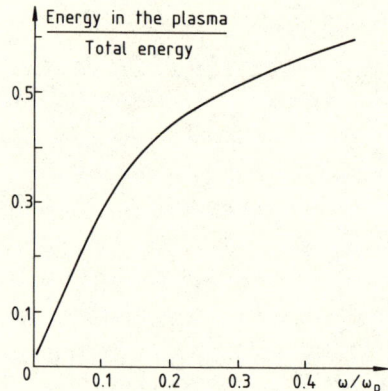

Fig. 18. Ratio of the energy inside the plasma over the total energy as a function of ω/ω_p.
$f = 915$ MHz. $- a = 7.5$ mm. $- b = 9$ mm. $- \varepsilon_g = 4$

incident power of the wave; otherwise, Θ is constant along the plasma column. The relation (28) may also be written

$$P_{abs}(\Delta z) = \Theta S \int_z^{z+\Delta z} n(z)dz \qquad (29)$$

where S is the plasma cross section and $n(z)$ the electron density. From (29), we may find the longitudinal variation of electron density.

$$\frac{dn}{dz} = -\frac{2\alpha n}{1 - \frac{n}{\alpha}\frac{d\alpha}{dn}} \qquad (30)$$

where α is the wave attenuation coefficient. We have shown [Glaude (1980)] that α depends also on the electron-neutral collision frequency through the relation

$$\alpha = \nu\, f(n). \qquad (31)$$

The value of $f(n)$ can be calculated from the surface wave dispersion equation for any value n. An interesting consequence of this relation (31) is the determination of the longitudinal profile of electron density without the knowledge of the incident power at the coupler. In most cases, the coefficient α can be written in the form

$$\alpha = \nu\, \gamma\, n^{-s} \qquad (32)$$

where γ and s are adjusted for a given density range.

From (30) and (31) we expect that n decreases nearly linearly with a slope proportional to the collision frequency ν, as shown in Fig. 19.

Although the assumption (28) seems very crude, it is in good agreement with results from computations of the longitudinal component of the electric field inside the plasma where it is the main component.

Effectively, as we can see in Fig. 20, the electric field E_z at the plasma boundary is nearly constant along the column (except near the plasma end where the electron density decreases very rapidly). This invariance is consistent with the independence of losses in the field's distribution.

Furthermore, the density measurements along the column give a linear profile whose slope is proportional to the collision frequency ν. As seen in Fig. 21, the experimental points, at a given pressure, are nearly on the same straight line, whatever the incident microwave power, although the initial density is higher for higher incident power.

To understand Fig. 21, we must note that the origin has been taken at the end of the plasma column while the symbol (→) represents the initial electron density and the plasma length obtained for every microwave power. Two measurement methods have been used to determine the electron density. First, it has been obtained by a classical interferogram method (Chaker et al., 1981). This shape is shown in Fig. 22. Second, for plasmas produced at low microwave frequency (about 200 MHz), we have used the properties of plasma surface waves (Nghiem et al., 1981). A test wave is excited at one point of the discharge and a moving antenna enables us to measure its phase β and its attenuation coefficient α along the column through a microwave receiver and a mixer. Since α and β depend on the ratio ω/ω_p, the electron density may be deduced.

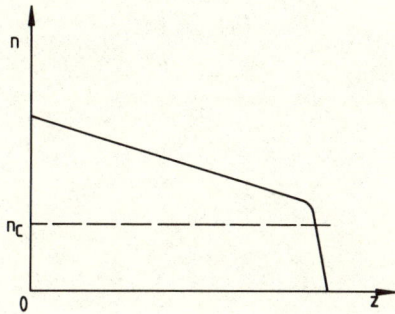

Fig. 19. Shape of the longitudinal profile of electron density.

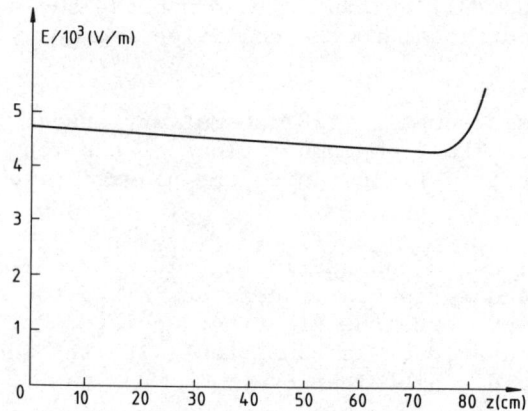

Fig. 20. Longitudinal profile of electric field inside the plasma E_z (r=a) f = 2,450 MHz,- a = 0.45 cm,- b = 0.6 cm,- ε_g = 4.5 - P_{in} = 100 w, ν = 10^9 s^{-1}.

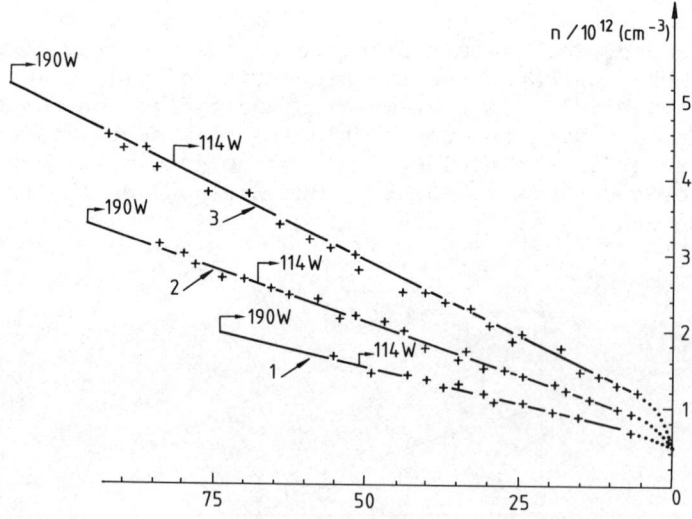

Fig. 21. Longitudinal profile of electron density in argon gas. + Experimental results f = 2,450 MHz,- a = 0.45 cm,- b = 0.6 cm. 1. p = 0.1 torr - 2. p = 0.2 torr - 3. p = 0.5 torr.

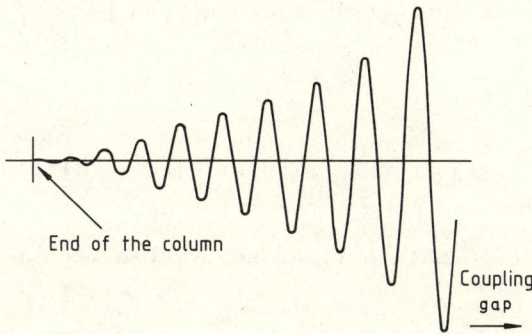

Fig. 22. Detected signal from the wave at 2,450 MHz. P_{in} = 114 W, - P_{ref} = 0 W. - p = 0.5 torr.

<u>Interest of Such Discharges.</u> As we have pointed out in previous sections, measurements are very difficult in most microwave discharges; hence very little information is available on laws governing the behavior of physical parameters. In the case of discharges sustained by surface waves, several measurements are possible.

- The electron density n can be determined from the interferogram (Fig. 22) as explained above.

- The collision frequency ν is measured from the slope of the axial density profile.

- The absorbed power P_{abs} is measured at each point by means of a moving antenna.

$$P_{abs}(\Delta s) = P_{in}(z) - P_{in}(z+\Delta z) \tag{33}$$

- The coefficient Θ is measured at each point along the discharge so that we have the relation

$$P_{abs}(\Delta z) = \Theta N \Delta z, \tag{34}$$

and the energy balance may be obtained through the relation

$$P_{in} = \Theta \int N dz \tag{35}$$

From those measurements, we can obtain some characteristics of the discharge:

- The electron temperature T_e can be deduced from the relation $\nu = f(T_e)$ if we assume a Maxwell distribution

- The effective electric field E_e can be also determined. It is defined by

$$E_e = \frac{\nu}{\omega} E(r=a) \qquad (36)$$

where $E(r=a)$ is the value of electric field at the plasma boundary.

- The mean ionization frequency $\bar{\nu}_i$ can be obtained through the relation

$$\bar{\nu}_i = f(E_e/p)$$

as shown by Chaker et al. (1981).

- The ionization energy w_i, i.e., the energy needed for the creation of an electron-ion pair is

$$w_i = \Theta/\bar{\nu}_i . \qquad (37)$$

More accurately, this can be written

$$w_i = eV_i + \frac{3m}{M} \frac{\nu}{\bar{\nu}_i} k T_e + \sum_j \frac{\bar{\nu}_j}{\bar{\nu}_i} e V_j \qquad (38)$$

where eV_i represents the losses by ionizing collisions, the second term of (38), the losses by elastic collisions, and the third term of (38) the losses by excitation. ν_j is the collision frequency leading to the j level whose potential is V_j.

Experimental Results. From the electron density profiles, the collision frequency ν can be deduced. An example of the variations of ν versus the pressure p is shown in Fig. 23 for argon gas with a discharge tube whose inner radius is a = 0.45 cm. It can be seen that the variation law is nearly linear from low pressures up to 2 torrs, with the classical proportionality coefficient of approximately 10^9.

If we assume, as mentioned above, that the electrons have a Maxwell distribution, we can obtain the variations of electron temperature T_e versus the product pa. In Fig. 24, we have plotted experimental results obtained with the same conditions as in Fig. 23. In this figure, we have also marked the variations of electron temperature in a positive column controlled by diffusion and having the same characteristics. It can be seen that the values of T_e that we find are of the same order. We have thus shown that in this type of discharge, we are able to determine the same quantities

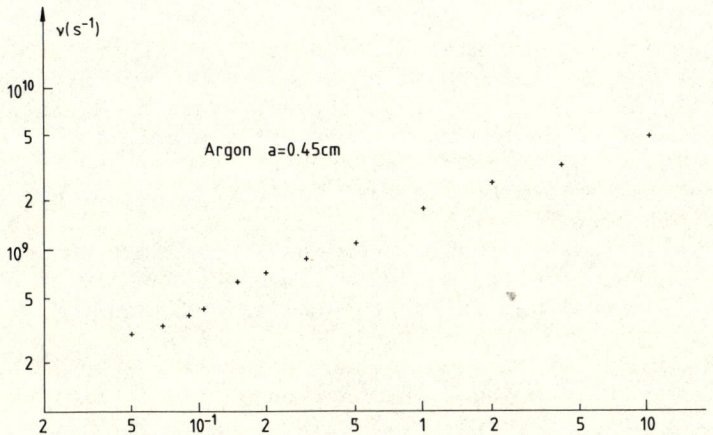

Fig. 23. Variation of the collision frequency versus the pressure in Argon. a = 0.45 cm. - b = 0.6 cm.

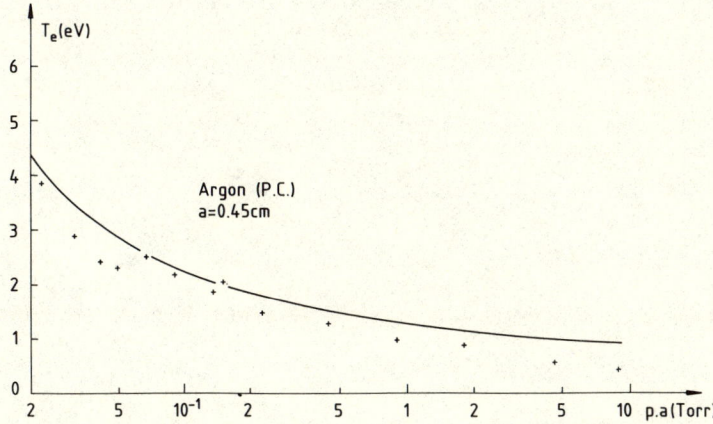

Fig. 24. Comparison of variations of the electron temperature in surface wave discharge and positive column in argon as a function of p.a. a = 0.45 cm. - b = 0.6 cm.

as in the classical glow discharges; that is not the case for most microwave discharges.

Furthermore, we are also able to write the energy balance equation for our discharge. We have shown that the proportionality coefficient Θ is a constant along the discharge at a given pressure and does not depend on the input power, the excitation frequency, and the radius. So, we have verified our assumption about the proportionality between the energy losses and the number of produced electrons. In Fig. 25, we have plotted variations of Θ versus the pressure with error bars which show that our results have good accuracy. In that figure, we see a linear dependence of Θ with pressure, below 2 torrs, similar to the collision frequency.

The knowledge of the coefficient Θ enables us to obtain a better energy balance through the determination of the ionization energy, w_i, i.e., the energy needed for the creation of an electron-ion pair. Results are plotted in Fig. 26, always with the same experimental conditions. For convenience, we have used the classical variable E_e/p as for the glow discharge, the effective electric field being given by relation (36). Three remarks can be made about this figure. First, the ionization energy decreases when the ratio E_e/p increases. Secondly, this decay of w_i is much slower for high values of E_e/p and w_i appears to approach a limit near 30 ev. We think this limit is the same as the energy lost by an ionizing beam in creating an electron-ion pair as it propagates through a gas. This energy has been found to be about 27 eV by Daugherty (1976).

Finally, we have also tried to estimate the losses by elastic and inelastic collisions, and we have defined three ratios:

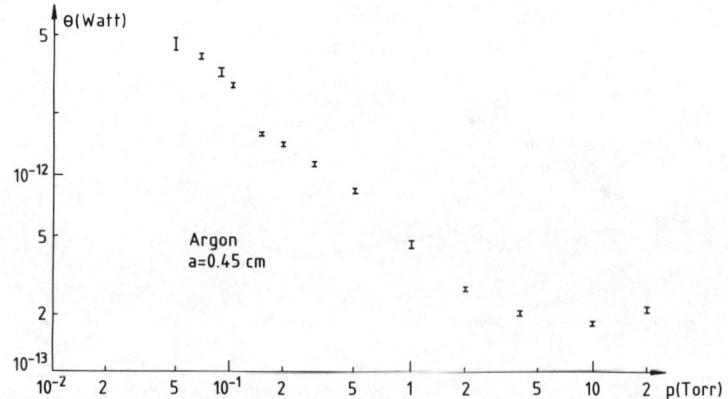

Fig. 25. Variations versus the pressure of the average power Θ absorbed by one electron from the microwave field. a = 0.45 cm. - b = 0.6 cm.

MICROWAVE DISCHARGES

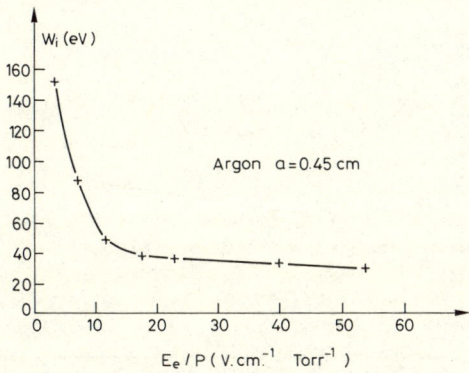

Fig. 26. Variations of the ionization energy versus the usual parameter E_e/p. a = 0.45 cm. - b = 0.6 cm.

(1) For losses by elastic collisions:

$$\delta_e = \frac{U_{el}}{\Theta} \quad (39)$$

where $U_{el} = \frac{3m}{M} \nu k \bar{T}_e \quad (40)$

(2) For losses by ionizing collisions

$$\delta_i = \frac{\bar{\nu}_i eV_i}{\Theta} \quad (41)$$

(3) For losses by excitation

$$\delta_{exc} = 1 - \delta_e - \delta_i \quad (42)$$

These three quantities have been plotted in Fig. 27 versus the usual parameter E_e/p.

At high values of E_e/p, the losses by ionization and excitation are of the same order while the losses by elastic collisions are negligible. Then, the losses by ionization decrease and those by excitation increase as do those by elastic collisions. However, when the losses by elastic processes remain lower than 10%, the losses by excitation become dominant.

This attempt to achieve energy balance gives only an approximated description of loss processes in the plasma. Nevertheless, one gains a better understanding of the behavior of loss mechanisms versus the effective electric field and the pressure.

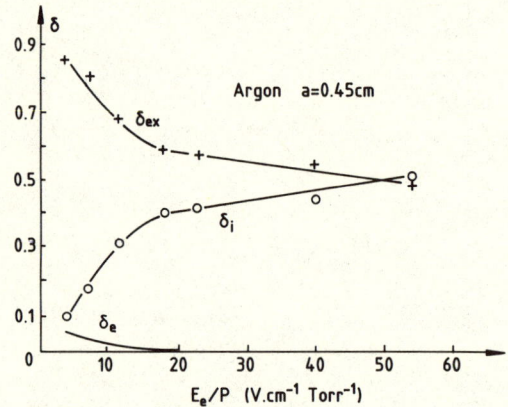

Fig. 27. Ratios of energy losses by elastic and inelastic collisions.
a = 0.45 cm. - b = 0.6 cm.

Conclusion

Concluding this part, let us note that recently there has been much interest in applications of microwave discharges. A good overview of applications to microwave systems, like switches or frequency converters, can be found in the review of MacDonald and Tetenbaum (1979). However, more promising applications seem to be be spectral radiation sources, ion sources, lasers, plasma-surface interactions, plasma chemistry, and chemical analysis.

The possibility of using microwave discharges as radiation sources is well known. A first lamp has been developed by Wilkinson and Tanaka (1955) in vacuum ultraviolet (VUV) region for absorption studies with xenon and krypton discharges. Also, many laboratories use the cavity lamps based on principle of Fehsenfeld et al. (1965). Since the work of Fehsenfeld, a waveguide cavity has been described by Beenakker (1976) for use at atmospheric pressure. A modified version has been used by Dalen et al. (1978). It has found application as a versatile selective potical emission detector in gas chromatography with applied power varying from 20 to 200 W. A new cavity shape called the slab-line cavity has been described by Hammond and Outred (1976), and its performance has been described recently by Outred and Hammond (1980).

More recently, a microwave discharge lamp at high pressure in krypton has been developed by Bloyet el al. (1979). A krypton-sealed tube (volume about 0.5 cm^3, pressure about 3 Kg/cm^2) is excited by a surfaguide (see earlier section) at 2.45 GHz. Power density as high as 2.5 KW/cm^3 has been fed in the discharge providing a strong line emission which could be used to pump a YAG laser.

A VUV source has been used by Pouey et al. (1980) with a surface system.

Also, a microwave-induced argon plasma has been produced at atmospheric pressure by Wong (1980). He uses a resonant cavity which is a cylindrical coaxial-line type operated at 2.450 MHz and 100 to 300 watts. The obtained plasma has the shape of a flame with a short tail outside the cavity. The electron temperature of the plasma is about 5.500°K and the electron density is about 10^{14} cm^{-3}, and gas translational temperatures about 4.000°K are measured. Application as a possible ion source for mass spectroscopy is mentioned.

Some attempts have been made to develop lasers using microwave discharges as an active medium. Ahmed and Kocher (1964) have used a 2.45 GHz cw magnetron to pump a He-Ne laser. A pulsed argon ion laser has been investigated by Paik and Creedon (1962), while Tuma (1970) studied plasma columns of xenon obtained form the dipolar mode. There has been interest in the possibility of chemical lasers, e.g., in the cw CO chemical laser of Stuart et al. (1972). However progress in practical microwave-excited lasers has been slow. More recently, an FH laser using discharges sustained by surface waves has been developed by Bertrand et al. (1977). Also a small CO_2-N_2He laser excited by a pulsed microwave source at 2.45 GHz has been operated at low powers by Handy and Brandelik (1978), the discharge structure being of the type described earlier with a tapered, rectangular waveguide. A pulsed microwave discharge has also been studied as a pump for the CO_2 laser by Vasyutinskii (1978) with an excitation structure of the type described in an earlier section. Muller (1979) has published a review paper on the use of a gas discharge in lasers and on their promising possibilities. Recently, a coaxial microwave structure has been studied by Kato et al. (1980) for argon-ion laser applications. We think, in fact, that it uses, a surface wave excitation.

Another interesting application is as a source of ions. Recent results have just been published by Henry et al. (1981). Using a surfatron as a plasma source, he has developed a simple extraction system which obtains ion currents of about 30 mA in argon.

At atmospheric pressure, some thermal applications based on plasma flames are possible. Interaction of microwaves with hydrocarbon fuel-air flame plasmas have been studied by Ward (1977, 1978, 1979) for applications of cylindrical geometry excited by modes independent of axial length. Another application involves stimulating combustion for accelerating the burning of lean air-fuel mixtures in internal combustion engines. We have also developed new techniques for welding and some other thermal applications with structures such as surface-wave couplers or microtorches. The possibilities of spectrochemical analysis with microwave induced plasmas at atmospheric pressure (CMP torch) were introduced by Mavrodineanu and Hughes (1963, 1964), and the first paper on the application of the CMP to analytical problem was published by Kessler and Gebhardt (1967). Then Goto et al. (1967), Yamamoto et al. (1967), Murayama et al. (1968), Kitagawa and Takenchi (1972, 1973) and Atsuya (1975) studied analytical properties and excitation mechanism of various CMP discharges. Their researches resulted in commercialization of a CMP spectrometer system by Hitachi (300 UHF Plasma Spectra Scan.). Successful analytical applications have been reported for determination of different materials like reagents and fine chemicals by Kessler (1971) and Dahmen (1978), minerals and geological samples by Govindaraju (1976) and Burman (1979). More recently a compilation of the most interesting results of CMP discharges has been published by Dahmen (1981). Also, a new CMP discharge used as excitation source for optical emission spectroscopy (OES) has been presented by Feuerbacher (1981). We must also mention that our group is working on that problem with a structure of the type described earlier (Patent 1980).

A final application which appears promising is in plasma chemistry. The reader will find a good discussion of the problem in the review paper of A. Goldman in these proceedings. We shall mention two applications of the interactions, between plasma and surfaces. Using microwave cavity discharges, Brodsky and Haller (1980) reported a new method of preparing hydrogenated amorphous silian, and etching systems have been developed by Hitachi (1980) and Toshiba Ltd (1980).

This short review of microwave discharge applications shows the recent interest manifested in these types of plasmas. It is justified by their many promising applications. Their main advantage is that they are electrode-less systems and, hence, they prevent most contamination. Also, in many cases, these systems are easier to operate than classical D.C. discharges. However, to extensively develop further applications, we must have a better understanding of the behavior of microwave discharges. We need to develop better models for microwave plasmas. We have seen that this is a very difficult problem in numerous cases and even an impossible one at times. That is why theoretical means must be actively employed. Our group is actively engaged in this activity.

REFERENCES

Ahmed, S. A. and Kocher, R., 1964, Proc. IEEE, 52: 1737.
Asmussen, J., Mallavarpu, R., Hamam, J. R., and Park, H. C., 1974, Proc. IEEE, 62: 109.
Atsuya, I., 1975, Anal. Chim. Acta, 74: 1.
Beenakker, C. I. M., 1977, Spectrochim. Acta, B, 32: 173.
Bertrand, L., Gagne, J. M., Mongeau, B., Lapointe, B., Conturie, Y. and Mosan, M., 1977, J. Appl. Phys., 48: 224.
Bloyet, E., Laprince, P., Marec, J., Bettini, J. P., Leroux, J. and Migne, J., 1979, in: Proceedings, Conference of Societe Francaise de physique, Toulouse, France."
Brodsky, M. H. and Haller, I., 1980, IBM Tech. Disclosure Bull., A, 22: (8), 3391.
Brown, S. C., 1959, "Basic Data of Plasma Physics," Wiley, New York.
Burman, J. O. and Boström, K., 1979, Anal. Chem., 51: 516.
Chaker, M., Bloyet, E., Leprince, P., Marec, J., and Nghiem, P., 1981, Int. Rept. LP190, Lab. Phys. Gaz et Plasmas, Orsay, France.
Chalker, M., Bloyet, E., Leprince, P., Marec, J., and Nghiem, P., 1981, in: "Proceedings, Conference on Surface waves in Plasmas, Blagoevgrad, Bulgaria."
Van Dalen, J. P. J., de Lezenne Coulander, P. A., and de Galan, L., 1978, Spectrochim. Acta, B, 33: 545.
Dahmen, J., 1978, in: "Procceedings, 6th Indo-German Seminars on Trace Element Analysis, Maria Laach, Austria.
Dahmen, J., 1981, ICP Inf. Newsletter 6: (11), 576.
Daugherty, J. D., 1976, in: "Principles of Laser Plasmas," G. Bekefi, ed., Wiley-Intersci., New York.
Fehsenfeld, F. C., Evenson, K. M., and Broida, H. P., 1965, Rev. Sci. Instrum., 36: 294.
Feuerbacher, H., 1981, ICP Inf. Newsletter, 6: (11) 571.
Glaude, V. M. M., Moisan, M., Pantel, R., Leprince, P., and Marec, J., 1980, J. Appl. Phys., 51: 5693.
Goto, H., Hirokawa, K., Suzuki, M., 1967, Z. Anal. Chem., 225: 130.
Govindaraju, K., Mevelle, G., and Chouard, C., 1976, Anal. Chem., 48: 1325.
Hammond, C. B. and Outred, M., 1976, Phys. Scripta, 14: 81.
Handy, K. G. and Brandelik, J. E., 1978, J. Appl. Phys., 49: 3753.
Henry, D., Hajlaoui, Y., Pantel, R., and Moisan, M., 1981, in: "Proceedings, Conference on Surface Waves in Plasmas, Blagoevgrad, Bulgaria."
Hitachi Ltd, 1980, Electronics, Nov. 20, p. 63.
Hitachi Ltd, 1980, Electronics, Dec. 4.
Kampmann, B., 1979, Z. Naturforsch., A, 34: 423.
Kato, I., Tsuchida, H., and Nagai, M., 1980, J. Appl. Phys., 51: 5312.
Kessler, W. and Gebhardt, F., 1967, Glastech. Ber., 40: 194.
Kessler, W., 1971, Glastech. Ber., 44: 479.
Kitagawa, K. and Takenchi, T., 1972, Anal. Chim. Acta, 60: 309.
Kitagawa, K. and Takenchi, T., 1973, Anal. Chim. Acta, 67: 453.
Leprince, P., 1968, Phys. Lett., A, 26: 431.

Leprince, P., Bloyet, E. and Milleon, H., 1973, J. Phys., 34: 185.
MacDonald, A. D. and Brown, S. C., 1949, Phys. Rev., 75: 411.
MacDonald, A. D. and Brown, S. C., 1949, Phys. Rev., 76: 1634.
MacDonald, A. D. and Betts, D. D., 1952, Can. J. Phys., 30: 565.
MacDonald, A. D., 1966, "Microwave Breakdown in Gases," Wiley, New York.
MacDonald, A. D. and Tetenbaum, S. J., 1978, in: "Gaseous Electronics," M. N. Hirsh and H. J. Oskam, eds., Vol. 1, Academic, New York.
Maksimov, A. I., 1974, Sov. Phys. Tech. Phys., 18: 1206.
Mavrodineanu, R. and Hughes, R. C., 1963, Spectrochim. Acta, 19: 1309.
Mavrodineanu, R. and Hughes, R. C., 1964, Develop. in Appl. Spectro., 3: 305.
Messiane, A. M. and Vandenplas, P. E., 1971, Appl. Phys. Lett., 18: 63.
Muller, Y. A., 1979, Radioel. and Comm. Syst., 22: 55.
Murayama, S., Matsumo, H., and Yamamoto, M., 1968, Spectrochim. Acta, B, 23: 513.
Nghiem, P., Bloyet, E., Chaker, M., Leprince, P., and Marec, J., 1981, Int. Rpt. LP 188, Lab. Phys. Gaz et Plasmas, Orsay, France.
Nghiem, P., Bloyet, E., Chaker, M., Leprince, P., and Marec, J., 1981, in: "Proceedings, Conference on Surface Waves in Plasmas, Blagoevgrad, Bulgaria,"
Poutred, M. and Hammond, C. B., 1980, J. Phys. D, 13: 1069.
Paik, S. F. and Creedon, J. E., 1968, Proc. IEEE, 56: 2086.
Pouey, M., Bloyet, E., Leprince, P., and Marec, J., 1980, in: "Proceedings, 6th International Conference on Vacuum Ultraviolet Radiation Physics Charlottesville, Ill," p. 65.
Rose, D. J. and Brown, S. C., 1955, Phys. Rev., 98: 310.
Stuart, R. D., Dawson, P. H., and Kimbell, G. H., 1972, J. Appl. Phys., 43: 1022.
Toshiba Ltd., 1980, Electronics, 53: 76.
Tuma, D. T., 1970, Rev. Sci. Instrum., 41: 1519.
Vasyutinshii, O. S., Kruzhalov, V. A., Perchanok, T. M., Tereklin, D. K., and Fridikhov, S. A., 1978, Sov. Phys. Tech. Phys. 23: 189.
Ward, M. A. V., 1977, J. Microwave Power.
Ward, M. A. V., and Wu, T. T., 1978, Combustion and Flame, 32: 57.
Ward, M. A. V., 1979, J. Microwave Power.
Wilkinson, P. G., 1955, J. Opt. Soc. Am., 45: 1044.
Wong, S. K., 1980, UTIAS Tech. Note CAN No. 225.
Yamamoto, M. and Murayama, S., 1967, Spectrochim. Acta, A, 23: 773.

PLASMA APPLICATIONS*

M. Kristiansen* and A. H. Guenther**

*Department of Electrical Engineering,
Texas Tech University, Lubbock, Texas 79409

**Air Force Weapons Laboratory, Kirtland Air
Force Base, Albuquerque, New Mexico 87117

INTRODUCTION

Plasmas have numerous applications for civilian as well as defense purposes. However, technical development is still in its infancy. Many new important applications depend only upon the imagination of engineers and scientists. In contrast to other develping technologies, applications from the fields of plasma science and engineering can only evolve through a multidisciplinary synergism. Research in plasma chemistry and physics together with gaseous electronics, fluid dynamics and thermodynamics, particularly mass and heat transfer, must be coupled with electro-chemistry and material science research particularly those aspects dealing with surfaces. In this paper we attempt to evaluate the importance of plasma applications. Obviously, it is impossible to do justice to all the important areas. The selection of topics is, therefore, influenced by the authors' interests and background. We will outline most of the applications rather briefly and concentrate in some detail on those areas in which we are interested.

Much of the information contained here was obtained directly or indirectly from the recent Workshop on Plasma Chemistry and Arc Technology sponsored by the National Science Foundation (Oskam and Pfender, 1980). Several of the participants in that workshop pro-

---

*This work was supported by the Air Force Office of Scientific Research, the Army Research Office, and the Air Force Weapons Lab.

vided us with useful material for this presentation. Whenever used, proper credit has been made in the reference section.

An important factor in a postulated continual increase in plasma applications stems from the fact that the ratio of electricity cost to fossil fuel cost has decreased steadily (Cheh et al., 1980), even accounting for the oil crisis as depicted in Fig. 1. This implies a gradual and continuing economic advantage for electric energy-intensive industrial processes. Additional benefits may be derived from environmental considerations. Electrically generated plasma processes can also produce conditions that cannot be easily duplicated by thermal means, such as the production of very energetic particles at a relatively low thermal loading. This last point is the consequence of so-called nonequilibrium plasma conditions. Nonequilibrium plasmas are characterized by low density, high electron temperature and low gas temperature, whereas thermal plasmas are generally hot and dense with an approximately similar electron and ion temperature. Advantages of plasma chemistry processes include relatively rapid turn-on and quenching, as well as precise control of the discharge conditions. Application must be looked at in terms of the unique characteristics of plasmas, and some examples of present plasma applications follow.

EXAMPLES OF SOME PLASMA APPLICATIONS

Controlled Thermonuclear Fusion

The most publicized plasma application field today is that of controlled thermonuclear fusion (CTF). This field is unique because of the special parameter conditions, such as an extremely high ion temperature ($T_i > 10^4$ eV). The topic is far too complicated and wide-ranging to discuss at any length in a review such as this. Suffice it to say that many different approaches to CTF are being

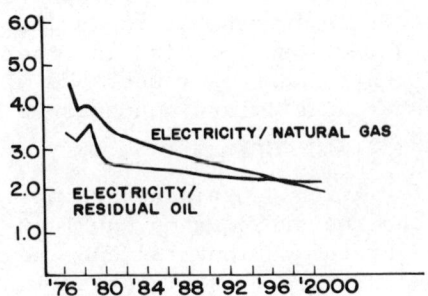

Figure 1. Ratio of average electricity cost to average fossil fuel cost in the industrial sector (adapted from Cheh et al., 1980).

pursued in major laboratories around the world. These approaches can be divided into two major categories: magnetic and intertial confinement.

In the magnetic confinement approach, the charged particle diffusion is slowed by an applied magnetic field so that the ions have many opportunities to undergo fusion collisions. Besides requiring a very high temperature so that some of the ions can overcome the Coulomb repulsive forces, the particles must also be confined for sufficiently long times so that they have the opportunity to undergo the fusion collisions (rather than just scatter). This is the famous "Lawson criterion" which states that the product of ion density and confinement time must exceed a given number in order for a fusion reactor to give a net power output. Under certain assumptions this criterion, for the $D + T \rightarrow He^4 + n +$ Energy reaction, is $N\tau > 10^{20}$ m^{-3}s where N is the ion density and τ is the confinement time. The D-T fusion reaction is the one with the largest cross section and, therefore, probably the one that will be demonstrated first.

The work in this field, since its declassification in 1958, has undergone large fluctuations of pessimism and optimism. Currently we are in a period of optimism. Several very clever approaches to magnetic confinement have been proposed, with exotic names such as: Scylla, Stellarator, Tokamak, Perhapsatron, Phoenix, Columbus, Tarantula, etc. The successful demonstration of "break-even" ($P_{in} = P_{out}$) has always seemed to be 20 years in the future. At the present time the Tokamak approach (a USSR invention) is the most favored one, and T > 2 keV, $N \tau > 10^{19}$ m^{-3}s has been achieved, simultaneously. It is expected that the large Tokamaks now under construction in the USA, USSR, Europe, and Japan will come very close to the break-even condition of T ~ 10 keV and $N \tau \sim 10^{20}$ m^{-3}s. A break-even demonstration is expected, maybe within this decade.

The inertial confinement fusion approach involves firing a large number of laser or charged particle beams simultaneously onto a small D - T target and thus compressing heating the target to ignition in a manner analogous to an H-bomb, in other words, causing a miniature explosion of the fuel particles while they are still confined by their own inertia. Because of the similarity with the H-bomb, much of this work has been classified, especially the advanced target design. The initial approach to this fusion scheme was to use high powered lasers. Recently the use of charged particle beams, especially light ions, for target heating and compression has achieved much attention. One of the main advantages of lasers is the ease of focusing the energy and thus of getting very high power densities. A chief advantage of the charged particle beam approach is the high beam production efficiency and the ease of producing large amounts of beam energy.

Many of the approaches to fusion are described on an elementary level, (Hagler and Kristiansen, 1977). It is still too early to say what may become the ultimate or even initially successful fusion scheme.

Radiation Sources

Energetic plasma discharges produce copius quantities of radiation, e.g., by carefully matching high power pulsed generators of a plasma load, the radiation spectrum can be tailored to simulate nuclear weapons effects at significant energy levels. The radiation can also be used in flash x-ray radiography for various diagnostic purposes, such as explosive or other dynamic events. The actual plasma load can have various initial forms, such as exploding wire arrays or imploding liners (Baker et al., 1977). Much of the work along these lines in the United States has been supported by the Defense Nuclear Agency. The physical installations required for generating the desired radiation spectrum and energy levels are quite impressive. An example of the magnitude of installation required is given by the Aurora facility which was oroginally designed as a 14 MV, 126 ns pulse length, 21 Ω generator (Bernstein and Smith, 1973).

Laser Discharges

In high power gas lasers, such as the CO_2 laser, it is important to obtain uniform (nonstriated) diffuse discharges with maximum energy density. Much work has been done towards understanding and achieving effective discharge conditions (Moeny et al., 1981; Filcoff, et al., 1981), but the understanding is still far from complete, especially at high pressures and for "unconventional" gases. This involves theoretical and experimental studies of discharge conditions under the influence of electrode geometry, rapid gas flow, and laser energy extraction. Large high performance computers are necessary to model the complex interplay between chemical kinetics, optical and electrical fields during excitation and extraction in a high speed flowing system, where one demands the most stringent uniformity conditions. Of great import is the "basic data" on transition probabilities, collision rates, etc., as a function of temperature, density, and slow rate for all important species present in the laser cavity plasma. Nowhere has today's plasma science been more rapidly applied than in the field of lasers.

Thermionic Energy Conversion

The basic principle of thermionic energy conversion has been known for over a century and is described in many textbooks (see Angrist, 1976). It has also been extensively investigated by NASA for possible use as an energy converter on various space systems for long duration missions. In principle, with a nuclear powered rocket, it could be possible to use the waste heat to operate a thermionic energy converter system surrounding the rocket to generate the needed on-board electric power. In its simplest form, the thermionic energy converter is just a heated cathode (electron emitter) and an anode (collector) connected to a load. The kinetic energy of the emitted electrons causes some of them to cross the interelectrode space and be collected by the anode, causing net current flow througy the load. However, this current is limited by space charge effects to low values, unless the interelectrode space is maintained at impractically close distances. In order to reduce the space charge effects, the interelectrode space is usually filled with a gas, such as Cesium which has an ionization potential less than the work function of the cathode material. The gas will then undergo contact-ionization upon impinging the heated cathode. Because of their larger mass and slower velocity, the ions then effectively neutralize the space charge by a factor of $(m_i/m_e)^{1/2}$ in number density. Among the practical problems preventing large scale application of thermionic energy conversion are the low output voltage per unit ($\gtrsim 1$ V), the low practical efficiency (~10%), and the high operating temperature (>1500°K) which results in severe materials problems. The output is also inherently DC.

Potentially important applications of thermionic energy conversion include topping units in large scale power production or in unconventional units such as those associated with space-based power. The topping unit application involves using the peak temperature in a high ΔT system with a conventional steam-cycle bottoming unit in order to achieve higher overall efficiency.

Magnetohydrodynamic Energy Conversion

The principle of magnetohydrodynamic (MHD) energy conversion is similar to that of a conventional, rotating generator in commercial power stations. A conducting medium is moved through a magnetic field causing a current to flow as a result of the $q(\bar{U}x\bar{B})$ force term. In the MHD generator, the moving conductor is an ionized gas rather than a metallic conductor as in a conventional generator. The basic system is depicted in Fig. 2. A hot, ionized gas is passed through a channel, with an imposed transverse magnetic field and two side electrodes, as shown. The interaction between the ionized gas with velocity $\bar{U}$ and the magnetic field $\bar{B}$ generates an electric field $\bar{E} = \bar{U}x\bar{B}$ which drives a current through the load. This basic principle is also described in many text books (see Angrist, 1976).

In order to increase the ionization fraction of the working gas, seed injection of materials with low ionization potentials (alkali metals) is commonly used. Besides applications of MHD generators as topping units in conventional power plants, there have also been many considerations of explosively driven MHD generators for various short duration, high peak power applications. The

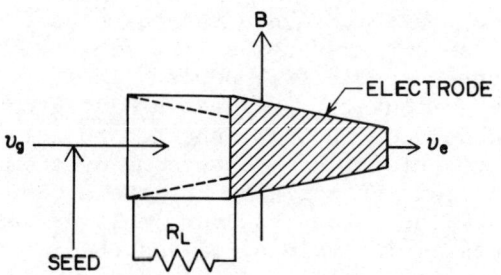

Figure 2. Basic MHD configuration.

largest MHD project for conventional power production is the U-25 installation at the Institute for High Temperature Physics near Moscow. This installation has reputedly been connected to the Moscow power grid and has produced in the excess of 10 MW for many hours. The next generation system is the U-500, which is still in the design phase.* Among the most serious operating problems with the U-25 has been electrode contamination and lifetime. Other inherent problems with MHD generators are: low output voltage (100's V) preferred DC operation, high temperatures (>2000° K), high magnet losses unless superconducting magnets are used, and need for seed material recovery because of both cost and environmental considerations.

Electrogasdynamic Energy Conversion

This is basically a flowing, ionized gas analog of the Van de Graaff generator. The basic arrangement is shown in Fig. 3. A high velocity gas stream between an ionizer and a collector performs the same function as the belt in the conventional Van de Graaff generator. The output power is characteristically in the form of high voltage (100's kV) and low current (100's μA). The efficiency has typically been found to be less than 10 percent (kinetic gas energy into electrical energy), although it should theoretically be much higher (>30%).

Communication

Reflection of radio signals off the ionosphere is, of course, what allows routine long distance radio communication around the

---

*The subnumbers -25 and -500 presumably indicate the MW power level of these generators. It appears, however, that it actually refers to the overall power station level where these generators serve as topping units. In other words, the U-500 would generate something like 200 MW in a 500 MW station where 300 MW would be generated by conventional, steam-cycle means.

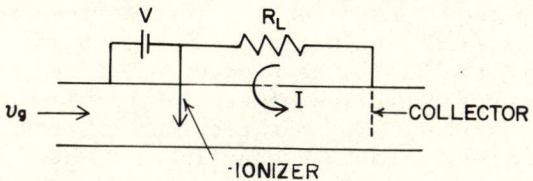

Figure 3. Basic electrogasdynamic energy converter.

world, not considered the introduction of communication satellites. Many books have been written about this subject and a classic is by Alpert et al., (1953).

Re-entry Blackout

The same phenomenon that makes long range radio propagation possible also creates communication problems during space vehicle re-entry. The shock waves and ablation products in front of the vehicle generate a plasma dense enough that communication signals are reflected ($\omega_p > \omega$) for part of the re-entry process ("blackout"). In principle, this can be overcome by going to higher carrier frequencies (optical lasers) or by generation of magnetic fields in the ionized gas layer, e.g., Alfvén wave propagation.

Ion Sources

The generation of intense ion beams is important to many applications, such as fusion plasma heating. In this case, an intense ion (H^+ or D^+) beam is accelerated and passed through a neutralizer cell, e.g., Cs vapor. The intense neutral beam is then injected into the fusion plasma, as shown in Fig. 4. Unfortunately, it is difficult to neutralize positive ion beams if their energy is too high. This energy is needed for the beam to penetrate to the center of a

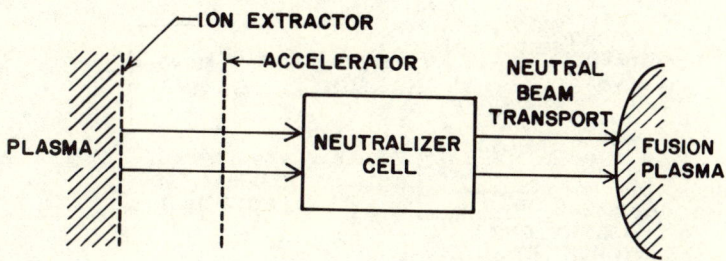

Figure 4. Neutral beam injection system for fusion reactor.

fusion reactor plasma. As a result, considerable research has been conducted in generating intense negative ion beams because they are much easier to neutralize. The production of these negative ion beams often occurs through the generation of plasmas which serve as the initial ion sources. The negative ion current density realized to date has been much less than what is needed for a successful system. Other technological problems are related to the beam transport and beam brightness.

Plasma Speakers

Electric discharge plasmas can also be modulated to act as "essentially" mass-less speakers which have very flat responde from 700 Hz to over 20 kHz (Plasmatronics, Inc., Albuquerque, New Mexico). In undergraduate laboratories, one often performs a rather simple experiment with very good results. The basic arrangement is shown in Fig. 5. The glass rod serves to seed the flame with sodium, thus increasing the percentage ionization. The sound fidelity from this simple experiment is quite surprising.

Microwave Generators

Since World War II, there has been a steady development of new, more efficient, and more powerful microwave generators. The "older" types include devices such as Klystrons, Magnetrons, Carcinotrons, and Travelling Wave Tubes. Among the newer devices are gyrotrons, gyrocons, and free electron lasers. Simple descriptions of the "older" devices can be found in Günther (1967). Some of the "newer" devices are described by Hirschfield and Granatstein (1967) and Mourier (1980). We will comment briefly upon the two most "successful" representatives of the "old" and "new" technology.

Klystrons work on the principle of velocity modulation and the resultant electron bunching in resonant cavity structures. At the present time, operating efficiencies of 35 to 40 % can be obtained, and as much as 70% may be ultimately achieved. Single sources of 1 MW continuous power at f~10 GHz and 3 MW at f~1 GHz have been demonstrated.

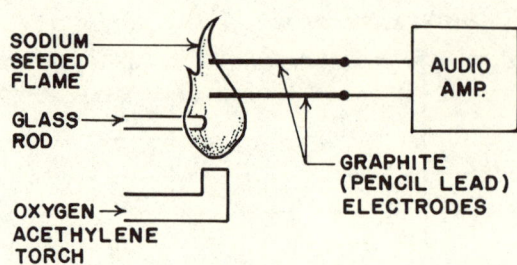

Figure 5. Simple plasma speaker.

Gyrotrons also work on the principle of electron bunching but in combination with $\bar{E} \times \bar{B}$ fields, resulting in a velocity loss or gain by the electrons because of a relativistic mass shift. The main advantage of gyrotrons is that the interaction region is approximately 10 times larger in linear dimensions than in other microwave power generating devices. In other words, their real advantages are at higher frequencies (>20 GHz) where fabrication tolerances and heat transfer will be easier to handle compared to the other more conventional devices. Gyrotrons presently exhibit a 30 to 35% efficiency with a 70% theoretical maximum. Among the demonstrated performance parameters for single units, are:

100 kW, continuous, at f=28 GHz (limited by rf window)

1.1 MW, pulsed, at f=100 GHz

25 MW, short pulse (30 ns), at f=34 GHz

Circuit Breakers, Switchgear, Interrupters

The range of commercial power industry "circuit breakers" span the voltage range from 110 V to 1.2 MV. Annually in the United States industry spends close to one billion dollars in this area as shown in Table 1 (Benenson et al., 1980). high voltage, high power circuit breakers may individually cost approximately one quarter million dollars.

At the present time, essentially all high voltage breakers are AC breakers where the current interruption occurs at "current

Table 1. Annual U.S. Industrial Expenditures for Circuit Breakers.

	Application	Annual U.S. Expenditures
1.	HV Transmission and Distribution	$ 2×10^8
2.	Medium Voltage Industrial Locations	3×10^8
3.	Contactors and Other Low Voltage Applications	2.5×10^8
4.	Homes	7×10^7
	TOTAL	$ 8×10^8

zero." However, there is a strong economic incentive for developing DC circuit breakers since DC transmission requires smaller transmission towers (less "peak" voltage) and less land for the right-of-way. Several studies of this technologically difficult topic have been conducted (Bowles et al., 1979). AC circuit breakers disconnect at a near zero current crossover and use powerful gas (air or SF_6) blasts, oil flow, and/or $\bar{J} \times \bar{B}$ forces to drive the arc into a quenching mode. Many phenomena in circuit breaker operation or development are poorly understood, including the gas nozzle design which controls the gas flow, turbulent gas flow, and B-field assisted vacuum interruption.

One of the better developed medium voltage breaker concepts is the vacuum interrupter — basically a vacuum container with two electrodes, one of which is movable through a mechanical bellows arrangement. The electrodes form solid metal-metal contact during the conduction phase and are then rapidly separated during a natural (AC) or forced (counterpulsed or commutated DC) current zero (Lafferty, 1980). Various clever electrode shapes and/or self-generated B-field systems help maintain a diffuse discharge. Severe electrode erosion may still occur during the arc or opening phase. This can lead to splatter debris which may form solid particles that move into highly stressed areas where they can be levitated and cause breakdown at levels far below normal. An example is shown in Fig. 6 where a 3 mm splatter particle has been welded to the edge of an electrode by electrical breakdown events caused by the particle itself (G. A. Garrall, General Electric Company, private communication).

In the general area of circuit breaker development, the United States is falling behind Europe and Japan. One reason may be that very few universities in the United States have research programs related to this important area.

Electric Discharge Lamps

The research in this field has largely been "Edisonian (Waymouth, 1980)" (lower empirical) in nature. High pressure arcs in metal or metal halide vapors can provide compact illumination sources with high efficacy [lumens (of light) per watt (LPW)] acceptable color rendition, and long life. It has been estimated that 30 million mercury lamps and 10 million high pressure sodium and metal halide lamps are in operation in the United States (Ingold and Zollweg, 1980).

Some interesting information about the performance of various light sources can be gleaned from Tables 2 and 3 (Ingold and Zollweg, 1980). A shocking piece of information to many is that the common incandescent light source is only 7% efficient in producing visible radiation; this is why the use of incandescent light for general purpose lighting has decreased in much of Europe.

PLASMA APPLICATIONS

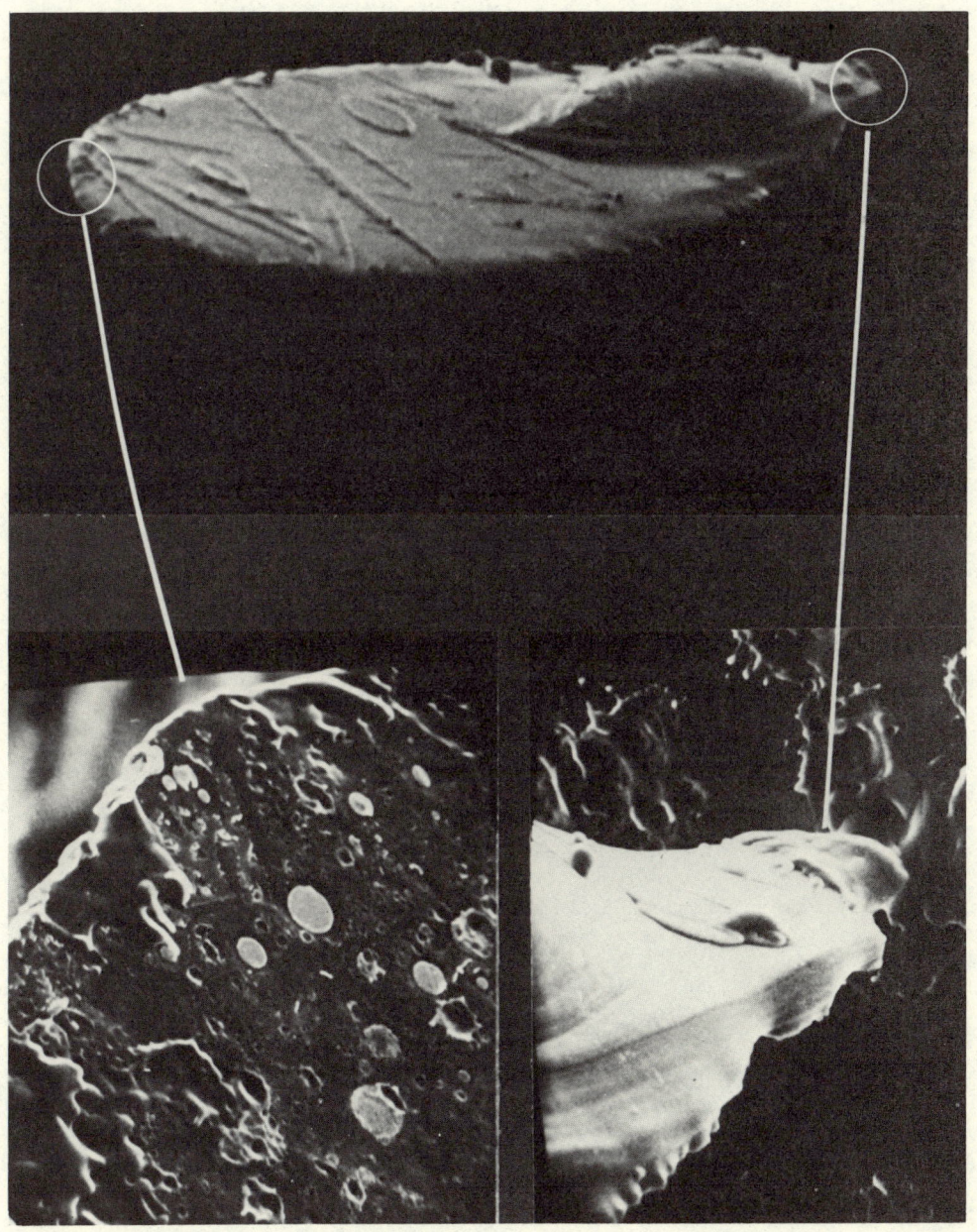

Figure 6. Splatter debris (~3mm long).

Table 2. Efficacy and Color Rendition (Ingold and Zollweg, 1980).

LIGHT SOURCE	LPW (TOTAL)	LPW (VISIBLE ONLY)	CRI
TUNGSTEN INCANDESCENCE	17	250	95
MERCURY ARC	52	347	22
SUN	95	250	100
METAL HALIDE ARC	100	294	85
SODIUM ARC	120	400	23
GREEN LINE (550 NM)	683	683	−40

LPW ≜ Lumens per watt
CRI ≜ Color rendition index

Table 3. Power Balance of Light Source (Ingold and Zollweg, 1980).

LIGHT SOURCE	RADIATED		NOT RADIATED	
	% VISIBLE	% NOT USED	% CONDUCTION	% ENDS
INCANDESCENT	7	80	12	1
MERCURY ARC	15	33	42	10
METAL HALIDE ARC	34	20	36	10
SODIUM ARC	30	20	40	10

The gases most often used in discharge lamps are mercury, sodium, rare gases, and metal halides such as thallium iodide. The electrodes are usually made of tungsten, with or without coatings, and the envelope is frequently made of fused silica (SiO_2) or polycrystalline alumina (A_2O_3). However, there are problems with the chemical compatibility of these materials. A new important market for these lamps is home lighting, but we must develop an inexpensive, reliable ignition system for these types of lamps.

Ion and Plasma Propulsion

Ion and/or plasma thrusters have been studied for space propulsion for a long time. Simple equations explaining the basic concept and defining the most important terminology are (Brewer et al., 1961):

$$\text{Thrust:} \quad T \triangleq \dot{m}_p v_{ex}, \quad (1)$$

where $\dot{m}_p$ = propellant mass flow rate and v_{ex} = propellant exhaust velocity

$$\text{Total Impulse:} \quad I_T \triangleq T\tau \quad (2)$$

where τ = thrust time duration.

$$\text{Specific Impulse:} \quad I_s = \frac{I_T}{W_p} = \frac{T}{\dot{W}_p}, \quad (3)$$

where W_p = propellant weight (gm_p) exhausted in time τ. Combining Eqs. (1) and (3) gives

$$I_s = \frac{v_{ex}}{g}, \quad (4)$$

where g = the gravitational acceleration.

In other words, I_s is a measure of the exhaust velocity v_{ex} and has units of seconds.

The equation of motion in free space for a vehicle with mass M and velocity v, which are functions of time, is:

$$M\,dv = -v_{ex}\,dM. \quad (5)$$

For an initial mass, M_i, starting from rest

$$v = I_s g \ln \left(\frac{M_i}{M}\right).$$

In other words, the effectiveness of mass expenditure to obtain a given velocity is a function of the specific impulse — one wants a high I_s.

Many clever propellant acceleration schemes have been tried. A key issue is that the expelled propellant must be neutral or the space vehicle will soon charge up to such a potential that further propellant expulsion becomes impossible. This implies that the propellant must be a neutral plasma or a nautral particle beam. In principle, the plasma can be accelerated by some fairly conventional $\vec{J} \times \vec{B}$ accelerator (e.g., a rail gun), and charged particles can first be accelerated in a conventional accelerator, then neutralized or stripped (positive or negative ion acceleration) in special cells.

<u>Plasma Torch</u>

The plasma torch is an application of plasmas which have utilized their unique properties to the extreme. Plasma torches are commercially available and in frequent use (Gross et al., 1969). The basic arrangement is shown schematically in Fig. 7. Plasma torches have higher temperatures than conventional gas torches because of the ionization and dissociation energy of the gas. They are used for such things as surface coating, or plasma spraying with metals and ceramics, and for spheroidization of metal particles. If the arc is transferred to a target, it can also be used for material processing such as cutting or drilling.

<u>Plasma Chemistry</u>

Plasma chemistry covers a wide range of subjects and applications. One way to clarify the field is to make a distinction between equilibrium and nonequilibrium conditions. In the use of equilibrium which involves arc processes such as those discussed in the previous section, local thermal equilibrium (LTE) exists; i.e., $T_e \sim T_{gas}$.

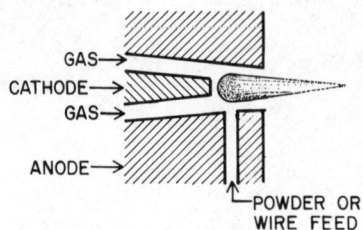

Figure 7. Plasma torch.

These are called thermal plasmas. In the case of nonequilibrium there is frequently a diffuse or glow discharge plasma where $T_e \gg T_{gas}$. These are called "cold" plasmas. This last case has the advantage of being able to produce highly active species while still keeping the plasma temperature quite low. The degree of coldness of hotness of a plasma generally deals with the gas or ion temperature value.

The general area of plasma chemistry appears to be poorly understood and is said (Mogab, 1980) to be one where "art is ahead of science." The complexity and multitude of the various processes which are of interest make the problem area a very difficult one, indeed. As a result, much of the industrial development has been rather empirical in nature. This has often been the case because urgent needs and potentially high pay-offs have not allowed time for careful, academic-like, basic research and a subsequent development program. The area is one of great, untapped potential provided the numerous fundamental processes and their interplay are better understood.

The following observation is taken from the introduction to the new Journal of Plasma Chemistry and Plasma Processing (Pfender and Veprek, 1981):

> Plasma chemistry and plasma processing make use of the plasma state also known as the "fourth state of matter." Although 99 percent of the universe is in the plasma state, this state of matter has been relatively little exploited by man.
>
> Over the past years there has been an increasing awareness of the potential which plasma chemistry and plasma processing and the associated technologies hold in our energy and raw material limited world. . .

In the following discussion, we only consider a limited number of plasma chemistry applications. Other cases were described in a separate paper at this NATO Study Institute by A. Goldman.

Thermal plasmas are used commercially in arc heaters and rf plasma generators at powers up to 50 MW and gas temperatures ranging from 4,000 - 10,000°K (Heberlein et al., 1980). Welding, cutting and spraying equipment at levels to 50 kW, and scrap melting equipment up to 1 - 2 MW are available at the present time and 10-20 MW becoming available very soon. Some examples of industrial furnaces are shown in Figs. 8 and 9.

The general arrangement of a nonequilibrium plasma reactor is shown in Fig. 10. (Much of the following information comes from

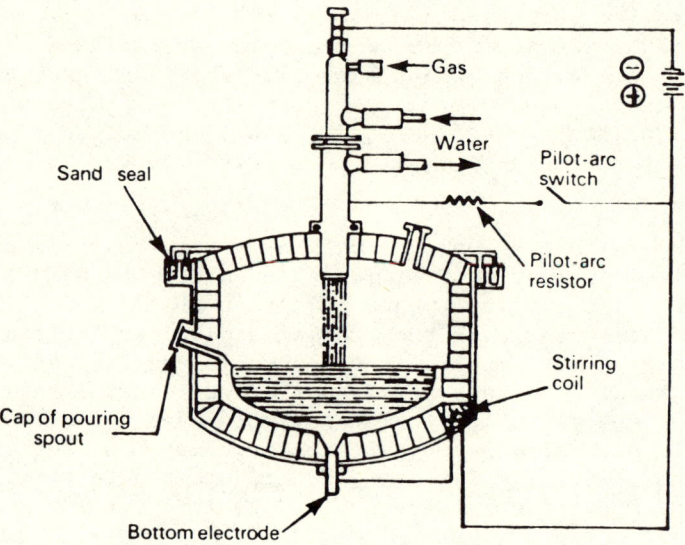

Figure 8. The 60 kW Linde plasmarc furnace (Courtesy of Professor E. Pfender, University of Minnesota).

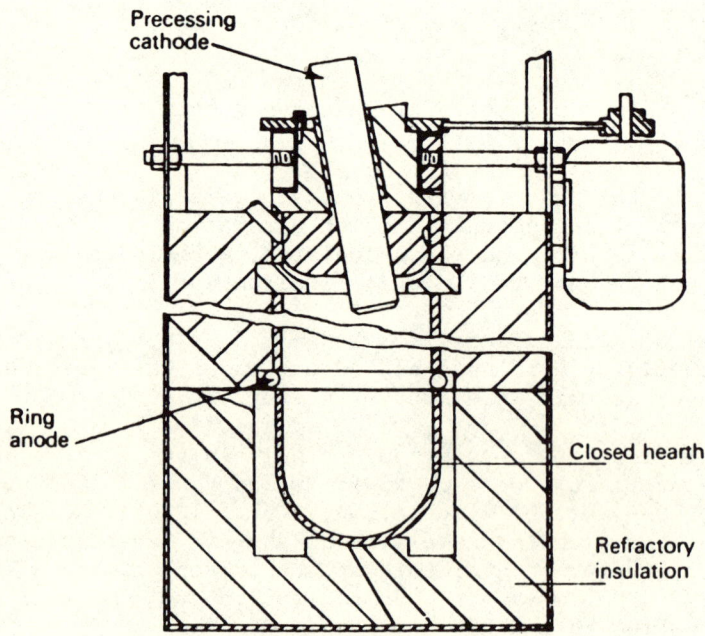

Figure 9. A 200 kW processing cathode, direct current plasma furnace (Courtesy of Professor E. Pfender, University of Minnesota).

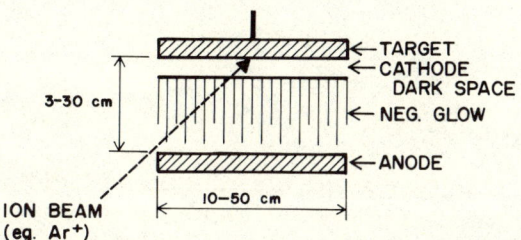

Figure 10. Nonequilibrium plasma reactor.

Oskam and Pfender, 1980; Winters, 1980; Coburn and Winters, 1979; Veprek, 1980; Vossen and Kern, 1978; Shen and Bell, 1979; and Hollahan and Bell, 1974).

The plasmas have an electron density of approximately $10^{10}/cm^3$ and operate in the abnormal glow region. Reactions generally follow the following sequence (Suhr, 1980):

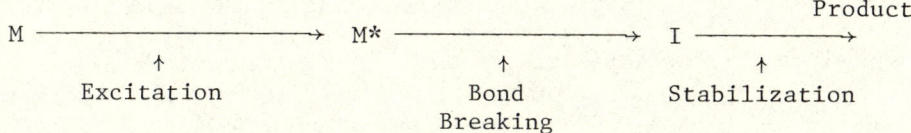

Whereas the energy transfer to a molecule in conventional chemistry is less than 5 eV and in photochemistry less than 7 eV, more than 300 eV can be transferred in plasma chemical processes. This situation often leads to the generation of super-excited molecules.

Depending upon the energy of the incident ions and on the details of the surface reactions, one can get sputtering and/or chemical reactions. The probability of ion neutralization at the surface is in excess of 0.99; e.g., 100 eV H_e^+ ion spends only 10^{-13} -10^{-14}s within 5 Å of a surface, but its probability of neutralization is 0.9983 (Winters, 1980). In most cases, these particle-surface interactions are poorly understood. The energy transfer can be electronic, vibrational rotational, or translational in nature. The resulting processes can be:

radiation-induced desorption

disassociation

displacement

valence excitation

ionization

sputtering

chemical etching

impurity and defect enhanced transport

absorbate-absorbate reactions

absorbate-surface reactions.

One of the most important applications of these reactions is in the semiconductor industry where plasma etching and sputtering are used to produce intricate electronic devices. Sputtering is a physical process in which energetic ions or neutrals bombard the surface and eject material because of the delivery of momentum to the surface. The sputtering yield is a function of the mass ratio between incident and sputtered particles, the angle of incident, and the incident energy. An allied field greatly benefitting from ion processes is optical thin-film coating. The advent of the laser has placed more stringent requirements on optical fabrication. No place is this more evident than in the preparation of impurity free (low absorption, low scattering), uniform, high quality dielectric or metallic coatings for reflection modification, environmental protection, wavelength control or polarization selection (Guenther and Glass, 1977). Plasma processes such as ion etching to clean substrates, ion-planing for polishing substrates or ion sputtering for deposition of materials are leading to great advances in high quality optical coatings.

In part, this is because all these steps can be accomplished without removing the materials from a vacuum chamber and exposing them to the deleterious influences of a generality adverse environment. Various ion sputtering schemes are given in an excellent recent review article (Jorgenson, 1981). In fact, because the plasma species are charged, they can be directed using electromagnetic deflection for removal or deposition at a given site. Their high energy insures excellent adhesion and dense films (Schroeder et al., 1971).

In the electronics industry great use is made of plasma etching. Etching is a chemical process in which chemically reactive species are generated in a glow discharge of relatively inert gases. The reactive species react with the surface and form volatile components that are pumped away. Different types of plasma etching are "reactive ion etching," "reactive sputter etching," and "plasma etching." However, in this discussion we will not make a clear distinction between the various processes. Figure 11 shows a basic etching arrangement. The figure also demonstrates how the addition

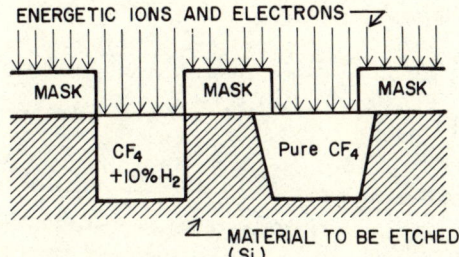

Figure 11. Importance of anisotropy in etching process.

of certain gases to the basic etch gas can be used to modify the lateral etch rate (vertical side walls are normally desired). In the example given of CF_4 + 10% H_2, only surfaces subjected to energetic particles bombardment will be etched (reactive ion etching) (Coburn and Winters, 1979). Figure 12 illustrates another important and poorly understood process: ion assisted gas-surface chemistry. When, for instance, an argon ion beam of a few hundred eV energy is incident upon the target, as indicated in Fig. 10, the etch rate can increase dramatically.

Examples of various plasma chemistry applications and the resulting produces are (Oskam and Pfender, 1980):

1) Plasma assisted chemical vapor deposition of a variety of materials including silicides, carbides, phosphides, and nitrides.

2) Plasma Surface Modification

    a) Surface Hardening (glow nitriding)

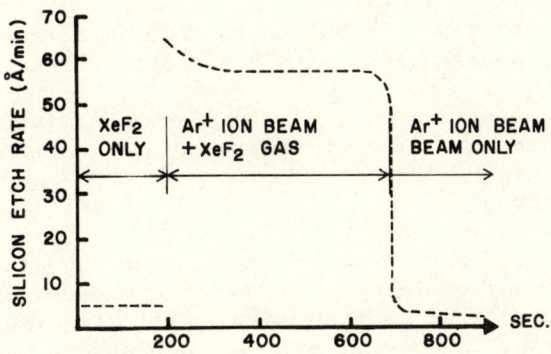

Figure 12. Ion-assisted, gas-surface chemistry (Winters, 1980).

b) Film Deposition

c) Anodization

d) Carburizing

e) Boriding

f) Oxidation

3) New Compounds or Known Materials with Unusual Properties

a) New: P_3N_5

b) Old: α-Si, dielectric films (coating at low temp.)

c) New: α-c, dielectric films (diamond-like coating)

4) Plasma Polymerization: Unique polymer properties

5) Powder Production: For instance, refractory compounds. At present the particle size tends to be too small (< 1 μm, whereas it needs to be > 1 μm).

6) Plasma Metallurgy

a) Scrap reclamation and reprocessing

b) New materials production

c) Metal refining

d) Novel processes that enable manufacturing of substitute

e) Alloys (e.g., N_2 for Ni in hardened steels)

Table 4 (Cheh et al., 1980) lists some important industrial plasma chemistry applications with a state-of-the-art assessment.

Some important research needs in plasma chemistry (Oskam and Pfender, 1980) are summarized in Table 5.

Pulsed Power Technology

Pulsed power is a relatively new engineering discipline. It is important in many emerging high technology endeavors, such as fusion power generation, nuclear weapons effects simulation, laser isotope separation, industrial material processing, and advanced weapons systems development. Most of the unclassified, recent

Table 4. Potential Industrial Processes of Thermal Plasmas (Chen et al., 1980).

	State-of-the-Art	Time to Commercialization (Years)
Metals Industry		
Blast Furnace Firing	Large Scale Demo	5-10
Reformers for Direct Reduction	Large Scale Demo	2-4
Remelting of Metallic Sponge/Scrap	Production	–
Reduction of Ferroalloy Ores	1 MW Pilot Demo	2-10
Preparation of Reactive Metals	Lab Scale Development	6-10
Preparation of Solar Silicon	Large Scale Demo	3-6
Reduction of Non-Ferrous Ores	Small Scale Production/Lab Demo	–
Firming of Metallurgical Slags	Lab Demo	?
Chemical Industry		
Oxygen Superheat for Titans or Manufacture	Production	–
Acetylene Manufacture	Production	5-15
Coal Pyrolysis	Large Scale Demo	5-15
Coal Desulfurization	Large Scale Demo	5-10
Cement Manufacture	Lab Scale Development	5-15
Nitrogen Fixation	Large Scale Demo	–
Ultrafine Silica, Carbides, etc.	Production	–
General		
Refractories/Glasses	Lab Scale Development	2-4
Plasma Welding, Cutting and Spraying	Production	–
Spheroidization	Production	–

Table 5. High Priority Research Needs Related to Plasma Chemistry.

1. Synergistic effects between beams
2. Charge exchange between particles and surface
3. Reactivity of atoms, ions & molecular fragments with surfaces
4. Selective sputtering and surface enrichment processes by electronic & ballistic processes
5. How is KE partitioned in a disassociating molecule?
6. Additional stopping mechanisms for ions
7. Reaction cross sections for beam-surface interactions
8. Characterization of macroscopic discharge conditions for various gases & pressures
9. Scaling & modeling
10. Plasma etching of organic material
11. Photon induced effects
12. Basic transport phenomena in plasma jets, including interaction with solid particles
13. Kinetics of absorption and desorption of plasma species by metal melts
14. Electrode erosion

work in pulsed power technology is summarized in Kristiansen (1976); Guenther and Kristiansen (1979); Martin and Guenther (1981); Creedon et al., (1979); Guenther and Kristiansen (1980); Burkes et al., (1978); Kristiansen and Schoenbach (1981).

Figure 13 shows a generalized pulsed power system and identifies some of the subareas in which plasma effects are important. The key plasma-dominated problem area is that related to high power, repetitive switching. At the present time, most energy storage systems

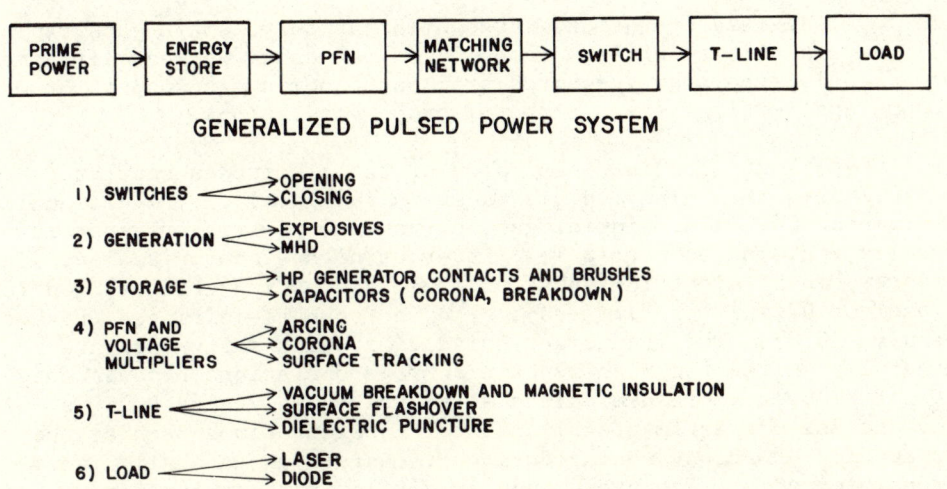

Figure 13. Examples of discharge problems and applications in pulsed power technology.

use capacitors and hence need closing switches to transfer stored energy to a load. Because of the much larger energy density possible with inductive storage, it is of great importance to develop repetitive opening switches which are necessary to break the storage loop and transfer the stored energy to the load. In this case, energy is stored as $1/2\ LI^2$ whereas in a capacitor, it is $1/2\ CV^2$. Simple calculations, based on fundamental materials limits, show that the inductive energy storage density is, conservatively, a factor of 100 larger than that of a capacitive system.

First, let us address closing switches. There are many candidate closing switches for high power repetitive operation, including spark gaps, ignitrons, thyratrons, crossed-field devices, and others. Chief among these are the spark gap and the thyratron.

The thyratron is a low pressure, gas-filled, three-electrode (triode) device that can switch over 100 kV and 10 kA at high repetition rates (>1000 Hz), and with a few ns jitter (reproducibility of discharge upon command of trigger). For even higher power applications, the spark gap has demonstrated impressive behavior at single shot or low repetition rate. In a single-shot mode, it has switched MV's and MA's with ns (or even sub ns) jitter. In a repetitive mode, it has switched 100's kV, 100's kA with ns jitter and more than 1000 Hz repetition rate in a short-burst mode.

Whereas most of the past development of spark gaps has been for single shot operation, much of the current interest is in repetitive, long-life operation. Present heavy duty, spark gaps typically have a utility of less than 10^4 to 10^6 shots whereas lifetimes of up to 10^9 shots are required from an economic standpoint for many future applications (e.g., laser isotope separation).

A spark gap consists basically of two electrodes separated by a dielectric switching medium, as shown in Fig. 14. The dielectric medium can be solid, liquid, gas, or vacuum. A trigger input is usually employed to obtain low jitter or high synchronization. The trigger can be electric (spark or field distortion trigger), optical (laser or UV initiated), x-ray, or electron-beam-initiated (Burkes, et al., 1978). The liquid and solid dielectric spark gaps are not generally suited for high repetition rate operations because they do not recover their dielectric strengths rapidly enough, if at all, and because of detrimental mechanical considerations such as the production of strong shocks during discharges in generally incompressible media. The vacuum gap is difficult to synchronize to better than a few 10's of ns, and the gas-filled gap has problems with heat transfer (gas flow). However, most of the work on repetitive operation is centered on the gas-filled gap. Some of the factors affecting repetitive operation of gas-filled gaps are summarized in Table 6.

Some of the factors affecting the gap life are summarized in Table 7, and some factors affecting the electrode erosion are summarized in Table 8. A schematic representation of a general switch operations sequence and some of the general research areas are shown in Fig. 15 (E E. Kunhardt, private communication).

The complicated, interrelated problem of plasma chemistry, electrode materials, dielectric properties, heat flow, breakdown initiation, and voltage hold-off recovery is being addressed in a United States Air Force supported coordinated research program at Texas Tech University. With Air Force and Army support, the even more difficult problem of repetitive opening switches is also being studied.

The only really successful high power opening switches to date are fuses. They have demonstrated the capability to interrupt MA's

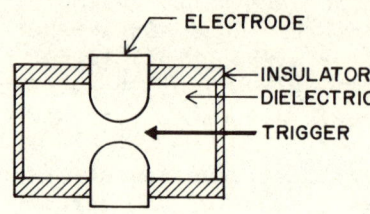

Figure 14. Spark gap.

Table 6. Factors Affecting Performance of Rep-Rated, Gas-Filled Spark Gaps.

- Shape of recharge pulse
- Gas deionization and deexcitation
- Heat removal (gas flow and electrode cooling)
- Plasma chemistry
- Change in gas composition (sealed gaps)
- Gas pressure

Table 7. Factors Affecting Gap Life.

Electrode Erosion ⟶ Excessive Jitter
Electrode Erosion ⟶ Prefire
Wall Damage ⟶ Tracking

Wall damage can be due to:
- plasma chemistry
- thermal shocks
- mechanical shocks
- metal vapor plating
- micro bullets
- UV, x-rays

New electrode materials of interest include:
- Poco Graphite
- Copper - infiltrated tungsten
- Tungsten-imbedded copper

Table 8. Factors Affecting Erosion.

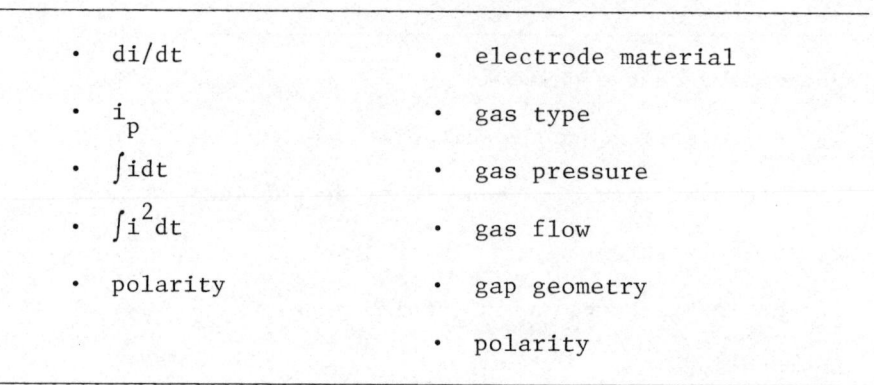

- di/dt
- i_p
- $\int i\,dt$
- $\int i^2\,dt$
- polarity

- electrode material
- gas type
- gas pressure
- gas flow
- gap geometry
- polarity

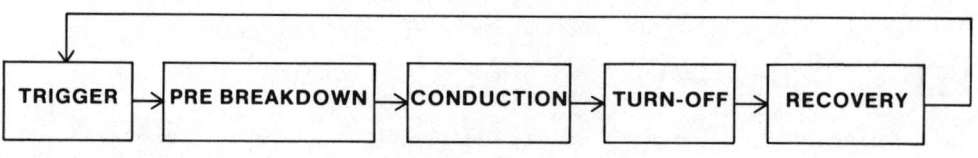

SYSTEMS ANALYSIS

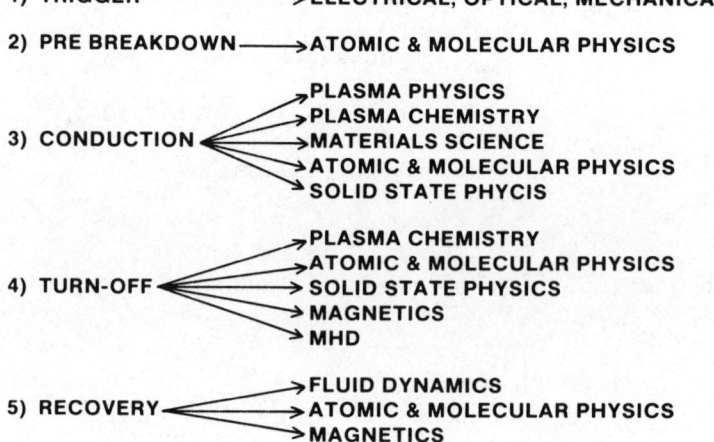

1) TRIGGER ⟶ ELECTRICAL, OPTICAL, MECHANICAL

2) PRE BREAKDOWN ⟶ ATOMIC & MOLECULAR PHYSICS

3) CONDUCTION ⟶ PLASMA PHYSICS
   PLASMA CHEMISTRY
   MATERIALS SCIENCE
   ATOMIC & MOLECULAR PHYSICS
   SOLID STATE PHYCIS

4) TURN-OFF ⟶ PLASMA CHEMISTRY
   ATOMIC & MOLECULAR PHYSICS
   SOLID STATE PHYSICS
   MAGNETICS
   MHD

5) RECOVERY ⟶ FLUID DYNAMICS
   ATOMIC & MOLECULAR PHYSICS
   MAGNETICS

Figure 15. Schematic representation of the switch operations sequence and some related research topics.

and hold off MV's in 100's ns. The issue of repetitive operation of opening switches was examined at two recent U.S. Army Research Office sponsored workshops (Kristiansen and Schoenbach, 1981; Schoenbach and Kristiansen, 1982). Many different switch concepts were considered. Much of the interest was devoted to diffuse discharge switches since it was felt that the task of interrupting an arc, on the time scale and with the precision needed, was much more difficult. Among the more speculative, but exciting, ideas was the optically controlled discharge. In principle, it should be possible to both increase or decrease the impedance of a diffuse discharge by illumination with light of the proper wavelength (e.g., the optogalvanic effect). Whether or not this can be developed to a workable switch is the subject of current research projects.

The basic research issues that were identified at the Workshop on Repetitive Opening Switches (Kristiansen and Schoenbach, 1981) are shown in Table 9. The suggested research priorities are given in Table 10.

Many other plasma related problems in pulsed power technology, as summarized in Fig. 13, are also of extreme importance. Many of them are discussed in cited references (Kristiansen, 1976; Guenther and Kristiansen, 1979; and Martin and Guenther, 1981).

Table 9. Basic Research Issues.

BASIC RESEARCH PROBLEMS	RESEARCH APPROACH	SWITCH CONCEPT
1. Discharge modeling and comparative experiments. Code development to produce circuit model. Inclusion of plasma chemistry.	Develop codes and carry out carefully planned experiments to check the code validity. Develop user oriented codes that enables the nonexpert to state problem in electrical circuit terms. Include plasma chemistry effects in the code.	Depending on the code (several will be needed) this applies to essentially all the gas discharge switch concepts. Particularly obvious ones are the electron beam and optically controlled, diffuse discharge switches.
2. Compile and measure (when needed) fundamental data such as rate coefficients, cross sections, etc.	Conduct literature search. Carry out basic measurements for gases and gas mixtures under conditions of interest to 1 & 3.	Essentially all gaseous switches, but especially those in 3.
3. Production of diffuse discharges. Establish conditions for arc development. Develop fundamental understanding to enable choice of gas mixtures with high conductivity in conducting phase, fast recovery, etc.	Carry out a series of comparable experiments, utilizing input from 2 above, to determine optimum gas mixtures, pressures, and excitation conditions for diffuse discharge switches.	Electron beam and optically controlled. Crossed field tube, spoiled electrostatic confinement, reflex discharge, plasma erosion, thyratrons.
4. Electrode phenomena. Surface physics of arc and diffuse discharges (e.g., sputtering) Photo-electric effect.	Conduct a careful experiment involving several gases, mixtures pressure, discharge conditions, etc., to establish the "best" electrode materials.	Essentially all gas discharge switches (including vacuum interrupter).
5. Motion of conducting plasma due to applied B-field. Effect of nonuniformities (asymmetries) in conduction channel on B field interaction.	Study interaction of high current plasma discharges with the vacuum plasma/field interfaces manipulated electromagnetically to explain such features as residual plasmas (outside main conduction region), propagation of electromagnetic energy in plasma/vacuum field environment, energy dissipation, low density-high current conduction, etc.	$\frac{dL}{dt}$, Dense plasma focus, Magneto-plasma-dynamic.
6. Plasma instabilities. Nonclassical transport phenomena (Generated beams, anamalous resistivity, etc.)	Identify conditions for "triggering" plasma instabilities. Utilize information developed in fusion research and adapt to partially ionized, low temperature discharges. Conduct comparative experiments.	Dense plasma focus. Magneto-plasma-dynamic. Macro- and micro-instabilities.

PLASMA APPLICATIONS

BASIC RESEARCH PROBLEMS	RESEARCH APPROACH	SWITCH CONCEPT
7. Effects of quenching media on inductive and resistive fields and the hydrodynamics of the media and their heating/cooling rates. Investigations of conductivity, diffusion mechanisms, breakdown energy requirements, etc.	Make systematic studies, under comparative conditions, of a wide range of quenching materials (solid, gas, liquid). Instrument for careful, V, I, T measurements. Interpret chemistry involved.	Fuses JxB and magneto-plasma-dynamic devices with quenching medium.
8. Develop superconducting materials with improved stability and higher transition temperatures.	Fundamental materials development.	Superconducting switches.
9. Flow dynamics in gas and liquid interrupters. Arc cooling rates. Recovery rate, effect of contamination.	Carry out measurements under controlled conditions with varying flow velocities, current densities, and electric fields.	Gas and liquid interrupters.
10. Magnetic switch material development.	Develop materials with high u_r, B_{sat}, low loss.	Magnetic (saturable inductor) switches.

Table 10. Suggested Research Priorities For Opening Switch Development.

1. a) Discharge Modeling (including plasma chemistry) and Comparative Experiments.

   b) Production of Diffuse Discharges with High Hold-off Voltage, High Conductivity, High Current Density, and Fast Recovery.

   c) Measurements of Basic Data Needed in a) and b) above, such as Cross Sections, Rate Coefficients, etc.

2. a) Motion of Gas Discharges in Magnetic Fields.

   b) Plasma Instabilities, Nonclassical Transport Phenomena.

3. Determine the Limits to Various Solid State Switching Schemes.

4. Electrode Surface Physics of Arc and Diffuse Discharges

5. Effects of Quenching Media on Discharges.

REFERENCES

Alpert, J. L., Ginzburg, V. L., and Feinberg, E. L., 1953, "Radio Wave Propagation," GITTL, Moscow.

Angrist, S. W., 1976, "Direct Energy Conversion," Allyn and Bacon, Boston.

Baker, W. L., Clark, M. C., Degnan, J. N., Kiuttu, G. S., McClenahan, C. R., and Reinovsky, R. E., 1976, J. Appl. Phys., 49: 4694.

Benenson, D. M., Frind, G., Frost, L. S., and Tuma, D. T., 1980, "Proceedings, NSF Workshop on Plasma Chemistry and Arc Technology, Minneapolis," Univ. Minn.

Bernstein, B., and Smith, I., 1973, IEEE Trans. Nucl. Sci., NS-20: 299.

Bowles, J. P., Turner, A. B., and Vaughan, R. L., 1979, EPRI Rept. EL-1260.

Brewer, G. R., Currie, M. R., and Knechtli, R. C., 1961, Proc. IRE, 49: 1789.

Burkes, T. R., Hagler, M. O. Kristiansen, M., Craig, J. P., Portnoy, W. M., and Kunhardt, E. E., 1978, Naval Surface Weapons Center Rept. NP 30/78.

Cheh, H., Fey, M., Sayce, G., and Wilks, P., 1980, in: "Proceedings NSF Workshop on Plasma Chemistry and Arc Technology, Minneapolis," H. J. Oskam and E. Pfender, eds., Univ. Minn.

Coburn, J. W., and Winters, J. F., 1979, J. Vac. Sci. Tech., 16(2): 391.

Creedon, J., Kristiansen, M., Gilmour, A. S., Jr., Vaughan, J. R. M., and Winslow, L. M., eds., 1979, Special Issue, IEEE Trans. Electron Devices, ED-26: 10.

Filcoff, J. A., Roache, P. J., and Moeny, W. M., 1981, in: "Digest of Technical Papers, Proceedings, 3rd. International Pulsed Power Conference, Albuquerque," T. H. Martin and A. H. Guenther, eds., p. 277.

Gross, B., Cryez, B. and Miklossy, K., 1969, "Plasma Technology," Illiffe, London; American Elsevier, New York.

Guenther, A., and Kristiansen, M., eds., 1980, Special Issue, IEEE Trans. Plasma Sci., PS-8.

Guenther, A., and Kristiansen, M., eds., 1979, "Digest of Technical Papers, 2nd IEEE International Pulsed Power Conference, Lubbock, Texas."

Guenther, A. H., and Glass, A. J., 1977, in: "Proceedings, NSF Workshop on Needs and Opportunities for Basic Research in Laser-Material Interactions, Univ. Southern California," p. 46.

Günther, W. A., 1967, "The Physics of Modern Electronics," Dover, New York.

Hagler, M. O., and Kristiansen, M., 1977, "An Introduction to Controlled Thermonuclear Fusion," Lexington Books, Lexington, Mass.

Heberlein, J. V. R., Toth, L. E., and Fauchais, P., 1980, "Proceedings, NFS Workshop on Plasma Chemistry and Arc Technology, Minneapolis," H. J. Oskam and E. Pfender, eds., Univ. Minn.
Hirschfield, J. L., and Granatstein, V. L., 1977, IEEE Trans. Microwave Theory and Tech., MTT-25: 522.
Hollahan, J. R., and Bell, A. T., eds., 1974, "Techniques and Applications of Plasma Chemistry," Wiley, N.Y.
Ingold, J. H., and Zollwen, R. J., 1980, "Proceedings, NSF Workshop on Plasma Chemistry and Arc Technology, Minneapolis," H. J. Oskam and E. Pfender, eds., Univ. Minn.
Jorgenson, G., 1981, Electro-Opt. Syst. Des., 13: (11)11.
Kristiansen, M., ed., 1976, "Proceedings, 1st. IEEE International Pulsed Power Conference, Lubbock, Texas."
Lafferty, J. M., 1980, "Vacuum Arcs, Theory and Applications," Wiley, New York.
Martin, T., and Guenther, A., eds., 1981, "Digest of Technical Papers, 3rd. IEEE International Pulsed Power Conference, Albuquerque, N. M." Lib. of Cong. No. 81-81007.
Moeny, W. M., Filcoff, J. A., and Roache, P. J., 1981, in: "Digest of Technical Papers, 3rd International Pulsed Power Conferences, Albuquerque, N. M." T. H. Martin and A. H. Gunther, eds., p. 273.
Mogab, C. J., 1980, "Proceedings, NSF Workshop on Plasma Chemistry and Arc Technology, Minneapolis," H. J. Oskam and E. Pfender, eds. Univ. Minn.
Mourier, G., 1980, Arch. Electron Übertragungstech., 40: 473.
Oskam, H. J. and Pfender, E., eds., 1980, "Proceedings, NSF Workshop on Plasma Chemistry and Arc Technology, Minneapolis," Univ. Minn.
Pfender, E. and Veprek, S., 1981, Plasma Chem. and Plasma Proc., 1: 1.
Schroeder, J., Dieselman, H. D., Douglass, J. W., 1971, Appl-Opt., 10: (2)295.
Shen, M., and Bell, A. T., eds., 1979, "Plasma Polymerization," Am. Chem. Soc. Symp. No. 108.
Suhr, H., 1980, "Proceedings, NSF Workshop on Plasma Chemistry and Arc Technology, Minneapolis," H. J. Oskam and E. Pfender, eds., Univ. Minn.
Waymouth, J. F., 1980, "Proceedings, NSF Workshop on Plasma Chemistry and Arc Technology, Minneapolis," H. J. Oskam and E. Pfender, eds., Univ. Minn.
Winters, H. F., 1980, Topics Cur. Chem., 94: 69.
Vossen, J. L. and Kern, W., eds. "Thin Film Processes," Academic, New York.

DIFFUSE DISCHARGE OPENING SWITCHES

K. Schoenbach, G. Schaefer, and M. Kristiansen
Electrical Engineering Department
Texas Tech University
Lubbock, Texas 79409, U.S.A.

L. L. Hatfield
Physics Department
Texas Tech University
Lubbock, Texas 79409, U.S.A.

and

A. H. Guenther
Air Force Weapons Laboratory
Kirtland Air Force Base, New Mexico 87117, U.S.A.

INTRODUCTION

Magnetic storage of energy for applications requiring large amounts of energy is preferable to capacitive storage because of its characteristically high energy density, some 10^2 to 10^3 times higher than electrostatic storage (Burton et al., 1979). Figure 1 shows an inductive energy storage circuit. While switch α is closed, the inductor L is charged by means of a current source; energy is thus stored in the magnetic field of the inductor. In order to transfer this energy into a load Z, switch β has to be interrupted; i.e., switch α has to be opened. For a certain time interval, determined by the opening speed of switch α, the voltage induced in the inductor L will be applied to Z at a relatively high load current level. The power transferred to Z depends on the rate with which switch α opens. The shorter the opening time τ_{op}, the higher the power transferred to the load.

Conditions for an opening switch suitable for a wide spectrum of pulsed power systems are (Kristiansen and Schoenbach, 1981):

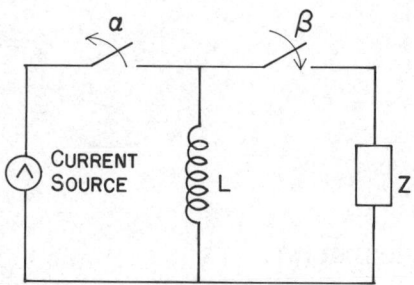

Fig. 1. Inductive energy storage circuit.

a) Opening time: $\tau_{op} < 1\ \mu s$

b) Switch conduction current: $I > 10^3 A$

c) Hold-off voltage after opening: $V_{oc} > 10^4 V$

d) Repetition rate: $f_r > 10$ pps

There exist single shot opening switches – fuses and explosively driven switches – which satisfy these current, voltage, and opening time conditions. However, at the present time, there is no repetitive opening switch which satisfies all the required conditions.

Recently several concepts have been discussed (Kristiansen and Schoenbach, 1981). One concept which seems to be particularly attractive for rep-rated opening switches is the diffuse discharge opening switch. There are two ways to use a diffuse discharge as an opening switch:

a) by turning off the ionization source in an externally sustained discharge,

b) by controlled reduction of the conductivity in a self-sustained discharge.

Whereas a controlled conductivity reduction seems to be possible only by using a laser, concept a) can be realized with either a laser or an electron beam as the ionization source. In such a switch, reduction of current carriers or carrier mobility after turn-off of the external source is determined by recombination and attachment processes, respectively.

EXTERNALLY SUSTAINED DIFFUSE DISCHARGES AS OPENING SWITCHES

Figure 2 shows experimental results (Beltzinger, 1981) with an e-beam as ionization source. Current traces of the externally sustained discharge in pure Argon are shown in the left picture for various e-beam currents. After turn-off of the electron beam, the current decreases because of recombination processes in the diffuse plasma. Time constants or "opening times" are 10 to 20 µs. The right picture again shows current traces of the externally sustained discharge for two different e-beam currents, but now the gas composition is slightly different: 0.02 torr SF_6 - an electronegative gas - has been added to the Argon. Because of attachment processes, the opening time is reduced by approximately one order of magnitude. However, this desirable effect - important for opening switches - has a side effect which limits the application of attachers in diffuse discharge switches. The faster the reduction of electron density due to attachment processes - the shorter the opening time - the higher is the e-beam or laser power necessary to keep the electron density and hence the conductivity of the plasma constant. In Bletzinger's experiment (1981) with the e-beam power kept constant, the switch current dropped from 14 A to 7.5 A after adding 0.02% SF_6 to the argon.

Using a simple balance equation, the power necessary to sustain a discharge in a diffuse discharge switch, is determined by:

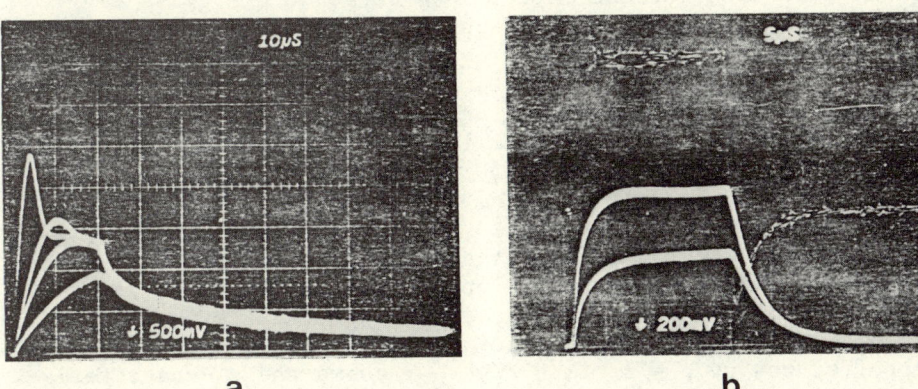

Fig. 2. Current decay in gas mixtures without (a) and with (b) attachers (Bletzinger, 1981). (a) Measured switch current in atmospheric pressure Argon (10 µs/div horizontal, 5A/div vertical). E-beam currents 20, 40, 80, 200 mA. (b) Measured switch current in atmospheric pressure Argon with 0.02 torr of SF_6 added (5 µs/div horizontal, 2A/div vertical). E-beam current 40, 80 mA.

$$P = V\, n_e\, E_{ion}\, \frac{1}{\tau_{op}}$$

where V is the volume of the discharge, n_e the electron density, E_{ion} the mean ionization energy, and τ_{op} the time necessary to replace all electrons lost due to recombination and attachment processes. $\tau_{op} = (k_A N_A)^{-1}$, where N_A is the concentration of the attacher and k_A is the attachment rate coefficient.

To overcome the conflict between short opening time and large energy consumption, the attachment rate coefficient should be kept small during charging to avoid large energy consumption. It should, however, have large values when the switch is to be opened to provide for short opening times. Because the reduced field strength E/N in the diffuse discharge is small during charging and starts growing after the ionization source is turned off, the required time dependence can be converted into a field-strength dependence for the attachment rate coefficient: k_A should be small for low values of E/N and have a maximum at high E/N.

To confirm this argument, computer calculations were performed for a diffuse discharge opening switch containing attachers with different E/N characteristics of k_A. The opening switch was assumed to be part of an inductive storage system (Fig. 1). The current source was a capacitor with a capacitance of 5 µF, charged to V = 20 kV. It is discharged through a diffuse discharge switch into an inductor with an inductance L = 1 µH. The electrons necessary to carry the discharge current in the 1 cm long, 25 cm² cross section discharge were assumed to be produced by an external ionization source, such as an e-beam or a laser. The source was turned on for t = 100 ns, corresponding to the time of maximum current flow in the charging circuit. After turning off the ionization source, switch β was closed and magnetically stored energy was transferred from the inductor to the load, a 50 Ω resistor.

The change in circuit voltages and currents was computed for times after turn off for different gas mixtures in the diffuse discharge switch. Three attachers, representative of electronegative gases, with falling, constant, and rising k_A-characteristics as a function of E/N were assumed to be additives in a gas mixture with N_2 as buffer gas at p = 1 atm. Figure 3 shows the k_A-characteristics (Nygaard, 1979) of F_2, NF_3, and I_2. For a reduced strength of E/N = 2 Td just before turn-off, F_2 represents attachers with declining attachment rate coefficient (SF_6 belongs to this group), NF_3 represents attachers with constant k_A, and I_2 represents those with strongly increasing k_A above the value of E/N which is characteristic for the conduction phase. Experimental results for I_2 exist up to E/N = 50 Td. A constant value of k_A was assumed for reduced field strengths above this value. The concentration of

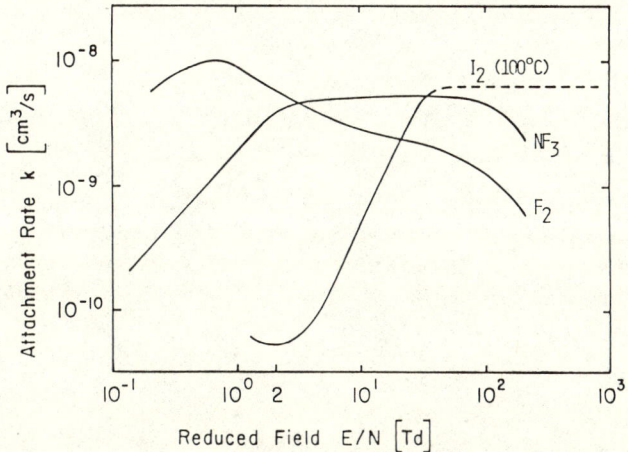

Fig. 3. Reduced field dependence of attachment rate (Nygaard, 1979).

the attaching gas was chosen so that the product of concentration and attachment rate coefficient at E/N - 2 Td was the same for all three attachers: $k_A N_A = 5 \cdot 10^7 \, s^{-1}$. Results of the current computation are shown in Fig. 4. Whereas the opening effect - decrease of current through the opening switch - is weak for F_2 and NF_3, it is pronounced for I_2. Because the attachment rate coefficient is rising with E/N, a feedback effect sets in after turning off the ionization source: the electron density is reduced because of attachment processes, causing an increase in the resistance of the diffuse discharge. The dynamics of the circuit mean that the voltage across the switch increases and thus the reduced field strength increases. This in turn gives rise to enhanced attachment, which causes a further increase in resistance, etc.

The use of this inherent attachment feedback effect requires an externally sustained diffuse discharge. There are two concepts for the externally controlled generation of electrons in a diffuse discharge:

a) the electron-beam sustained discharge.

b) the optically (laser) sustained discharge.

E-BEAM SUSTAINED DIFFUSE DISCHARGE

In the first concept (Kovaltchuk and Mesyats, 1976; Hunter, 1976; Fernsler et al., 1980; Dzimiansky and Kline, 1980), secondary electrons are produced by high energy electrons with an efficiency up to 50%. To test the applicability of such a system as part of

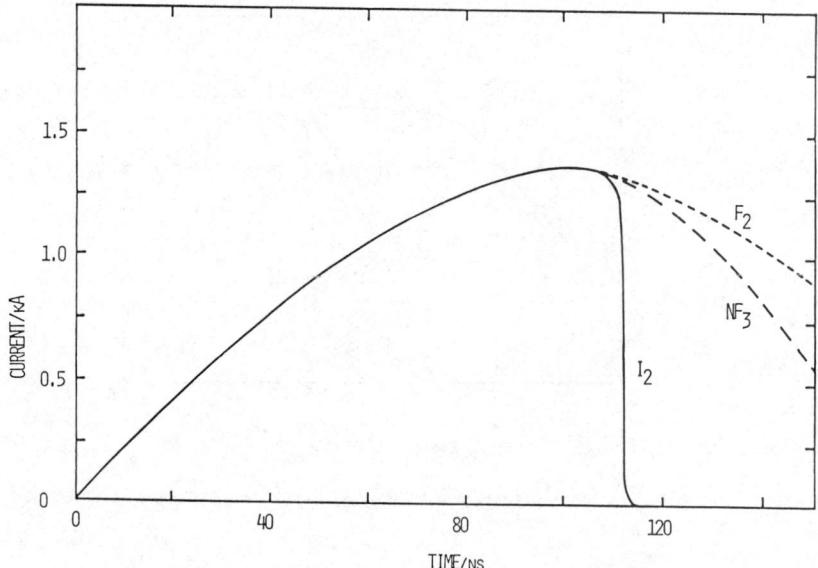

Fig. 4. Time-dependence of current through opening switch in a divertor circuit.

an inductive energy storage circuit, an electron-beam controlled discharge experiment is under construction in our laboratory. The arrangement is shown in Fig. 5. The power supply for the electron beam, a pulsed voltage source, consists of two sets of capacitors, each charged to 130 kV. By means of a switch the capacitors are discharged in series, so that a total voltage of 260 kV appears at the electrodes of the electron gun. The voltage pulse has a rise-time of 10 ns and a 1/e - fall time of 1 μs. It can be clipped by means of a crowbar switch or by a control grid (not shown) between the cathode and the foil.

A heated cathode is used in the electron gun to achieve reliable rep-rated operation and long, variable pulse lengths with good electron emission after a number of shots. To avoid insulation problems in the heater circuit, lead batteries are used as the power supply. The electrons accelerated across an evacuated space penetrate a 1 mil Titanium window which serves as the anode for the electron gun and the cathode for the gas-filled discharge system. The electrons ionize the gas in this chamber, changing its conductivity. After turn-off of the electron beam, the electrons in the discharge chamber are removed by attachment processes. The objective of the investigation is to obtain guidelines for operating an electron-beam controlled switch in an attachment-dominated gas. Closing and opening times, the switch conductivity, the switch hold-off voltage, and the current gain (discharge current/electron beam current), will be measured.

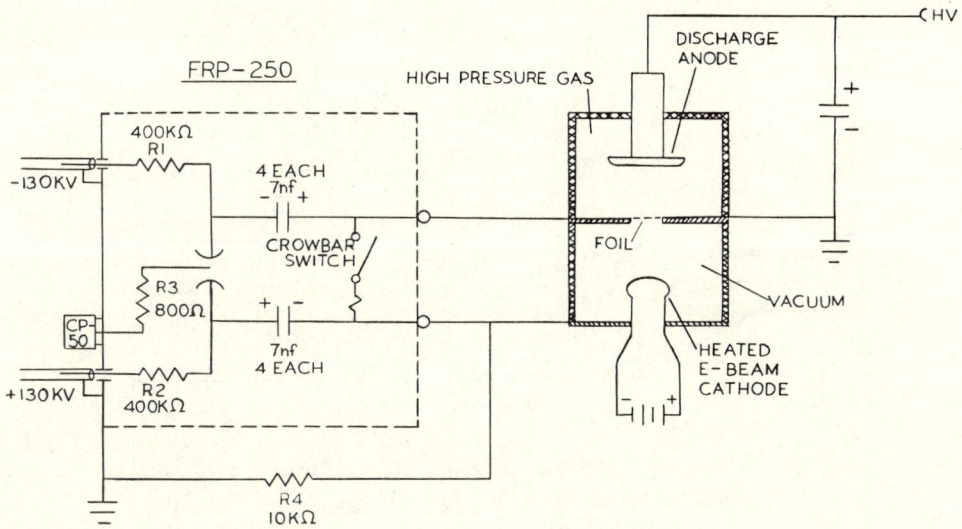

Fig. 5. Functional circuit of electron-beam controlled switch.

LASER-CONTROLLED DIFFUSE DISCHARGE

The second concept, the use of a laser to sustain or control a diffuse high pressure discharge, has not been considered previously for opening switches. The main reason is that direct ionization, as in the case of electron beam sustained discharges, is possible only in some alkali metal vapors using UV-lasers. However, it is possible to use a combination of collisional and photoprocesses to ionize commonly used gases even with visible laser radiation. Laser control has the advantage that resonance processes can be used to ionize the gas. This may be important for high rep-rate opening switches, where excessive heating of the filling gas has to be avoided.

To check the feasibility of the second concept - control of a diffuse discharge by a laser - a highly resistive diffuse discharge will be initiated by overvolting a gap. The laser serves to change the conductivity of the plasma from a highly resistive to a low resistance state by photoionization from an excited state in one of the filling gases. The experimental arrangement for testing this concept is shown in Fig. 6. A coaxial line is charged to a voltage greater than the breakdown voltage of the opening switch O. The voltage is applied to the opening switch through a laser-triggered (Harjes et al., 1980) spark gap C. A few nanoseconds before the voltage is applied, the cathode of switch O is irradiated by means of a flash lamp through a metallic mesh which serves as the anode. The photoelectrons released from the cathode form a homogeneous layer, guaranteeing the development of a homogeneous broad discharge (Koppitz, 1973). To avoid instabilities extending from the cathode,

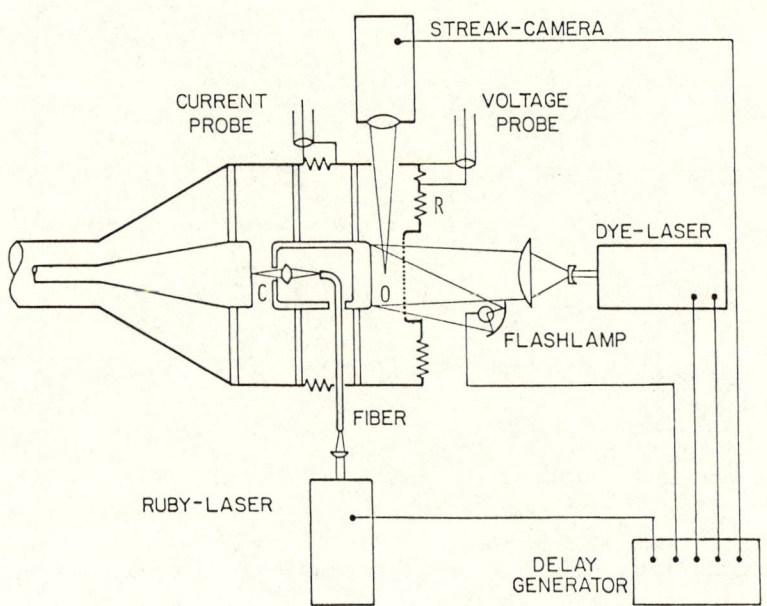

Fig. 6. Experimental set-up for optically controlled switch.

a Rogowski-type electrode shape was designed, according to the boundary conditions of the discharge chamber, using a field plotting computer code. The discharge current can be controlled by means of a variable water resistor R. Measurements of current and voltage across the gap and optical measurements by means of a streak camera are planned to get information on the stability of the diffuse discharge and its resistance.

After a diffuse discharge is initiated, a dye-laser will be used to enhance the plasma conductivity. We have planned to use, in a first experiment, a combination of N_2 (p = 1 atm) as buffer gas, a small percentage of NO, and a small percentage of I_2 or another attacher with the desired E/N dependence of the attachment rate coefficient k_A.

Figure 7 shows an energy level diagram with the pertinent transitions in the system N_2, NO and an attacher (AB). The most important process is the two-step photoionization from an excited NO* ($A^2\Sigma^+$) via an intermediate state ($E^2\Sigma^+$). This process is resonant, which means that almost all the laser energy is used for ionization. N_2 as the buffer gas has, besides its good dielectric strength, the advantage that the large population of the collisionally excited metastable state N_2 ($A^3\Sigma_u^+$) serves as an energy reservoir for the ionization process. Collisions of the second kind with a very high

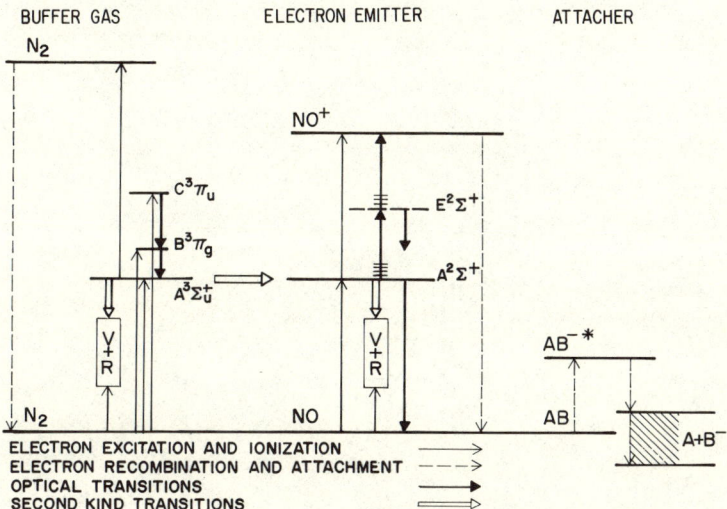

Fig. 7. Energy level diagram with transition mechanisms for a laser-controlled, opening switch gas mixture.

transition rate provide the population of the NO* state necessary for a high laser-induced ionization rate.

The electrons produced by two-step photoionization are continuously removed by attachment processes. To keep the system in a highly conductive equilibrium state (the "on" state) laser energy has to be supplied continuously. Once the laser is turned off, the electron density decreases, and, with the appropriate attacher, a "feedback" effect sets in which provides the desired short opening times.

LASER-STIMULATED ATTACHMENT

A second possibility using a diffuse discharge as an opening switch is to reduce its conductivity by stimulating loss processes in the plasma when the switch is to be opened. It is possible to get higher attachment cross sections by vibrationally exciting certain electronegative gases. The mechanism can be understood by considering the potential energy diagram of attaching diatomic molecules (Fig. 8). The potential energy curve of a neutral diatomic molecule, AB, is crossed at an energy E_v above the ground state by a repulsive potential energy curve of the negative ion AB^-. One way to produce negative ions is resonant dissociative electron attachment (Christophorou, 1971). A second way is the excitation of vibrational states of the molecule AB near the curve-crossing point. The probability of attachment and succeeding dis-

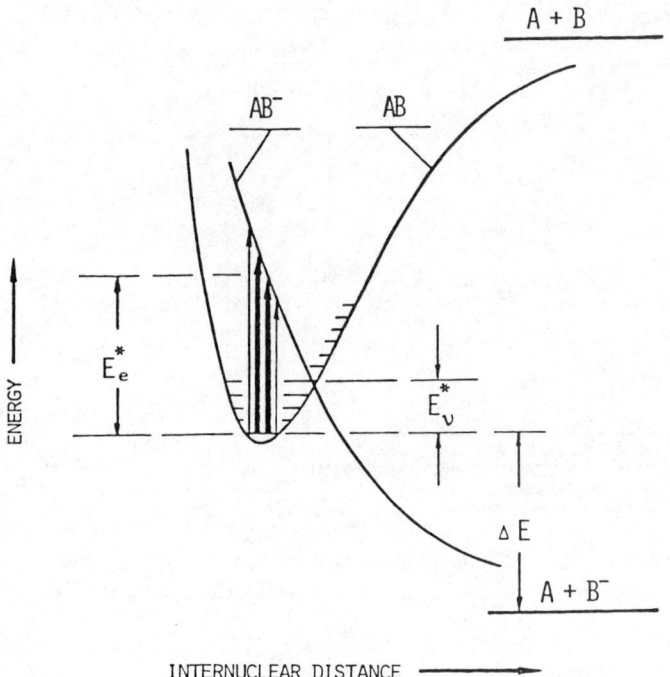

Fig. 8. Resonance dissociative electron attachment.

sociation is higher when the energy state of the vibrationally excited molecules is near the curve-crossing.

A higher population of vibrational levels above the ground state can be obtained by raising the temperature of the gas. This explains the strong temperature dependence of the attachment rates for gases like I_2, where the curve-crossing is near the 5th vibrational level of the neutral molecule (Brooks, et al., 1979). It is important that the same result can be obtained by optical vibrational excitation using an IR laser. Using this method an increase in attachment rate of several orders of magnitude can be obtained. For HCl molecules, calculations have been performed (Morgan and Pound, 1980), based on measurements of attachment rates as a function of temperature (Allan and Wing, 1981), which show an increase of the maximum value of the cross section from 10^{-17} cm^2 for the $v = 0$ vibrational level to 10^{-14} cm^2 for $v = 2$ (Fig. 9).

Figure 10 shows a schematic energy level diagram with transition mechanisms for a gas system, which may be used for an attachment-controlled opening switch. N_2 serves as the buffer gas and is also the gas that is ionized. The attacher is "triggered" by means of an IR-laser when the switch is to be opened. The discharge

DIFFUSE DISCHARGE OPENING SWITCHES

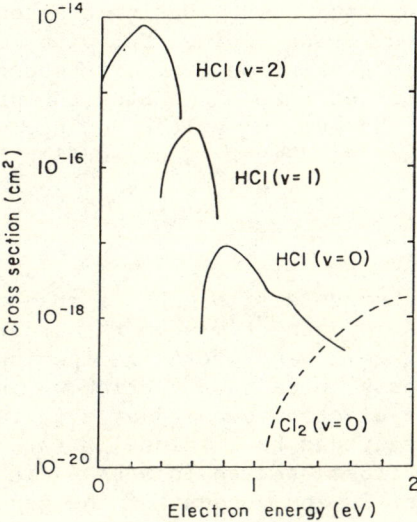

Fig. 9. Dissociative attachment cross sections for vibrationally excited HCL molecules (Morgan and Pound, 1980).

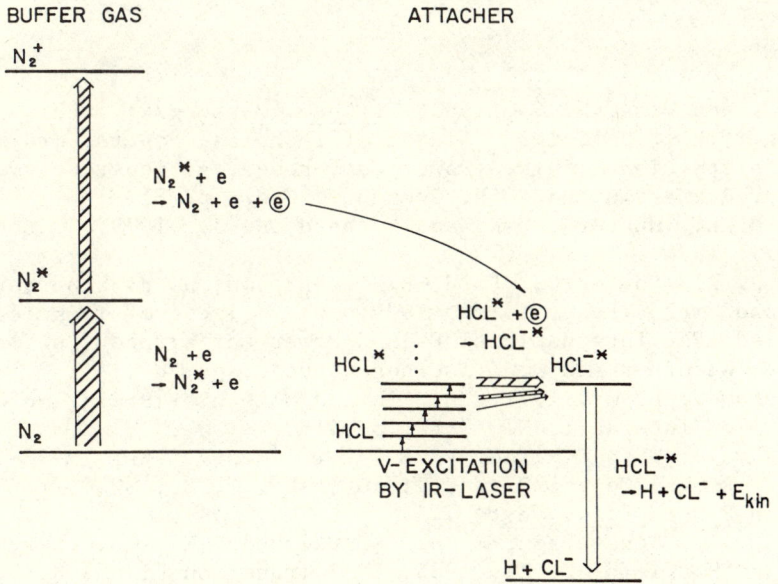

Fig. 10. Energy level diagram with transition mechanisms for a gas mixture with optically controlled attachers.

again has to be diffuse and cold. During the "on" state, the discharge can be either self-sustaining, but with low resistance, or sustained by means of an electron beam or a second laser. The experimental arrangement described in one of the previous sections can also be used to test the opening switch concept which is based on optical control of the attachment cross section.

SUMMARY

Using attachment processes to enhance the resistance of a diffuse discharge switch, opening times on the order of 10 ns should be possible. Opening switches, where the opening effect is based on attachment, are lossy, because a continuous energy supply is necessary to keep the electron density at a certain level. Improvement of their efficiency can be attained if attachers are used for which the attachment cross section increases with reduced field strength. This effect is independent of the generation mechanism for the electrons. It can be used in electron-beam controlled as well as in laser-controlled switches. A concept which can only be realized by means of lasers is that of enhancement of the attachment cross section by optical vibrational excitation of attachers. The efficiency of opening switches based on this mechanism should be high because the laser energy is used only during the opening phase and not to sustain the discharge.

REFERENCES

Allan, M. and Wong, S. F., 1981, J. Chem. Phys., 74: 1687.
Bletzinger, P., 1981, in: "Digest of Technical Papers, 3rd IEEE International Pulsed Power Conference, Albuquerque, N.M.," T. H. Martin and A. H. Guenther, eds., p. 81.
Brooks, H. L., Hunter, S. R. and Nygaard, K. J., 1979, J. Chem. Phys., 71: 1870.
Burton, J. K., Conte, D., Ford, R. D., Lupton, W. H. Scherrer, V. E., and Vitkovitsky, I. M., 1979, in: "Digest of Technical Papers, 2nd IEEE International Pulsed Power Conference, Lubbock, Texas." A. Guenther and M. Kristiansen, eds., p. 284.
Christophorou, L. G., 1971, "Atomic and Molecular Radiation Physics" Wiley Intersci., New York, p. 416.
Dzimiansky, J. W. and Kline, L. E., 1980, Rept. AFWAL-TR-80-2041.
Fernsler, R. F., Conte, D. and Vitkovitsky, I. M., 1980, IEEE Trans. on Plasma Sci., PS-8, 176.
Harjes, H. C., Schoenbach, K. H., Kristiansen, M., Guenther, A. H., and Hatfield, L. L., 1980, IEEE Trans. on Plasma Sci., 8: 1494.
Hunter, R. O., 1976, in: "Proceedings, 1st International Pulsed Power Conference, Lubbock, Texas," M. Kristiansen, ed., Paper 1C-8.
Koppitz, J., 1973, J. Phys. D., 6: 1454.

Kovaltchuk, B. M. and Mesyats, G. A., 1976, in: "Proceedings, 1st International Pulsed Power Conference, Lubbock, Texas," M. Kristiansen, ed., Paper 1C-7.

Kristiansen, M. and Schoenbach, K., eds., "Proceedings, Army Research Office Workshop on Repetitive Opening Switches, Tamarron, Colorado, January, 1981," Texas Tech University, Lubbock, Texas.

Morgan, W. L. and Pound, M. J., 1980, in: "Abstracts of 33rd Gaseous Electronics Conference, University of Oklahoma, Norman," Abstract FB-3.

Nygaard, K. J., 1979, IEEE J. Quantum Electron., 15: 1716.

THE ELECTROLUMINESCENCE OF PURE NOBLE GASES BELOW THE

THRESHOLD FOR ELECTRON AVALANCHE*

C. A. N. Conde,[+] M. F. A. Ferreira,[+] T. H. V. T. Dias and
A. D. Stauffer[++]

[+]Physics Department, University of Coimbra
Coimbra, Portugal

[++]Physics Department, York University
Toronto, Ontario, Canada M3J1P3

INTRODUCTION

The study of the light produced when electrons are released in a noble gas by radiation, the so-called primary scintillation, has been for a long time the subject of research (Sayres and Wu, 1957; Bennett, 1962; Leite, 1980) and has been applied to nuclear radiation detection among other fields.

Some years ago it was found that the intensity of the light produced can be increased by a few orders of magnitude if the primary electrons are allowed to drift under the influence of an electric field (Conde and Policarpo, 1967; Policarpo et al. 1968). This electroluminescence phenomenon occurs even for fields well below the threshold for charge multiplication, and it is called secondary or proportional scintillation. This has increased the range of applications of noble gases to radiation spectrometry and has led to the development of a new type of radiation detector: the Gas Proportional Scintillation Counter. This development has stimulated much research on the basic physics of the processes involved which is also relevant for gaseous electronics. The characteristics and main conditions for the production of the secondary scintillation are the following:

---

*Work partially supported by INIC (Instituto Nacional de Investigacão Científica) and by NATO under Research Grant RG 133-80.

1) Very pure noble gases (Xe, Kr and Ar) are used at pressures of around 1 atmosphere.

2) The reduced electric field intensities range from about 1 to 5 V cm^{-1} torr^{-1} (3 x 15x10^{-21} Vm^2). For stronger fields, there is charge multiplication besides light production.

3) The light is emitted in the VUV region and its intensity is so large that a single electron drifting along the field lines can lead to the production of a few hundred photons.

4) The intensity of the secondary scintillation does not depend on the influence of magnetic fields even for field intensities as strong as a few kGauss.

The understanding of these phenomena requires the use of the theory of the transport of electrons in gases. Recently an effort (Ferreira et al., 1980; Feio and Policarpo, 1981 and Dias et al. 1981) has been made towards that understanding using the Boltzmann transport equation and Monte Carlo methods.

In the present work, we review the experimental results obtained (mainly those with interest for gaseous electronics), discuss the application of the Boltzmann equation and Monte Carlo methods to these phenomana, and present as well some unpublished results of our research.

THE ELECTROLUMINESCENCE OF NOBLE GASES AT PRESSURES OF THE ORDER OF ONE ATMOSPHERE

The experimental study of the electroluminescence phenomena referred to before is generally carried out with a cell like the one shown in Fig. 1, filled with the appropriate noble gas.

Radiation entering the window produces a swarm of primary electrons in the region, a few centimeters long, that extends down to grid 2. An electric field not too strong (below the threshold for light production) produced by the application of a potential to grid 2 and to the field rings, drifts those electrons towards the region between grids 1 and 2. Once they reach this region they meet a fairly strong electric field (a few kV/cm) that accelerates them. The energy distribution function of the electrons in the swarm is thus changed, and some may reach enough kinetic energy to excite the noble gas atoms. As a result of the de-excitation processes, light is produced in the VUV region. This is the secondary scintillation or electroluminescence which is detected with a photomultiplier. A thin wavelength shifter deposit may be required to convert the VUV photons to longer wavelengths if the photomultiplier has not the appropriate VUV sensitivity. The experimental fact

ELECTROLUMINESCENCE OF PURE NOBLE GASES

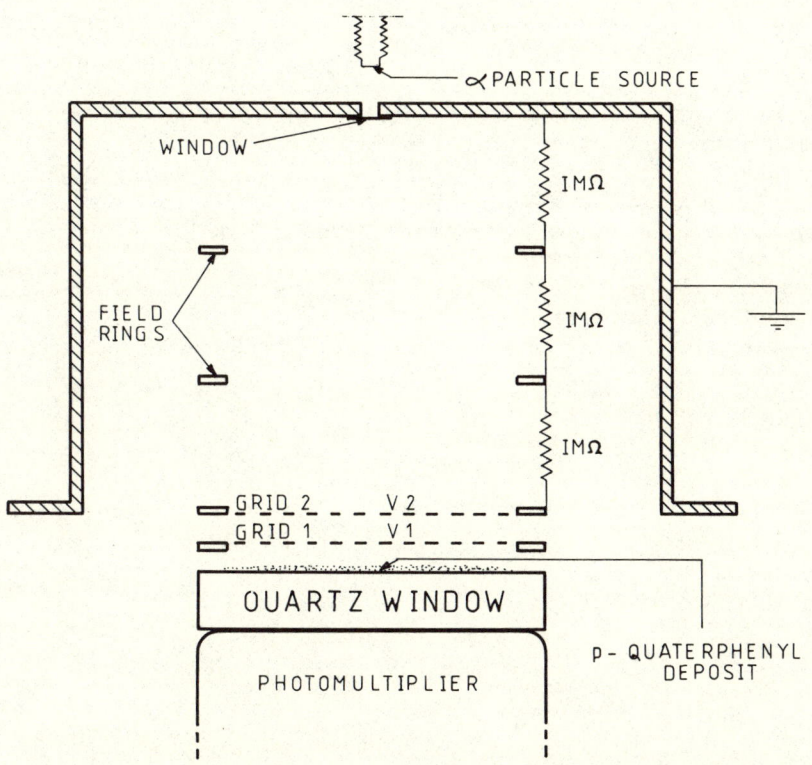

Fig. 1. Schematic diagram showing the principle of the uniform-field, gas-proportional scintillation counter (Conde et al., 1975).

(Conde et al., 1975) that the intensity of the secondary scintillation depends only on the difference of potential between the grids shows that indeed the light is originated in the strong field region and not, as Braglia et al. (1966) initially proposed, in the weak field one.

The intensity of the light is expected to be proportional to the number of primary electrons for the weak ionization densities normally used for this kind of work, and also proportional to the length of the path travelled by the electrons in the strong field region. This is supported by experimental data. If excitation of the noble gas atoms is carried out directly by electrons in two-body collision processes the dependence of the intensity of light on the gas pressure is expected to obey an appropriate scaling law

$$\frac{1}{p} \frac{dn}{dx} = f\left(\frac{E}{p}\right) \tag{1}$$

where n is the number of photons produced by a single electron travelling along the distance x in the gas at pressure p, and E is the electric field intensity. It is assumed that the de-excitation mechanisms do not depend on the gas pressure. This is, of course, an approximation. The experimental results (Conde et al., 1977) are in good agreement with this interpretation and show (Fig. 2) that the function $f(E/p)$ is a straight line that, for Xe, crosses the E/p axis at about 1 Vcm^{-1} $torr^{-1}$, the threshold for light production.

Suzuki and Kubota (1979) were the first to observe the optical spectrum of the proportional scintillation. It consists of a continuum peaked at the 173nm for Xe, 147nm for Kr, and 128nm for Ar

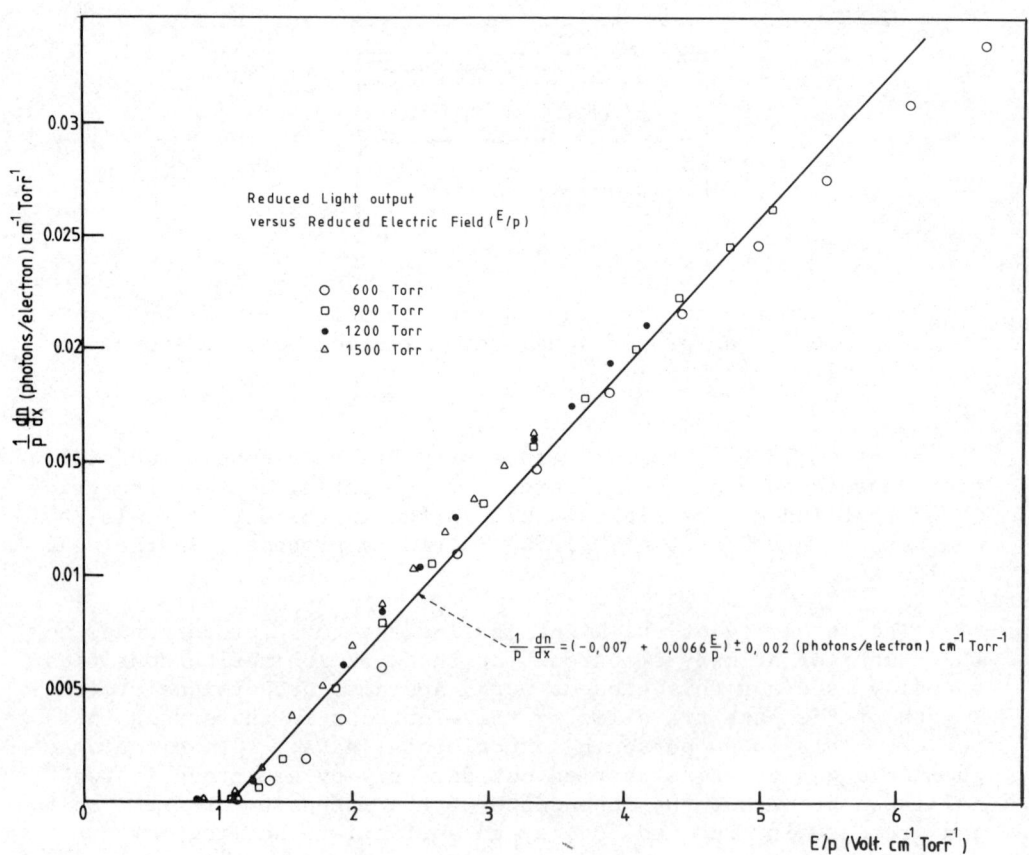

Fig. 2. Reduced secondary light output for a xenon uniform-field, gas-proportional scintillation counter with p-terphenyl wavelength-shifter (Conde et al., 1977).

with FWHM of 14,12, and 10nm, respectively. Unlike the spectra for RF discharges and primary scintillation (Bennett, 1962), no photons were detected in the visible region. Their observations clearly show that only the lower excited states of the noble gases are populated, at around 9 eV for Xe, 10.2 eV for Kr, and 11.7 eV for Ar. This means that, as expected, electron energy is lower for proportional scintillation than for primary scintillation or discharges which require stronger electric fields.

A noble gas atom, R, once excited to a low state can lead through collisions to the formation of a molecular excimer, R_2^*, which breaks up as

$$R_2^* \rightarrow 2R + h\nu , \qquad (2)$$

leading to the production of VUV photons. This is the secondary or proportional scintillation of the noble gas. The visible light observed in discharges is attributed to energetic electrons which excite higher states that decay afterwards to one of the lower states. As noble gas atoms have only electronic states and as the gases used are very pure, the drifting electrons cannot lose energy through excitation of low lying rotational on vibrational states which exist only for impurity molecules. Therefore, electrons with energy below the first excited state can only lose energy through the recoil of the noble gas atom that suffered an elastic collision. This is a relatively small amount even if we consider the large number of elastic collisions an electron makes before it gains from the electric field enough energy to excite a noble gas atom. Thus, except for relatively low electric fields, we expect a fairly high efficiency for conversion of electrical energy gained from the field into optical energy observed as VUV photons. Experimental values above 70% have been measured for Xe and Kr (Policarpo, 1981). Thus one single electron travelling through a difference of potential of 7 kV can produce a number of photons of the order of 500 if the E/p values are optimized.

Up to now the most important application of the proportional scintillation phenomena has been to radiation spectrometry since the development of gas proportional scintillation counters for X-rays. Energy resolutions for soft X-rays are unmatched by any other type of radiation detector: 85 eV FWHM for 140 eV X-rays (Hamilton et al., 1980) and 100 eV for 283 eV X-rays (Alves et al., 1975). For large area detectors, the superiority of the gas proportional scintillation counter is extended far into the keV region.

The high efficiency for conversion of electrical into optical energy is also of interest for laser applications.

THE ELECTRON ENERGY DISTRIBUTION FUNCTION AND THE ELECTROLUMINESCENCE OF NOBLE GASES

The understanding of the electroluminescence of noble gases is simplified if we consider separately two main steps:

1) excitation of noble gas atoms with electrons.

2) de-excitation mechanisms leading to light production.

While a lot of research has been already carried out concerning step 2) for the primary scintillation [for a review see the article by Leite (1980)], not much is known about the secondary scintillation de-excitation mechanisms besides the work by Suzuki and Kubota (1979), which we have already mentioned. Thus in this section we only discuss step 1).

For the reasons explained before, we assume that noble gas atoms are excited directly by electrons. In a steady-state situation, where there is a continuous flow of electrons drifting under a uniform electric field, the number of excited species will be a function of the number of electrons reaching a kinetic energy larger than that of the first excited state. Feio and Policarpo (1981) assumed that that number is proportional to the number of electrons in the tail of the electron distribution with an energy between the first excited state of the noble gas and the threshold for ionization. The electron distribution function they used was obtained neglecting inelastic collisions. The results of their calculations are shown in Fig. 3 where the percentage of electrons with an energy between those two limits is plotted as a function of the electric field intensity at atmospheric pressure for the five noble gases. These results are in reasonably good agreement with the experimental 1V cm^{-1} $torr^{-1}$ threshold for proportional scintillation in Ar and Xe and the lower limit for Kr (Leite et al., 1981).

It is known that magnetic fields affect the energy distribution function for electrons not only by changing its shape, but also by destroying the cylindrical symmetry around the electric field direction. A consequence of this is that electrons drift along a direction that makes an angle with the electric field which can be of the order of 10° to 20° for magnetic field intensities of a few kGauss, as experimentally observed by Schultz (1976). Another expected consequence of magnetic fields would be the changing of the proportional scintillation intensity. However, experimental results for field intensities as strong as 4.5 kG (Ferreira et al., 1980a) showed that the secondary scintillation of argon and xenon remains constant within an experimental accuracy of $\pm$ 1%. Similar results were obtained for krypton (Ferreira et al., 1980b).

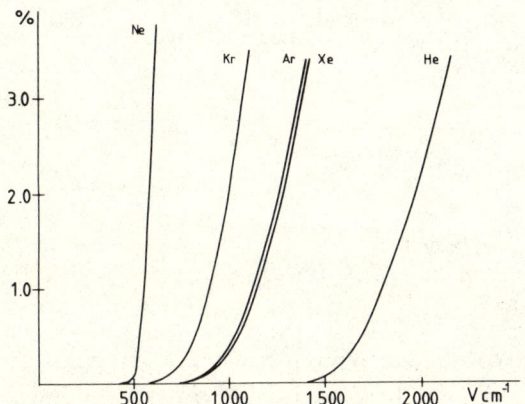

Fig. 3. Theoretical curves for the percentage of the number of electrons with energy between the first excited state and the threshold for ionization as a function of the electric field intensity for the various noble gases at atmospheric pressure (Feio and Policarpo, 1981) with permission from the *Japanese Journal of Applied Physics*.

To interpret those results, we carried out an investigation of Boltzmann transport equation for electrons in the presence of a magnetic field $\vec{B}$, perpendicular to the electric field $\vec{E}$, as shown in Fig. 4. The distribution function $F(\varepsilon, \cos\Theta, \phi)$, in terms of the electron energy ε, the cosine of the polar angle $\cos\Theta$, and the azimuthal angle ϕ, can be developed using a spherical harmonics expansion.

In a first order approximation, we can write:

$$(F(\varepsilon, \cos\Theta, \phi) = F_0(\varepsilon) + \cos\Theta F_1(\varepsilon) + \sin\Theta \cos\phi\, F_2(\varepsilon) + \\ + \sin\Theta \sin\phi\, F_3(\varepsilon) + \ldots) \quad (3)$$

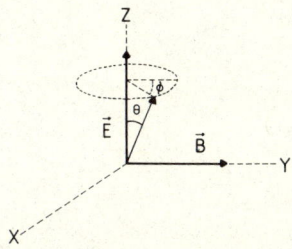

Fig. 4. Geometry of the electric and magnetic fields used for proportional scintillation experiments and for the calculation of the electron energy distribution function.

If we consider both elastic and inelastic collisions, $F_o(\varepsilon)$ obeys the differential equation (Ferreira, 1979)

$$\frac{2}{3} \frac{(eE)^2}{m} \frac{\partial}{\partial \varepsilon} \left[\frac{\varepsilon\, l_e(\varepsilon)}{1+e^2 B^2 l_e^2(\varepsilon)} \frac{\partial}{\partial \varepsilon} \left(\frac{F_o}{v}\right) \right] + \frac{2m}{M} \frac{\partial}{\partial \varepsilon} \left(\frac{\varepsilon v\, F_o}{l_e(\varepsilon)}\right) +$$

$$+ \sum_h \left[\frac{\sqrt{2/m}\, \sqrt{\varepsilon+\varepsilon_h}}{l_h(\varepsilon+\varepsilon_h)} F_o(\varepsilon+\varepsilon_h) - \frac{\sqrt{2/m}\, \sqrt{\varepsilon}}{l_h(\varepsilon)} F_o(\varepsilon) \right] = 0 \quad (4)$$

where $l_e(\varepsilon)$ is the mean free path referred to the momentum transfer cross section for an electron with energy ε, ε_h the energy of the excited level h, and $l_h(\varepsilon+\varepsilon_h)$ the mean free path for an electron with energy $\varepsilon+\varepsilon_h$. This equation is similar to that deduced by Schultz (1976) except that it includes the case where electrons have enough energy to excite the atoms or molecules in the gas mixtures. Whenever

$$\frac{e^2 B^2 l_e^2(\varepsilon)}{2m\varepsilon} \ll 1$$

is a valid approximation, that equation reduces to the one without magnetic fields (Palladino and Sadoulet, 1974).

The remaining coefficients of the expansion obey the equations

$$F_1(\varepsilon) = - \frac{e E v\, l_e(\varepsilon)}{1 + \frac{e^2 B^2 l_e^2(\varepsilon)}{2m\varepsilon}} \frac{\partial}{\partial \varepsilon} \left(\frac{F_o}{v}\right) \quad (5a)$$

$$F_2(\varepsilon) = - \frac{eB}{mv} l_e(\varepsilon)\, F_1(\varepsilon) \quad (5b)$$

$$F_3(\varepsilon) = 0 \quad (5c)$$

Thus, the influence of magnetic fields on the electron energy distribution function is only relevant if the term

$$\frac{e^2 B^2 l_e^2(\varepsilon)}{2 m \varepsilon}$$

cannot be neglected when compared with 1. The calculation of this term was carried out for the different noble gases under various magnetic field intensities and for an electron energy ε equal to that of the first excited state of the noble gas considered. The calculated values are shown in Table I. As can be seen, for the stronger magnetic field used in the proportional scintillation experiments referred to before (Ferreira et al., 1980), the values that term takes are much smaller than 1 (of the order of 10^{-4} or 10^{-5}).

The apparent discrepancy of the influence of magnetic fields on the proportional scintillation experiments and on the electron drifting experiments in noble gases remains to be explained. Actually, if we consider that in most drifting experiments the electric fields used are weaker so that most electrons travel with an energy near the Ramsauer minimum, their average energy ε is lower. Thus, the electron mean free path is larger, and the term $e^2 B^2 l_e^2 / 2m\varepsilon$ cannot be neglected. On the other hand, for the proportional scintillation, the term referred to above can now be neglected because the average electron energies involved in the excitation processes are higher and the corresponding mean free path smaller. This means that only magnetic fields stronger than a few Tesla can affect the proportional scintillation of the gases normally used: Ar, Kr and Xe. Those fields are not found in most experimental situations except in high energy physics work.

Table I. Calculated values for the $e^2 B^2 l_e^2 / 2m\varepsilon$ term for the various noble gases and different magnetic field intensities.

B(Tesla)	He	Ne	Ar	Kr	Xe
0.4	1.8×10^{-3}	1.6×10^{-3}	7.6×10^{-5}	6.3×10^{-5}	2.9×10^{-5}
1	1.1×10^{-2}	1.0×10^{-2}	4.7×10^{-4}	3.9×10^{-4}	1.8×10^{-4}
2	4.5×10^{-2}	4.1×10^{-2}	1.9×10^{-3}	1.6×10^{-3}	7.2×10^{-4}
5	2.8×10^{-1}	2.6×10^{-1}	1.8×10^{-2}	9.8×10^{-3}	4.5×10^{-3}
10	1.13	1.03	4.7×10^{-2}	3.9×10^{-2}	1.8×10^{-2}

MONTE CARLO SIMULATION OF THE TRANSPORT OF ELECTRONS
AND THE ELECTROLUMINESCENCE OF NOBLE GASES

While the Boltzmann transport equation can be used to interpret the proportional scintillation phenomena of noble gases, as we have shown, much more information can be obtained with Monte-Carlo simulation techniques. Thus we have engaged ourselves in a research program (Dias et al., 1981) along these lines. We present here some preliminary results obtained with a one dimensional Monte Carlo model for the transport of electrons in Xe under the influence of electric fields. An analytical expression for the total cross section was obtained with a polynomial fitting. As we were dealing with a one dimensional model, only forward and backward collisions were considered whose relative probabilities were obtained by intergrating the differential 3- dimensional cross section over the forward and backward hemispheres. The differential cross sections were calculated using theoretical phase shifts, δ_ℓ, up to $\ell = 50$. Figure 5 shows the calculated probability for forward scattering as a function of the electron energy. It is interesting to notice that near the Ramsauer minimum the scattering direction is almost 100% forward.

One dimensional Maxwellian velocity distributions were considered both for the atom and the starting electron, enabling us to study the effect of temperature.

Under the influence of the electric field, each electron was allowed to collide with a moving atom after a path length l according to

$$\ln R = - N \int_0^l \sigma(l') \, dl' \qquad (6)$$

where σ is the total cross section, N the number density and R a random number uniformly distributed between 0 and 1, generated by a subroutine. Depending on the probability of backward/forward scattering, the electron direction was modified after each collision, using a random number technique. The average energy transferred to the recoiling atom was evaluated using momentum-transfer cross sections integrated over the forward and backward hemispheres. Along the next collision path, the electron may gain or lose energy from the electric field, depending on its direction of movement. This simulation process was then repeated until the electron reached an energy equal to that of the first excited state of the noble gas atom. It is assumed that soon after that, the electron loses almost all its energy through an inelastic collision leading to excitation of the noble gas. Light is then produced by de-excitation mechanisms described before.

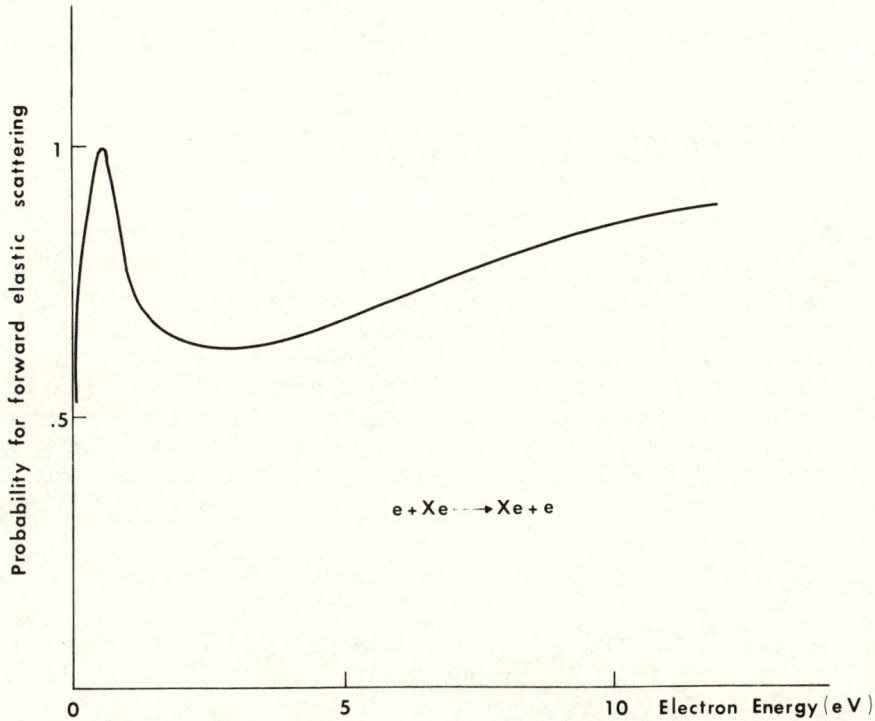

Fig. 5. Calculated probability for forward elastic scattering of electrons in Xe.

The total path length travelled by a single electron and the total number of collisions from the starting position until excitation is expected to decrease with the electric field intensity. The computer simulation for Xe shows that indeed this is the case. Path lengths of the order of 100 μm with drift times of the order of 10 nsec were obtained, which are within the expected values. The large number of collisions (of the order of 10^4) prior to photon production suggests that the gas purity is a critical factor for the success of a proportional scintillation experiment; even fairly small concentrations of impurity atoms are likely to make the drifting electron lose its energy in one of those collisions before it excites a noble gas atom. The large number of collisions also means that if mixtures of noble gases are used, the atoms with lower energy levels will be preferentially excited even if they are in small concentrations. We calculated as well the efficiencies for conversion of electrical into optical energy, which are of the order of 85% assuming that the 8.3 eV Xenon excited state decays through the 7.3 eV (average energy) state of the Xenon excimer Xe_2^*. In our simulation, we also studied the effect of changing the temperature of the gas and found that it can be neglected for constant number density.

To improve the accuracy of our model, we are now carrying out three-dimensional Monte Carlo simulations.

CONCLUSIONS

The VUV electroluminescence of pure noble gases at pressures of around one atmosphere, with small electron concentrations and weak electric fields (below the threshold for ionization), is a phenomenon that was essential in the development of radiation detectors such as the gas proportional scintillation counter. Transport theory techniques, either Boltzmann equation analysis or Monte Carlo simulations, seem to be successful in the understanding of the light-production mechanisms.

REFERENCES

Alves, M., Alice, F., Policarpo, A. J. P. L., Leite, M., and Salete, S. C. P., 1975, IEEE Trans. Nucl. Sci., NS-22, 109.
Bennett, W. R., 1962, Ann. Phys., 18: 367.
Braglia, G. L., Munari, G. M., and Mambriani, G., 1966, Il Nuovo Cimento, B, 43: 130.
Conde, C. A. N. and Policarpo, A. J. P. L., 1967, Nucl. Instrum. Methods, 53: 7.
Conde, C. A. N., Santos, M. C. M., Ferreira, M., Fatima, A., and Sousa, Celia A., 1975, IEEE Trans. Nucl. Sci., NS-22, 104.
Conde, C. A. N., Requicha Ferreira, L., and Ferreira, M. F. A., 1977, IEEE Trans. Nucl. Sci., NS-24, 221.
Dias, T. H. V. T., Stauffer, A. D., and Conde, C. A. N., 1981, to be published.
Feio, M. A. and Policarpo, A. J. P. L., 1981, submitted to Jap. J. Appl. Phys.
Ferreira, M. F. A., Conde, C. A. N., Ayres De Campos, N., and Gill, J. M. S. C., 1980, IEEE Trans. Nucl. Sci., NS-27, 208.
Ferreira, M. F. A., Dias, T. H. V. T., Campos, A. J., and Conde, C. A. N., 1980, "Proceedings of the 5th Europhysics Sectional Conference on the Atomic and Molecular Physics of Ionized Gases, Dubrovnik, Yugoslavia," 4D, P. 120.
Ferreira, M. F. A., 1979, Post-Graduation Course Report, Physics Department, University of Coimbra, Portugal.
Hamilton, T. T., Hailey, C. J., Ku, W. H. -M., and Novick, R., 1980, IEEE Trans. Nucl. Sci., NS-27, 190.
Leite, M. and Salete, S. C. P., 1980, Portgal. Phys., 11: 53.
Leite, M. Salete, S. C. P., Policarpo, A. J. P. L., Feio, M. A., and Alves, M. A. F., 1981, Nucl. Instrum. Methods, 179: 295.

Palladino, V. and Sadoulet, B., 1974, Rep. LBL-3013, Lawrence Berkeley Laboratory, University of California, Berkeley.
Policarpo, A. J. P. L., Conde, C. A. N., and Alves, M. A. F., 1968, Nucl. Instrum. Methods, 58: 151.
Policarpo, A. J. P. L., 1981, Physica Scripta, 23: 539.
Sayres, A. and Wu, C. S., 1957, Rev. Sci. Instrum., 28: 752.
Schultz, G., 1976, Thesis, Université Louis Pasteur, Strasbourg, France.
Suzuki, M. and Kubota, S., 1979, Nucl. Instrum. Methods, 164: 197.

PLASMA FORMATION IN A TOKAMAK

R. D. Bengtson and J. F. Benesch

Physics Department and Fusion Research Center
University of Texas at Austin
Austin, Texas 78712

INTRODUCTION

In fusion energy research the goal is to contain a hot plasma for a sufficiently long period of time so that the reacting species (deuterium and tritium nuclei) will release more energy in exothermic fusion reactions than is required to heat and contain the plasma. One of the promising concepts in magnetic fusion is the tokamak, a toroidal magnetic containment vessel first developed by Artsimovich (1972) and described in a recent survey article by Sheffield (1981). The magnetic field in a tokamak is composed of three parts: a strong toroidal magnetic field which is the main constraining field for particle motion, a poloidal field which is produced by a current in the plasma in the toroidal direction, and a vertical field to counteract the hoop force generated by the currents in the toroidal direction. The plasma current is driven by a transformer, and so the duration of a plasma discharge is limited by the volt-second capability of this transformer. One of the major problems in tokamak research is the impurities that are found in the hot plasma, causing a large radiative power loss and the development of possibly catastrophic instabilities.

The initial phases of plasma formation in a tokamak may be the source for many of the impurities observed later in the discharge and of the high energy runaway electrons observed in many tokamaks. The relevance of the plasma formation period to successful tokamak operation has been demonstrated in several experiments. Estimates of Z_{eff} on the ST tokamak (Dimmock et al., 1973) suggested that the lower molecular weight impurities were present at the beginning of the discharge and did not increase substantially during the discharge. This is in contrast to the heavy impurities which were injected throughout the discharge. Measurement of hard X-rays produced in normal ORMAK discharges by Knoepfel and Zewben (1975) provides direct evidence that runaway electrons can be produced at early times and can be contained stably during the entire discharge time. Other investigations (Spong et al., 1974) have noted runaways at low density which were correlated with a poor preionization and/ or high impurity content in the filling gas. In spite of the consequences of runaway electrons formed in the early part of the discharge and of low molecular weight impurities injected into the plasma when it is formed, comparatively little work has been focused on the initial phases of operation. Hutchinson and Strachan (1974) and Hutchinson and Morton (1976) have studied the formation of plasmas in a small tokamak. The TFR group (1975) recently has investigated the formation processes in their tokamak. In 1976, Papoular surveyed the extant work on tokamak startup.

The formation period of the plasma has relevance in the design of a large machine. The portion of the volt-second capability of the ohmic heating transformer which must be used to form the plasma and arrive at a stable equilibrium has significant consequences in the design of the transformer and core. Of greater importance is the loop voltage required for breakdown of the neutral gas to form a plasma. The loop voltage required for breakdown grows linearly with the major radius. Present day machines such as PLT (R = 1.0 m) require a loop voltage of 60 V while TFTR (R = 2.7 m) will require about 150 V, and a reactor will need 300-500 V. These voltages are secondary voltages, and the primary contains a large number of turns, of order 100. These voltages then become higher than is convenient for the power supply to drive in the primary.

DESCRIPTION OF PLASMA FORMATION

Following Papoular (1976), we will divide the start of a tokamak discharge into three phases: breakdown, plasma formation, and current rise. Breakdown is indicated by the first appearance of neutral hydrogen light, indicating significant ionization. In this first phase, electron neutral collisions dominate the physics, so that Townsend avalanche and gaseous electronics provide the framework for the examination of this phase. When the number of free electrons is such that the electron-ion collision frequency is higher than the electron-neutral collision frequency, the long range electromagnetic interactions become dominant, and the resistivity may be viewed as depending only on the electron-ion collision and density. This happens when the electron density reaches about 10% of the original neutral density which is of the order of 10^{13} cm^{-3} and marks the start of the plasma formation phase. The plasma temperature is clamped during this period by radiation and ionization losses until the neutral fraction becomes small. Finally, after the plasma is substantially ionized, the current rise phase begins.

The breakdown phase of a tokamak is usually viewed in light of the Townsend avalanche theory in which the important physical parameters are the ratio of E_a/p and pd. E_a is the applied electric field in V/cm, p the gas pressure in torr, and d is the electrode separation. Plate separation is generally one to ten millimeters, voltages range from 0.1 to a few kV, and pressures from 0.1 to 10 torr. The data available from such work covers the E_a/p range 10-1000. Tokamaks are generally run towards the high end of this range. Thus it may be possible to gain some insight into the physics of tokamak breakdown by examining results from linear discharges. Conversely, study of plasma formation in a tokamak may contribute to the field of gaseous electronics. One must be careful, however, because the two discharges are quite different. The difference is that the parameter pd which is essential to the Townsend theory is not well defined in a tokamak geometry. A parameter, as yet undefined, associated with the confinement time will play a similar role as pd in Townsend theory.

There are many physical processes which may affect the course of Townsend breakdown. Among these, the ones which may be relevant to this experiment are: gas ionization by electrons emitted from the filament in the shadow of the limiter; ionization by (secondary) electrons from the gas; ionization by ions; fast neutral formation via charge exchange with subsequent Penning ionization; ionization by fast neutrals; photoionization; ionization by electrons or ions produced by electron ion or neutral collisions with the wall or limiter; particle drifts in the toroidal field; particle diffusion due to collisions with neutrals; and the effects of the stray fields due to coil misalignments, principally those associated with the toroidal field coils.

In the second phase of plasma formation, plasma effects become important, and the electron distribution function becomes more thermal, although runaway electrons may still be present in small numbers. A zero-dimensional model of the electron-density increase in a hydrogen gas was developed by Kaplan and Bengtson (1981) by assuming thermal equilibrium for the plasma. The time evolution of the number densities of various chemical species was calculated from a set of rate equations using Maxwellian rate coefficients, based on data in the literature. The breakdown was found to be basically a two-step process: dissociation of the molecules overlapped and followed by ionization of the atoms. The H_2^+ reactions were found to be necessary for modeling the rapid breakdown of the gas. The model was compared to experimental data from the formation phase of Texas Turbulent Torus (Kaplan, 1976), and reasonable agreement was found when gain and loss terms were included to account for plasma wall interactions.

The stage of tokamak startup which has been most thoroughly studied, and which is least understood, is the current-penetration phase. One starts with the cold, fully ionized plasma which is present at the end of the plasma-formation phase and finishes with a hot plasma with centrally peaked temperature distribution. The expected result is that the electrons on the outside will increase in temperature, thereby decreasing in effective resistivity because the Spitzer resistivity goes as $T_e^{-3/2}$ (Spitzer, 1962), and will shield the interior from the electric field as in any other conductor. Inward diffusion of the electric field and reorganization of the current from a skin to a centrally peaked distribution, should be very slow. Experimentally, the observations have shown a rapid penetration with little indication of a skin effect (Hawryluk et al., 1980). Several explanations of current penetration have been proposed, but experimental results are inadequate to confirm or eliminate any of them.

EXPERIMENTAL RESULTS ON BREAKDOWN

In the PRETEXT tokamak, we apply a sinusoidal loop voltage pulse to control the frequency of the applied voltage by changing the capacitors in the driving circuit. We define the breakdown voltage V_{br} as that voltage at which the secondary loop voltage first departs from its unloaded, sinusoidal form. Within experimental error, this was the time at which the current departed from zero and the H_β light signal sharply rose. Typical loop voltages for the breakdown were 10 volts. In Fig. 1 we display breakdown voltage vs. pressure curves taken with three values of capacitance and thereby three distinct half-periods or frequencies. For comparison,

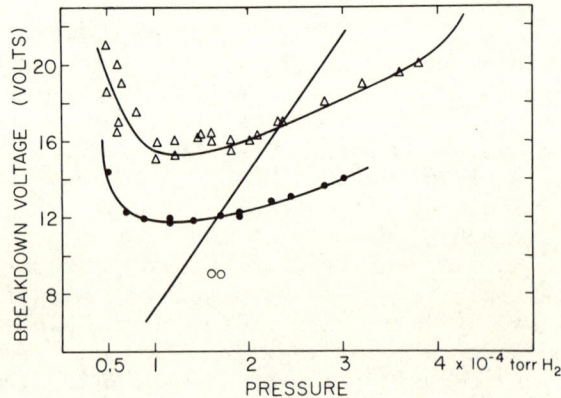

Fig. 1. Effect of frequency on breakdown voltage as a function of pressure. The straight line represents the minimum of the V vs. pd curve for H_2 taken with Pt electrodes, 295V @ 1.25 torr-cm. ○: 26 Hz, half period 19 ms; ●: 50 Hz, half period 10 ms; Δ: 71 hz, half period 7 ms.

we show a line which corresponds to the minimun of the V vs. pd
(applied voltage vs. pressure in torr electrode distance in cm)
curve taken in a linear unmagnetized discharge in hydrogen, 295 V
@ 1.25 torr-cm (von Engel, 1965). The frequency dependence in the
breakdown voltage is obvious. We also checked for frequency dependent
effects by changing the charging voltage and thereby changing
dV/dt. Graphing V_{br} vs. dV/dt at breakdown time leads to equations
of the form V_{br} = C dV/dt + D as may be seen in Fig. 2. It
should be mentioned that the frequency dependent results were identical
whether they were taken by varying the charging voltage or
the frequency. Figure 2 shows the influence of small vertical fields
on the breakdown. The direction of this field is such that it compensates
for the stray vertical field due to toroidal field misalignments.
The slopes of the lines are clearly strong functions of the
compensating field. Radial fields of similar strength have a much
smaller effect on the slope. The radial-field error of the TF
coil system causes charge separation which will move particles upward,
interacting in a manner hard to predict with the charge separation
caused by the stronger VF errors. The effect of the compensating
vertical field (Fig. 2) is much stronger on the slopes of the lines
than it is on the breakdown voltage.

It is possible to obtain information about the role of ions
and neutrals by changing the working gas. We have taken breakdown
voltage curves with the method described above for hydrogen, deuterium,
and nitrogen. Nitrogen was chosen because it is diatomic
like hydrogen, and has a molecular ionization potential only 1%

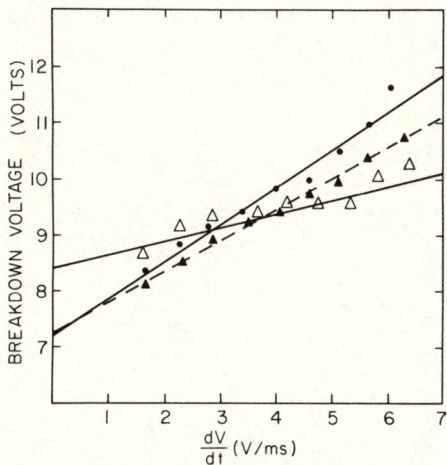

Fig. 2. Effect of compensating vertical field on breakdown. All
points taken with 8kG TF, 3.5G RF and P = 2×10^{-4} torr H_2.
●: 5G V = .67dV/dt + 7.2; ▲: 8.6G V = .55dV/dt + 7.3; △:
12G V = .28dV/dt + 8.4.

higher. Results are shown in Fig. 3. The ratio of slopes H_2: D_2: N_2 is 1: 1.3: 2.3, clearly showing a mass dependence. Note that H_2 and D_2 have the same electronic properties. The difference in intercepts for hydrogen and deuterium is probably due to vacuum conditions — the walls were dirtier during the run with deuterium, and the gas was not as pure.

Figure 4 demonstrates the effect of pressure on breakdown. At lower pressures the average electron energy and drift velocity will increase. At higher pressures and fixed voltage, the ionization gain drops with E/p producing a lower net gain or even a net loss. Our experimental results suggest that the critical parameter, previously mentioned, is the confinement time due to field errors in the magnetic field (Benesch et al., 1981). We reached this conclusion from results obtained from experiments which varied the vertical and horizontal fields and thereby pointed out any effects due to errors in coil alignment. This changed the particle loss time and directly affected breakdown (Benesch et al., 1981).

ACKNOWLEDGEMENTS

We would like to thank J. Dutton, L. Frommhold, M. Oakes, S. Mahajan, and L. Pitchford for helpful comments. This work was supported by the U.S. Department of Energy.

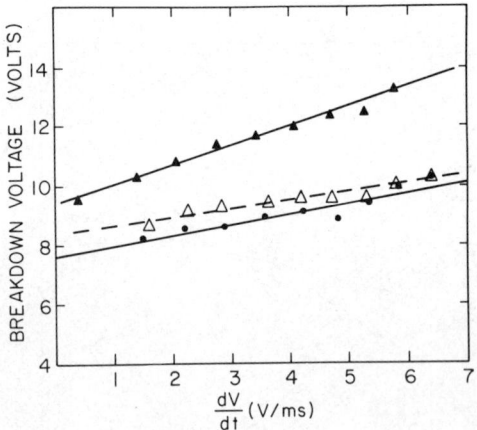

Fig. 3. Effect of working gas on breakdown. All points taken with 8 kG toroidal field (TF), 12 G compensating vertical field (VF) and 3.5 G compensating radial field (RF) at P = 2×10^{-4} torr. $\triangle$: H_2 V = .28 dV/dt + 8.4; $\bullet$: D_2 V = .37 dV/dt + 7.6; $\blacktriangle$: N_2 V = .65 dV/dt + 9.4.

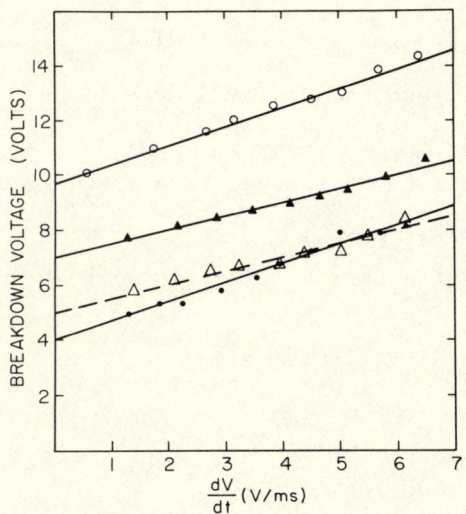

Fig. 4. Effect of pressure on breakdown. All curves taken with 8kG TF, 8G VF and 7G RF in H_2. ○: $P = 4 \times 10^{-4}$ torr, $V = .72 dV/dt + 9.7$; ▲: $P = 2 \times 10^{-4}$ torr, $V = 5.1 dV/dt + 7.0$; △: $P = 1 \times 10^{-4}$ torr, $V = .49 dV/dt + 5.1$; ●: $P = 5 \times 10^{-5}$ torr, $V = .71 dV/dt + 4.0$.

REFERENCES

Artsimovich, L. A., 1972, Nucl. Fus. 12: 215.
Benesch, J. F., Bengtson, R. D., Oakes, M. E., and Mahajan, S. M., submitted to Nucl. Fus.; also FRCR 236, Oct. 1981.
Dimock, D. L., Eubank, H. P., Hinnov, E., Johnson, L. C., and Meservey, E. B., 1973, Nucl. Fus. 13: 271.
Hawryluk, R. J., Bretz, N., Dimock, D., Hinnov, E., Johnson, D., Monticello, D., McCune, D., and Suckewer, 1980, Princeton Plasma Physics Laboratory Report PPPL-1572.
Hutchinson, I. H. and Morton, A. H., 1976, Nucl. Fus. 16: 447.
Hutchinson, I. H. and Strachan, J. D., 1974, Nucl. Fus. 14: 649.
Kaplan, D. J., 1976, Doctoral Dissertation, The University of Texas at Austin.
Kaplan, D. J. and Bengtson, R. D., 1981, J. Phys. B: At. Mol. Phys. 14: 1893.
Knoepfel, H. and Zweben, S. J., 1975, Phys. Rev. Lett., 35: 1340.
Papoular, R., 1976, Nucl. Fus., 16: 37.
Sheffield, J., 1981, Proc. IEEE, 69: 885.
Spong, R. A., Clarke, J. F., Rowe, J. A., and Kammash, T., 1974, Nucl. Fus., 14: 397.
TFR Group, 1975, IAEA CN/A6-1.
Von Engel, A., 1965, "Ionized Gases," Clarendon, Oxford.

NOTES ON SYMBOLS, UNITS AND NOMENCLATURE IN GAS DISCHARGE PHYSICS

R. S. Sigmond

Electron and Ion Physics Group
Norwegian Institute of Technology
7034 Trondheim NT, Norway

INTRODUCTION

At this NATO ASI the usual confusion regarding Symbols, Units and Nomenclature (S.U.N) was not absent. These notes were written on the request of the ASI Scientific Director, but represent only my personal, quite strong opinions on the subject.

Our problems in the S.U.N. field can be divided into three classes:

(1) A certain, easily rectified confusion in the use of symbols and nomenclature (gas number densities, secondary ionization coefficients).

(2) A stubborn, but still correctable tendency to stick to the use of gas pressure p when density n is the relevant physical quantity (as using E/p instead of E/n).

(3) A complete incompatibility between our needs for units or prefixes x units, and the SI units and prefix rules. A handy unit like the Townsend is not in the SI, nor can it legally be formed by SI units with prefixes.

The general guidelines are found in a booklet entitled <u>Symbols, Units and Nomenclature in Physics</u>, Document U.I.P. 20 (1978), reprinted from <u>Physica</u>, 93A (1978) 1-60, prepared by the S.U.N. Commission of the International Union of Pure and Applied Physics (IUPAP). The recommendations contained in this document are in general agreement with those of the ISO and IEC and have been sent to physicists in all IUPAP member countries for review and comment. Two paragraphs from this document of special relevance to us are reproduced in Table 1 and 2.

SYMBOLS AND NOMENCLATURE

As seen from Table 1, gas discharge and plasma physicists should have few, if any, difficulties complying with the recommended list of symbols. The most notable and still current deviations seem to be:

(1) Using N for number densities, while the recommended usage is N for numbers (e.g. of particles) and n for number densities, and

(2) Using ω/α for the secondary electron emission coefficient, instead of the recommended γ.

Personally, I would like to advocate some additions:

(3) Using γ as the total secondary ionization coefficient, defined as the number of secondary electrons produced per primary electron by electron-gas collisions. According to the secondary electron producing agent, γ can be subdivided into:

γ_i Positive ion feedback

γ_p Photon feedback

γ_m Metastable feedback

Furthermore, if necessary, feedback to the cathode can be designated by the subscript C and to the gas by the subscript G.

Thus, γ_{Cm} is the secondary emission coefficient due to metastables impinging on the cathode, while γ_{Gi} (= Townsend's β) is the secondary emission coefficient due to positive ions ionizing the gas molecules.

(4) Apart from the γ, reserving the name "coefficient" (without qualifier) for electron and ion reactions per unit "drift length" (ionization coefficient, attachment coefficient, ion conversion coefficients). For reactions per unit "time," the word "rate" as a qualifier should always be included. "Rate coefficient" is to be preferred to "rate constant," as these so-called "constants" vary rapidly with parameters like E/n.

SYMBOLS, UNITS AND NOMENCLATURE

PREFIXES OR POWERS OF TEN?

Table 2 shows the SI rules. Note that the prefixes hecto, deca, deci and centi are to be avoided. We are thus left with a set of recommended prefixes differing by factors 10^3. This is adequate and convenient for all <u>linear</u> problems: you cover nearly all relavant distances using the range 1 am - 1 Em, etc. The use of, e.g., nm in spectroscopy and μm - m for Debye lengths is very acceptable.

The rule for prefixes in higher dimensions is less satisfactory for our needs: a prefix is to be raised to the same power as the unit symbol it preceeds. Thus 1 km^2 is (naturally, you would say) 10^6m^2. However, this leads in three dimensions to the completely impractical minimum prefix "ratio" of 10^9, though it must be admitted that the prefix "range" is assuredly adequate! Note that any combined (not compound!) prefix is allowed: =1 A/m^2 = 1 MA/m^2

It is very illuminating to experiment with the use of SI units and prefixes to express gas discharge quantities:

Linear Units (Length, Velocities Etc.), 10^3 Prefix Ratios

No problems, adequate prefix ratios and range.

Area-Containing Units, 10^6 Prefix Ratios

Cross-sections: $10^{-18} - 10^{-30}$ m^2 = 1 nm^2 - 1 fm^2

Current densities: $10^{-15} - 10^{15}$ A/m^2

$= 1$ fA/m^2 - 1 PA/m^2

$= 1$ mA/Mm^2 - 1 kA/μm^2

$= 1$ kA/Gm^2 - 1 mA/nm^2

Mobilities: $10^{-8} - 1$ m^2/Vs = 10 m^2/GVs - 1 km^2/μVs

Reduced fields: E/n = $10^{-17} - 10^{-24}$ Vm^2

$= 10$ Vnm^2 - 1 Vpm^2

(= 10 aVm^2 - 1 aVmm^2 ??)

Volume-Containing Units. 10^9 Prefix Ratios

Number densities: $10^9 - 10^{27}$ m^{-3}

$= 1$ mm^{-3} - 1 nm^{-3}

$$1 \text{ atm} = 2{,}687 \cdot 10^{25} \text{ m}^{-3} = 26870000 \text{ μm}^{-3}$$
$$= 0.02687 \text{ nm}^{-3}$$

It is readily seen that:

- Pure prefixes to area and volume units have minimum ratios too large.

- Inverse prefixes convey an inverted picture of size: $1 \text{ μm}^{-3} \gg 1 \text{ m}^{-3}$.

- Combined prefixes tend to be strange, bordering on ridiculous, and one can write many combinations for the same power of 10 : $1 \text{ pA/m}^2 = 1 \text{ A/Mm}^2$.

Now, the reasons for using suitably sized units or units with prefixes are (a) to get numbers that are easier to write, print, and read than numbers with an excessive number of digits or with power-of-ten multipliers, and (b) in this way to get number scales that become familiar to workers in a particular field. A compromise between (a) and (b) is sometimes necessary: one often accepts inconveniently large or small numbers if they fit in a commonly used scale (e.g., one talks about 3000 Townsends, not 3 kiloTownsends). Within the recommended SI, however, none of these benefits may be reaped by gas discharge workers. The Townsend unit for E/n, for instance, does not belong within the SI framework.

My own recommendations for the gas discharge field are:

1. <u>Avoid prefixes, except in simple cases where the prefixes have a direct bearing to the problem at hand.</u> As an example, I would say that the quantity 6.8 pA/mm^2 is well accepted, and perhaps recommended, when it expresses the result of an electrometer measurement of the current to a small Langmuir probe. But it is deplorable if it expresses the current density 6.8 A/km^2 to earth from the ionosphere (both equal to $6.8 \cdot 10^{-6} \text{ A/m}^2$).

2. Reasonably sized numbers and easily familiarized scales are obtained by <u>expressing all quantities involving the gas density n to the power k, in SI units with $10^{21 \cdot k}$ multipliers.</u> Cross sections can also be expressed on this type of scale. Personally, I simply use the free letters ℓ and L to represent the numbers 10^{-21} and 10^{21}, respectively. Thus, ℓ and L are not prefixes, but pure numbers, and are exempted from the prefix power rule: e.g. a cross section of $15 \text{ }\ell\text{m}^2 = 15 \cdot 10^{-21} \text{ m}^2 = 1.5 \cdot 10^{-16} \text{ cm}^2$.

With this convention we obtain:

SYMBOLS, UNITS AND NOMENCLATURE

<u>Densities (n)</u> 1 atm = 101.32 kPa, 20°C: n=26870 L m^{-3}

$\qquad$ 1 Pa 0°C: $\qquad$ n = 0.265 L m^{-3}

$\qquad$ 1 Pa 16.4°C: $\qquad$ n = 0.250 L m^{-3}

$\qquad$ 1 Pa 20°C: $\qquad$ n = 0.247 L m^{-3}

$\qquad$ 1 torr 0°C: $\qquad$ n = 35.36 L m^{-3}

$\qquad$ 1 torr 20°C: $\qquad$ n = 32.94 L m^{-3}

$\qquad$ 10^{15} electrons/cm^3 = 1.0 L m^{-3}

<u>"Reduced" fields (E/n)</u> 1 Td = 10^{-21} Vm^2 = 1 ℓVm^2

$\qquad$ 1 ℓ Vm^2 = 0.3536 V/torr cm $\qquad$ 0°C

$\qquad$ 1 ℓ Vm^2 = 0.3294 V/torr cm $\qquad$ 20°C

<u>"Reduced" mobilities (μn)</u> μn = 2 - 10 L (Vm s)$^{-1}$ (light ions)

$\qquad$ 1 L (Vm s)$^{-1}$ = 0.372 cm^2 atm/Vs 0°C

<u>Conversion Coefficients (α/n, η/n, χ/n etc.)</u>

$\qquad$ Reactions·m^3/m => ℓ m^2

$\qquad$ 1 ℓ m^2 = 0.3536 reactions/torr cm $\qquad$ 0°C

$\qquad$ 1 ℓ m^2 = 0.3294 reactions/torr cm $\qquad$ 20°C

<u>Three-body coefficients (χ/n^2)</u>

$\qquad$ Reactions·m^6/m => ℓ^{2}m^5

$\qquad$ 1 ℓ^{2}m^5 = 0.1250 reactions/torr^{2}cm $\qquad$ 0°C

$\qquad$ 1 ℓ^{2}m^5 = 0.1085 reactions/torr^{2}cm $\qquad$ 20°C

<u>Rate coefficients (k_b, k_t)</u>

$\qquad$ 1 ℓ m^3/s = 10^{-15} cm^3/s

<u>Cross sections</u>

$\qquad$ 1 ℓ m^2 = 10^{-17} cm^2

In publications the L, ℓ nomenclature could easily be used, as it can be defined in the one sentence: $L \equiv 10^{21}$; $\ell \equiv 10^{-21}$. I would then not mix in the Townsend (Td).

SYMBOLS, UNITS AND NOMENCLATURE

Table 1. Recommended symbols for plasma physical quantities. Reproduced from Document U.I.P. 20 (1978) of the IUPAP S.U.N. Commission.

Remark: In general no special attention is paid to the name of the quantity.

7.13 Plasma physics

energy of particle: *énergie d'une particule*	ε
dissociation energy (e.g. of molecule X): *énergie de dissociation (par ex., d'une molécule X)*	$E_d, E_d(X)$
electron affinity: *affinité électronique*	E_{ea}
ionization energy: *énergie d'ionisation*	E_i
degree of ionization: *degré d'ionisation*	x
charge number of ion (positive or negative): *charge ionique (positif ou négatif)*	z
number density of ions of charge number z: *densité ionique des ions de charge z†)*	n_z
degree of ionization for charge number $z \geq 1$: *degré d'ionisation pour un nombre de charge $z \geq 1$*: $x_z = n_z/(n_z + n_1)$	
neutral particle temperature; *température des neutres*	T_n
ion temperature: *température ionique*	T_i
electron temperature: *température électronique*	T_e
electron number density: *densité électronique*	n_e
electron plasma circular frequency: *fréquence de plasma*: $\omega_{pe}^2 = n_e e^2/\varepsilon_0 m_e$	ω_{pe}
Debye length: *longueur de Debye*	λ_D
charge of particle; *charge d'une particule*	q
electron cyclotron circular frequency; *fréquence cyclotron électronique*: $\omega_{ce} = (e/m_e)B$	ω_{ce}
ion cyclotron circular frequency: *fréquence cyclotron ionique*: $\omega_{ci} = (ze/m_i)B$	ω_{ci}
reduced mass: *masse réduite*: $\mu = m_1 m_2/(m_1 + m_2)$	μ, m_r
impact parameter: *paramètre d'impact*	b
mean free path: *libre parcours moyen*	l, λ
collision frequency: *fréquence de collision*	ν_{coll}, ν_c
mean time interval between collisions: *intervalle de temps moyenne entre collisions*: $\tau_{coll} = 1/\nu_{coll}$	τ_{coll}, τ_c
cross section: *section efficace*: $\sigma = 1/l n$	σ
(electron) ionization efficiency: *efficacité d'ionisation (électronique)*: $s_e = (p_0/p)dN/dx$	s_e
(dN: number of ion pairs formed by an ionizing electron travelling through dx in the plasma at gas density p; p_0: gas density at $p_0 = 1$ Torr, $T_0 = 273.15$ K)	
rate coefficient: *taux de réaction*	k
one-body rate coefficient: *taux de réaction unimoléculaire*: $-dn_x/dt = k_m n_x$	k_m
relaxation time; *temps de relaxation*: (e.g. $\tau = 1/k_m$)	τ
two-body rate coefficient, binary rate coefficient: *taux de réaction binaire*: (e.g. $X + Y \rightarrow XY + h\nu$) $dn_{XY}/dt = k_b n_X n_Y$	k_b
three-body rate coefficient, ternary rate coefficient: *taux de réaction ternaire*: (e.g. $X + Y + M \rightarrow XY + M^*$) $dn_{XY}/dt = k_t n_X n_Y n_M$	k_t
Townsend (electron) ionization coefficient; *coefficient de Townsend†)*	α
Townsend (ion) ionization coefficient: *coefficient ionique de Townsend*	β
secondary electron emission coefficient: *taux d'émission secondaire*	γ
drift velocity: *vitesse de mouvement*	v_{dr}
mobility: *mobilité*: $\mu = v_{dr}/E$	μ
positive or negative ion diffusion coefficient: *coefficient de diffusion des ions*	D_+, D_-
electron diffusion coefficient; *coefficient de diffusion des électrons*	D_e
ambipolar (ion-electron) diffusion coefficient: *coefficient de diffusion ambipolaire*: $D_a = (D_- \mu_e + D_e \mu_-)/(\mu_- + \mu_e)$	D_a, D_{amb}
characteristic diffusion length: *longueur caractéristique de diffusion*	L_D, Λ
ionization frequency: *fréquence d'ionisation*	ν_i
ion-ion recombination coefficient; *coefficient de recombinaison ion-ion*: $dn_-/dt = -\alpha_i n_- n_+$	α_i
electron-ion recombination coefficient: *coefficient de recombinaison électron-ion*: $dn_e/dt = -\alpha_e n_e n_+$	α_e
plasma pressure; *pression cinétique du plasma*	p
magnetic pressure: *pression magnétique*: $p_m = B^2/2\mu$ (μ: permeability)	p_m
magnetic pressure ratio; *coefficient* β: $\beta = p/p_m$ (p_m: magnetic pressure outside the plasma)	β
magnetic diffusivity: *diffusivité magnétique*: $\nu_m = 1/\mu\sigma$ (σ: electric conductivity; μ: permeability)	ν_m, η_m
Alfvén speed: *vitesse d'Alfvén*: $v_A = B/(\mu\rho)^{1/2}$ (ρ: (mass) density; μ: permeability)	v_A

†) When only singly charged ions need to be considered, n_- and n_+ may be represented by n and n_i.

†) The same name is also used for $\eta = \alpha/E$, where E denotes electric field strength.

Table 2. Prefixes in the SI (Systéme International) Reproduced from Document U.I.P. 20 (1978) of the IUPAP S.U.N. Commission.

2.2 Prefixes–General rules

1. The *prefixes* which should be used to indicate decimal multiples or submultiples of a unit are given in table 1.

Table 1

deci; *déci*	($= 10^{-1}$)	d	deca; *déca*	($= 10^{1}$)	da
centi; *centi*	($= 10^{-2}$)	c	hecto; *hecto*	($= 10^{2}$)	h
milli; *milli*	($= 10^{-3}$)	m	kilo; *kilo*	($= 10^{3}$)	k
micro; *micro*	($= 10^{-6}$)	μ	mega; *méga*	($= 10^{6}$)	M
nano; *nano*	($= 10^{-9}$)	n	giga; *giga*	($= 10^{9}$)	G
pico; *pico*	($= 10^{-12}$)	p	tera; *téra*	($= 10^{12}$)	T
femto; *femto*	($= 10^{-15}$)	f	peta; *peta*	($= 10^{15}$)	P
atto; *atto*	($= 10^{-18}$)	a	exa; *exa*	($= 10^{18}$)	E

2. *Compound prefixes*, formed by juxtaposition of two or more prefixes, are not to be used.

Not: mμs,	*but:* ns	(nanosecond)
Not: kMW,	*but:* GW	(gigawatt)
Not: μμF,	*but:* pF	(picofarad)

3. When the symbol of a prefix is placed before the symbol of a unit, the *combination of the two symbols* should be considered as *one new symbol*, which can be raised to a positive or negative power without using brackets.

Examples: cm^3 mA^2 μs^{-1}.

Remark:

cm^3 means always $(0,01\ m)^3$ *but never* $0,01\ m^3$
μs^{-1} means always $(10^{-6}\ s)^{-1}$ *but never* $10^{-6}\ s^{-1}$.

PARTICIPANTS

Allen, J. E.
Department of Engineering Science
University of Oxford
Parks Road, Oxford
OXI 3PJ
England

Allis, W. P.
33 Reservoir St.
Cambridge, MA 02138
USA

Aschwanden, T.
ETH
Swiss Fed. Inst. of Technology
High Voltage Laboratories
Physikstr 3
CH - 8092 Zuerich
Switzerland

Baldo, G.
Instituto di Electrotecnica
ed Elettronica
Via Grademigo 6A
Universita di Padova
Padova
Italy

Bardsley, J. N.
Physics Department
University of Pittsburgh
Pittsburgh, PA 15260
USA

Barrault, M. R.
University of Liverpool
Liverpool L69 3BX
England

Bastien, F.
Laboratoire de Physique
des Decharges
Ecole Superieure d'Electricite
Plateau du Moulon
9110 Gif sur Yvette
France

Bayrakceken, F.
O.D.T.U.
Cevre Muh Bolumu
Ankara
Turkey

Bengston, R. D.
Department of Physics
University of Texas at Austin
Austin, TX 78712
USA

Bicknell, J. A.
University of Manchester
Inst. of Science and Technology
Physics Department
Sackville Street
Manchester M60 1QD
England

Boeuf, J. P.
Laboratoire de Physique
des Decharges
Ecole Superieure d'Electricite
91190 Gif sur Yvette
France

Bradley, L. P.
LLNL
7000 East Avenue
L-464
Livermore, CA 94550

Braithwaite, N. St. J.
Department Engineering Science
Parks Road
Oxford, OXI 3PJ
England

Burgmans, A. L. J.
Philips Research Labs
5600 MD Eindhoven
The Netherlands

Chang, J. S.
McMaster University
Dept. of Engineering Physics
Hamilton, Ontario
Canada L8S 4M1

Christophorou, L. G.
Oak Ridge National Laboratory
Building 4500S, H-158
P.O. Box X
Oak Ridge, TN 37830
USA

Cole, H. R.
1161 West 37th Street
Los Angeles, CA 90007
USA

Conde, C. A. N.
Physics Department
University of Coimbra
3000 Coimbra
Portugal

Danzmann, K.
Institute fur Plasmaphysik
der Universitat Hannover
Callinstr, 38 D-3000
Hannover 1
West Germany

Davidson, H.
DAMA-AR
Washington, D.C. 20310
USA

Desoppere, E.
State University
Rozier 44
B-900
Belgium

DeWitt, R. N.
Code F12
Naval Surface Weapons Center
Dahlgren, VA 22448
USA

Dias, T. T.
Physics Department
University of Coimbra
3000 Coimbra
Portugal

Dorelon, A. A.
Laboratoire de Spectrometrie
BP 53X
38041 Grenoble Cedex
France

Dupuy, J.
Universite de PAU
Laboratoire d'Electricite
Avenue Phillippou
64000 PAU
France

Dutton, J.
University College of Swansea
Singleton Park
Swansea, Glamorgan
SA 2 8PP, England

Ecker, G.
Ruhr-Universitat Bochum
423 Bochum
Universitatsstrasse
Geb. NB7
West Germany

PARTICIPANTS

Eliasson, B.
Brown Boveri
Research Center
5405 Baden - Daettwil
Switzerland

Fang, D. Y.
Lehrstuhl fur Technische
Elektrophysik der TUM
Arcisstr. 21
8000 Munchen 2
West Germany

Fernsler, R. F.
Naval Research Laboratory
Code 4770
Washington, D.C. 20375
USA

Ferreira, C. M.
Centro de Electrodinamica
Instituto Superior Tecnico
Av. Rovisco Dais
1000 Lisboa
Portugal

Ferreira, M. de F.
Departamento Fisica
Universidade Coimbra
3000 Coimbra
Portugal

Gayet, P.
Electricite de France
Direction des Etudes et Recherches
B.P. no. 1
77250 Moret/S/Loring
France

Germain, G.
Commissariat a l'Energie Atomiques
Centre d'Etudes Scientifiques
et Techniques D'Aquitaine
Boite Postale no. 2.
Le Barp
33830-Belin-Beliet, France

Goldman, A.
Laboratoire de Physique
des Decharges
Ecole Superieure d'Electricite
Plateau du Moulon
91190 Gif sur Yvette
France

Goldman, M.
Laboratoire de Physique
des Decharges
Ecole Superieure d'Electricite
Plateau du Moulon
9119 Gif sur Yvette
France

Golightly, D.
U.S. Geological Survey
957 National Center
Reston, VA 22092
USA

Gripshover, R. J.
RR2, Box 360 B
King George, VA 22485
USA

Guenther, A.
Chief Scientist (Code CA)
Air Force Weapons Lab
Kirtland AFB, NM 87117
USA

Hackmann, J.
Phys. Inst. II
Der Universitat Dusseldorf
4000 Dusseldorf
Universitatsstrasse 1
West Germany

Hamden, M.
University College of Wales
Department of Physics
Aberystwth SY23 3BZ
United Kingdom

Harjes, C.
Department of Elec. Eng.
Texas Tech University
P.O. Box 4439
Lubbock, TX 79439
USA

Kline, L. E.
1540 Shady Avenue
Pittsburgh, PA 15217
USA

Kristiansen, M.
Plasma and Switching Laboratory
Texas Tech University
P.O. Box 4439
Lubbock, TX 79409
USA

Kunhardt, E. E.
Ionized Gas Laboratory
Texas Tech University
P.O. Box 4439
Lubbock, TX 79409
USA

Lakdawala, V. K.
Department of Elec. Eng.
University of Liverpool
P.O. Box 147
Liverpool, L69 3BX
England

Lawler, J.
Physics Department
University of Wisconsin
Madison, WI 53706
USA

Lea, L. M.
University of Oxford
Dept. of Engineering Science
Parks Road
Oxford, OXI 3 PJ
England

LeNy, R.
Lab. Phys. Exp.
Institut de Physique
Universite de Nantes
2, Rue de la Houssiniere
44072 Nantes
France

Llewellyn-Jones, F. P.
University College of Swansea
Singleton Park
Swansea, Glamorgan
SA 2 8PP, United Kingdom

Leussen, L. H.
Naval Surface Weapons Center
Code F12
Dahlgren, VA 22448
USA

Marec, J.
Laboratoire de Physique
des Plasmas
Universite de Paris Sud
Batiment 212
91405 Orsay
France

Marode, E.
Laboratoire de Physique
des Decharges
Ecole Superieure d'Electricite
Plateau du Moulon
91190 Gif sur Yvette
France

Mentzoni, M.
Physics Department
University of Oslo
Box 1038, Blindern
Norway

Molen, G. M.
Dept. of Electrical Engineering
Old Dominion University
Norfolk, VA 23508
USA

PARTICIPANTS

Nolting, E. E.
6142 Camelback Lane
Columbia, MD 21045
USA

Penven, H. Le
Societe Protel
BP7 ZI La Roche de Glun
26600 La Roche de Glun
France

Phelps, A. V.
Campus Box 440
JILA
University of Colorado
Boulder, CO 80309
USA

Pinnekamp, F.
BBC - Research Center
CH 5400 Baden - Daettwil
Switzerland

Pitchford, L.
Division 4211
Sandia National Laboratories
Albuquerque, NM 87117
USA

Plueksawan, W.
Laboratoire de Genie Electrique
U.P.S.
118 Route de Narbonne
31062 Toulousse
France

Pointu, A. M.
Lab Physique des Plasmas
Batiment 212
Univ. de Paris Sud
91405 Orsay
France

Proud, Jr., J. M.
GTE Laboratories, Inc.
40 Sylvan Road
Waltham, MA 02154
USA

Rees, J. A.
Dept. of Electrical Engineering
University of Liverpool
Brownlow Hill, P.O. Box 147
Liverpool, L69 3BX
England

Robledo, A.
Instituto De Investigaciones
    Electricas
Div. de Equipos
Shakespear 6
Mexico 5 D.F.

Rodrigo, H.
Dept. of Electrical Engineering
University of Liverpool
P.O. Box 147
Liverpool, L69 3BX
England

Rose, M. F.
Code F-04
Naval Surface Weapons Center
Dahlgren, VA 22448
USA

Savic, P.
National Research Council
Ottawa, Ontario
Canada K1AOR6

Schoenbach, K. H.
Dept. of Electrical Engineering
Texas Tech University
P.O. Box 4439
Lubbock, TX 79409
USA

Schwirzke, F.
Dept. of Physics
Naval Postgraduate School
Monterey, CA 93940
USA

P. Segur
Universite de Paul Sabatieur
118 Route de Narbonne
31062 Toulousse
France

Sigmond, R. S.
Dept. of Physics
University of Trondheim
Norwegian Institute of Tech.
N7034 Trondheim, Norway

Skullerud, H. R.
Electron and Ion Physics Group
Dept. of Physics
University of Trondheim
Norwegian Institute of Tech.
N7034 Trondheim, Norway

Spiga, G.
Via dei Colli, 16
Laboratorio di Ingegneria Nucleare
I - 40136
Bologna
Italy

Szabo, I.
Chemical Center
P.O. Box - 40
S-22007 Lund 7/
Sweden

Tang, T. M.
Dept. of Physics
University of Manchester
Inst. of Science and Tech.
Sackville Street
Manchester M60 1QD, England

Teich, T. H.
Swiss Fed. Inst. of Tech.
ETH, High Voltage Labs
CH - 8092 Zuerich
Switzerland

Thompson, J. E.
College of Engineering
University of South Carolina
Columbia, SC 29208
USA

Tredicce, J.
National Inst. of Optics
Largo Enrico Fermi 6
50125 Florence
Italy

Twiddy, N. D.
University College of Wales
Department of Physics
Aberystwyth SY23 3BZ
United Kingdom

VanBrunt, R. J.
Electro Systems Div., 722
National Bureau of Standards
Washington, D.C. 20234
USA

Vandevender, J. P.
Org 4252
Sandia National Laboratories
P.O. Box 5800
Albuquerque, NM 87185
USA

Vriens, L.
Philips Research Labs, WAG
Eindhoven
The Netherlands

Vitkovitsky, I. M.
10406 Burnt Ember Drive
Silver Spring, MD 20903
USA

Walter, W.
Polytechnic Institute of NY
Microwave Research Institute
Route 110
Farmingdale, NY 11735
USA

Waters, R. T.
Inst. of Science and Technology
University of Wales
Cardiff CF1 3NU
United Kingdom

Williams, F.
Dept. of Electrical Engineering
Texas Tech University
P.O. Box 4439
Lubbock, TX 79409
USA

PARTICIPANTS

Wilson, R. M.
Department of Chemistry
University of Durham
South Road
Durham DH1 3LE
England

Wyatt, K.
Apt. D-4
West Gate
Cambridge, MA 02139
USA

INDEX

Arc, 167-180, 181-202
  free burning, 185, 211
  in a magnetic field, 182
  modeling of, 169, 182, 193
  wall-stabilized, 183
Avalanches, 35, 124

Breakdown, 26, 119, 166
  in microwave discharges, 351
  in tokamaks, 446
Broadening, 225
  stark-, 225, 285

Catalysis, 315, 319
Cathode
  fall, 122
  spot, 168, 176-178
  sputtering, 35, 36, 400
Conductivity, 197
Confinement, 385
Conservation equations, see
  Transport equations
Corona, 1-64
  applications, 55
  continuous glow-, 39, 43, 47
  negative, 4, 8, 9, 24, 28
  point-to-plane, 40, 45, 49
  positive, 4, 8, 9, 12, 40
  streamer, 43, 47

Debye length, 294
Diagnostics, 203-266
Diffuse discharge, 415, 428
Diffusion
  ambipolar, 91, 145
  length, 197
  neutral, 91, 194

Discharge
  arc, 167-180, 181-202, 266
  corona, 1-66
  diffuse, 415-428
  glow, 65-118, 121
  hollow cathode, 236
  laser, 386
  microwave, 347-382
Dopler anemometry, 215

Electroluminescence, 429-442
Electron distribution function
  in glow discharges, 77-79
  in microwave discharges, 353
  in noble gases, 434
Energy balance, see Transport
  equations
Energy conversion, 386-388
Equilibrium, 68, 267, 296
Etching, 326, 329, 343, 400

Field measurement, 243, 260
  (see also Probes)

Glow discharge, 65-118, 121
  metal-doped, 66, 72, 73
  modeling, 65, 105
  positive column, 68

Heating, 82, 146, 172
Hydrodynamic equations, see
  Transport equations

Instabilities
  attachment, 14
  in corona, 13-14
Ionization growth, 125, 139, 140

Ionization waves, 154, 161

Laser
  discharges, 386, 421
  probes, 216
  stimulated attachment, 423
Lightning, 227

Microwave
  discharge, 367-382
  breakdown, 348
  generator, 390
  modeling, 369
  structure, 360
Modeling
  arc discharge, 169
  glow discharge, 65, 105
  microwave discharge, 369
  streamer, 57
Monte-Carlo method, 438

Nonequilibrium, 68

Ozone synthesis, 330

Paschen curve, 125-127
Plasma
  applications, 383-414
  catalysis, 315, 319
  chemistry, 293-346, 396
  diagnostics, 203-266
  etching, 326, 329, 363, 400
  generation, 299, 360
  metallurgy, 402
  propulsion, 395
  reactors, 302, 320, 399
  torch, 396
Point-to-plane corona, 40, 45, 49
Positive column, 68, 129
Probes
  laser, 116
  field-filter, 245
  for field measurements, 243
  langmuir, 232, 236
  rotating, 256
  shock wave, 213
Pulsed power techniques, 402

Radiations
  transport, 103
  detection, 429
Radiative lifetime, 91
Rate coefficients, 85-88
Resonance, 363

Schlieren interferometry, 209
Schlieren photography, 205
Scintillation, 429
Spark
  breakdown, 26, 27
  formation, 123
  long-, 226
Spectroscopy, 221, 267-292
  molecular, 271
Sputtering, 35, 36, 400
Spark broadening, 225, 285
Streamer, 119-166
  mode, 125
  modeling, 57
  preonset-, 42
  primary, 48
  secondary, 48
  -to-spark transition, 118
Surface effects, 176 (see also
  Sputtering)
Surface treatment, 334
Surface wave, 348
Switch, 391, 415-428

Temperature
  electron-, 270, 295
  gas-, 270, 295
  rotational, 269, 275, 284
  vibrational, 269, 270, 283, 146
Thermodynamic equilibrium, 267,
  296
Thermonuclear fusion, 384
Thomson scattering, 217
Time lag, 9
Tokamak, 443-450
  breakdown in, 446
Townsend coefficients, 125
Townsend formula, 16
Transition
  glow to arc, 119-166, 229
  Trichel to glow, 33-35
  Trichel continuous glow, 35, 39

Trichel to spark, 33, 38
  streamer to spark, 33, 34
Transport equations, 93, 132, 142, 146, 156, 187, 195, 307, 349, 377
Trichel pulse, 9, 35, 37
Trichel to spark transition, 33, 34
Trichel to glow transition, 33-35

Vibrational energy, 146

Vibrational temperature, 269, 270, 283

Warburg distribution, 15, 18

Wave
  ionizing, 154, 161
  antiforce, proforce, 163
  surface, 348
Wave guide, 365